Workbook To Accompany
Anderson, Sweeney and Williams'

STATISTICS FOR BUSINESS AND ECONOMICS

SIXTH EDITION

By:

Mohammad Ahmadi

The University of Tennessee at Chattanooga

WEST'S COMMITMENT TO THE ENVIRONMENT

In 1906, West Publishing Company began recycling materials left over from the production of books. This began a tradition of efficient and responsible use of resources. Today, 100% of our legal bound volumes are printed on acid-free, recycled paper consisting of 50% new paper pulp and 50% paper that has undergone a de-inking process. We also use vegetable-based inks to print all of our books. West recycles nearly 27,700,000 pounds of scrap paper annually—the equivalent of 229,300 trees. Since the 1960s, West has devised ways to capture and recycle waste inks, solvents, oils, and vapors created in the printing process. We also recycle plastics of all kinds, wood, glass, corrugated cardboard, and batteries, and have eliminated the use of polystyrene book packaging. We at West are proud of the longevity and the scope of our commitment to the environment.

West pocket parts and advance sheets are printed on recyclable paper and can be collected and recycled with newspapers. Staples do not have to be removed. Bound volumes can be recycled after removing the cover.

Production, Prepress, Printing and Binding by West Publishing Company.

Printed with Printwise
Environmentally Advanced Water Washable Ink

610 Opperman Drive
P.O. Box 64526
St. Paul, MN 55164–0526

Printed in the United States of America
03 02 01 00 99 98 97 96 8 7 6 5 4 3 2 1 0

ISBN 0–314–08459–2

CONTENTS

CHAPTER		PAGE
13.	Analysis of Variance and Experimental Design	439
14.	Simple Linear Regression	495
15.	Multiple Regression	535
16.	Regression Analysis: Model Building	567
17.	Index Numbers	593
18.	Forecasting	609
19.	Nonparametric Methods	633
20.	Statistical Methods for Quality Control	659
21	Sample Survey	677
22.	Decision Analysis	723

TO THE READER

This study guide is designed to reinforce and elaborate on the main concepts and ideas of the text by Anderson, Sweeney and Williams. The purpose of the guide is to help the reader more easily understand the material covered in the text. The chapters in the study guide parallel the chapters of the text, and the same notations are used in both. In this sixth edition, each chapter contains four distinct parts: (1) chapter outline and review, (2) chapter formulas, (3) exercises, and (4) self-testing questions. In part one (chapter outline and review), the material covered in the text is outlined and briefly reviewed. Some of the explanations are phrased differently than the text; but of course, the meaning is the same. In the second part of each chapter (chapter formulas), the formulas which have been introduced in the text are organized in a concise form for easy reference. The same numbering system is used in both so that cross reference can be easily achieved. Part three (exercises) consists of an extensive exercise section. In this part, two forms of exercises are presented. First, for each topic which is covered in the text, an exercise is given; and then, a step-by-step procedure for the solution of the exercise is presented. The purpose of this section is to show the reader how she/he should approach the problems in order to solve them. For easy reference, all the solved exercises are marked by an asterisk (*). These exercises are followed by a series of additional exercises, which the reader must solve. The answers to these latter exercises are provided in the back of each chapter. In part four, I have designed a series of self-testing questions where the reader can test her/his understanding of the concepts covered in each chapter. Following the self-testing questions, an answer key is provided so the reader can evaluate her/his test results. To use this study guide, the reader must first read the material in the text. Then, to better understand the material, the reader should review the chapter outline of the study guide; review the problems which are solved completely in order to understand the solution procedure; and then, work as many problems as possible and check the answers for correctness with those provided at the end of each chapter. Finally, the reader should check her/his understanding of the materials covered in the chapter by answering the self-testing questions and checking the answers with the key. I trust the workbook will enhance the reader's understanding of the material covered in the text.

I would like to express my great appreciation to many people who made this project possible. First, I would like to thank my lovely wife Nancy, who typed this manuscript. My special thanks to Tammy Bastien and Michelle Russell who edited, proofread, and meticulously checked the accuracy of the manuscript. Finally, I would like to thank all my students who encouraged me to write this sixth edition.

Mohammad Ahmadi
Chattanooga, Tennessee

CHAPTER ONE

DATA AND STATISTICS

CHAPTER OUTLINE AND REVIEW

In this chapter you have learned that *data* are the raw materials of statistics. Data can be qualitative or quantitative. Qualitative data are labels used to identify attributes of elements, whereas quantitative data indicate the value of a variable; i.e., how much or how many. Furthermore, you were introduced to the concept of statistics and were given a few examples of how statistics is applied in economics and various functional areas of business, such as accounting, finance, marketing, and production. You were also informed that statistics can be used to make inferences about the population characteristics from sample information. The key terms which you should have learned from this chapter are as follows.

A. **Data:** Factual information which is collected, analyzed, presented, and interpreted. Data may be either numeric or nonnumeric.

B. **Data Set:** All of the data collected in a particular study.

C. **Elements:** The entities on which data are collected.

D. **Variable:** A characteristic of interest for the elements.

E. **Observations:** The set of measurements or data obtained for a particular element.

F.	**Qualitative Data:**	Data that provide labels or names for a characteristic of an element. Example: Labels of *male* and *female* . Qualitative data may be numeric or nonnumeric.
G.	**Qualitative Variable:**	A variable with qualitative data.
H.	**Quantitative Data:**	Data which indicate the quantity of a variable in terms of how much or how many. Example: Price of an automobile. Quantitative data are always numerical.
I.	**Quantitative Variable:**	A variable with quantitative data.
J.	**Crossectional Data:**	Data gathered at the same time or approximately the same time.
K.	**Time Series Data:**	Data collected at several successive periods of time.
L.	**Statistics:**	A body of principles which deals with collection, analysis, interpretation and presentation of numerical facts or data.
M.	**Population:**	The aggregate of all elements of interest in a particular study.
N.	**Sample:**	A portion of the population selected to represent the whole population.
O.	**Descriptive Statistics:**	The study of the methods of organization, summarization, and presentation of statistical data.
P.	**Statistical Inference:**	Making inferences about the characteristics of the population from the information which is provided by the sample.

EXERCISES

*1. The following data set provides information about five college professors.

Professor Name	Specialty	Sex	Age	Rank
Able	Economics	M	40	Associate Professor
Bastien	Accounting	F	33	Full Professor
Martin	Marketing	F	52	Associate Professor
Russell	Management	F	30	Assistant Professor
White	Finance	M	38	Assistant Professor

(a) How many elements are in this data set?

Answer: Elements are entities on which data are collected. In this case, each professor represents an element. Therefore, in this data set there are five elements.

(b) How many variables are in this data set?

Answer: Variable refers to a characteristic of interest for an element. In this case, four characteristics are being observed. Thus, there are four variables *(specialty, sex, age, and rank)* in this data set.

(c) How many observations are in the above data set?

Answer: The set of data gathered for an element is an observation. For instance, professor Able (the first element) has the following observation: *Economics, M, 40, and Associate Professor*. Since there are five professors in this data set, we say that there are five observations.

(d) Which variables are qualitative and which are quantitative variables?

Answer: *Specialty, Sex,* and *Rank* are simply labels, therefore, these three variables are qualitative variables. The variable *Age* is a numerical measure and it is a quantitative variable.

2. In many universities, students evaluate their professors by means of answering a questionnaire. Assume a questionnaire is distributed to a class of 45 students. Students are asked to answer the following questions:

1. Sex
2. Race (Black, White, Other)
3. Age
4. Number of hours completed
5. Grade point average
6. My instructor is a very effective teacher

1	2	3	4	5
strongly agree	moderately agree	neutral	moderately disagree	strongly disagree

(a) How many elements are in the above data set?

(b) How many variables are in this data set?

6

(c) How many observations are in this data set?

(d) Which variables are qualitative and which are quantitative variables?

3. The following information regarding the top eight Fortune 500 companies was presented in the April 18, 1994 issue of *Fortune Magazine.*

Company	Sales $ Millions	Sales Rank	Profits $ Millions	Profits Rank
General Motors	133,622	1	2,466	6
Ford Motor	108,521	2	2,529	5
Exxon	97,825	3	5,280	1
Int'l Business Machines	62,716	4	(8,101)	485
General Electric	60,823	5	4,315	2
Mobil	56,576	6	2,084	10
Philip Morris	50,621	7	3,091	4
Chrysler	43,600	8	(2,551)	484

(a) How many elements are in the above data set?

8

(b) How many variables are in this data set?

4

(c) How many observations are in this data set?

8

(d) Which variables are qualitative and which are quantitative variables?

qualitative Sales Rank Profit rank

quantitative Sales million Profit millions

4. The following national weather report gives the temperatures and weather conditions on the previous day in cities across the nation.

City	Hi	Lo	Condition
Albany, N.Y.	88	60	cloudy
Chicago	92	64	clear
Dallas-Ft.Worth	89	72	cloudy
Denver	75	54	clear
Hartford	88	61	cloudy
Honolulu	86	70	clear
Kansas City	93	74	clear
Los Angeles	80	62	cloudy
Nashville	94	72	rain
New York City	90	69	rain
Philadelphia	90	67	rain

(a) How many elements are there in this data set?

11

(b) How many variables are in this data set?

3

(c) How many observations are there in the above data set?

11

(d) Which variables are qualitative and which are quantitative variables?

Temp is quantitative
Condition is qualitative

*5. The July 1995 issue of a national magazine reported that in a national public opinion survey conducted among 1200 registered voters, 36 percent favored Candidate A, 34 percent favored Candidate B, and 23 percent were in favor of Candidate C for president.

(a) What constitutes the population?

Answer: The population is the aggregate of all elements of interest in a study. In the above example, the population will be all the registered voters.

(b) What is the sample?

Answer: The sample is a portion of the population which represents the population. Hence, the 1200 people who were surveyed represent the sample.

(c) Based on the sample, what percentage of the population would you think favors none of the candidates?

Answer: Based on this sample, our best estimate is that 36 percent favored Candidate A, 34 percent favored Candidate B, and 23 percent were in favor of Candidate C, or a total of 93 percent indicated a favorite candidate, which means 7 percent did not indicate any preference. Thus, our best estimate is that 7 percent of all registered voters favor none of the three candidates.

(d) Based on the sample, what percentage of the population would you think favors Candidate C?

Answer: We assume that the sample represents the population. Based on this sample, our best estimate is that 23 percent of all registered voters are in favor of Candidate C. In statistics, this type of estimation is called a "point estimate." That is, determining a measure from a sample and inferring that the best estimate for the population's measure is that of the sample. We shall study the concept of estimation at length in Chapter 8.

***6.** The following table shows the age distribution of a sample of 180 students at a local college.

AGE DISTRIBUTION OF
180 STUDENTS AT A LOCAL COLLEGE

Age of Students	Number of Students
15 - 19	36
20 - 24	44
25 - 29	60
30 - 34	38
35 - 39	2
Total	180

(a) Of the students in the sample, what percentage is younger than 20 years of age?

Answer: There are 36 students whose ages are less than 20. Therefore, the percentage of the students whose ages are less than 20 is calculated as

(36/180) x 100% = 20%

(b) What percentage is at least 30 years of age?

Answer: There are 40 students who are at least 30 (38 in the "30 - 34" category and 2 in the "35 - 39" category). Therefore, the percentage of the students who are 30 or older is calculated as

(40/180) x 100% = 22.22%

(c) Based on this sample, what percentage of the students at the college do you estimate to be younger than 25 years of age?

Answer: There are 80 students who are younger than 25 (36 in the first category and 44 in the second category). Therefore, the percentage of the students (in the sample) who are younger than 25 is

(80/180) x 100% = 44.44%

Since we assume the sample represents the population, our best estimate (based on the sample) is that 44.44% of all the students in the college are younger than 25.

7. In the shrimping industry, the fishermen bring their boats to the packing house pier and unload their catch into larger holding tanks. The price that the packing house will pay the fishermen is based on the average size of the shrimp. To determine the average size, the shrimp are thoroughly mixed in the holding tank, and then a bucket of shrimp is taken out. The shrimp in the bucket are then sized.

(a) What is the population of the shrimp?

(b) What constitutes the sample?

(c) If the contents of the bucket showed a count of 50 shrimp per pound, what would be an estimate for the shrimp per pound in the population?

8. Ryan, Inc., a manufacturer of solar panels, is a small firm with 80 employees. The table below shows the hourly wage distribution of the employees:

Hourly Wages (in dollars)	Number of Employees
2 - 5	3
6 - 9	12
10 - 13	18
14 - 17	20
18 - 21	15
22 - 25	10
26 - 29	2
Total	80

(a) How many employees receive hourly wages of at least $18?

(b) What percentage of the employees have hourly wages of at least $18?

(c) What percentage of the employees have hourly wages of $13 or less?

(d) How many variables are presented in the above data set?

(e) This data set represents the results of how many observations?

(f) Which variables are qualitative and which are quantitative variables?

9. To determine the average typing speed of 700 students who just finished Typing 101, a group of 20 students is randomly selected. It is determined that the average typing speed is 47 words per minute.

(a) What is the population for this study?

(b) What constitutes the sample?

(c) Based on the sample, what is an estimate for the average typing speed of the population?

(d) If you allow "plus or minus" 5 words per minute "margin for error," what range will include the average speed of the population?

(e) If you wanted the sample size to be 10% of the population, how many students would be in the sample?

10. The highway patrol is interested in determining the average speed of automobiles traveling on I-75 between Chattanooga and Atlanta. To accomplish this task, the speed of every tenth car passing a particular point is recorded.

(a) What is the population for this study?

(b) What constitutes the sample?

(c) Is speed a qualitative or a quantitative variable?

11. In order to determine the delinquent loans in a credit union, every single account is audited. Is the credit union using the sampling concept?

12. In order to assure the quality of wine, the quality control departments of wineries determine the acidity, the correctness of taste, and the aroma. This task is accomplished by opening the seal, determining the quality of the wine, and then discarding the bottle. Do you think the quality control departments use a sampling concept?

13. Briefly explain the meaning of statistics and distinguish between descriptive statistics and statistical inference.

14. Explain what is meant by quantitative data and qualitative data.

15. The following data show the age distribution of a sample of employees of Research Inc.

Age	Number of Employees
20 - 24	2
25 - 29	48
30 - 34	60
35 - 39	80
40 - 44	10

(a) What percentage of employees are at least 35 years of age?

(b) Is the figure (percentage) that you computed in Part a an example of statistical inference? If no, what kind of statistics does it represent?

(c) Based on this sample, the president of the company said that "45% of our employees are 35 or older." The president's statement represents what kind of statistics?

(d) What percentage of the employees are 29 years or younger?

SELF-TESTING QUESTIONS

In the following multiple choice questions, circle the correct answer. An answer key is provided following the questions.

1. The sample size

a) can be larger than the population size
b) is always smaller than the population size
c) can be larger or smaller than the population
d) is always equal to the size of the population

2. A population is

a) the same as a sample
b) the selection of a random sample
c) the collection of all items of interest in a particular study
d) none of the above

3. A portion of the population selected to represent the population is called

a) descriptive statistics
b) inferential statistics
c) a statistic
d) a sample

4. The study of the methods of organization, summarization, and presentation of statistical data is referred to as

a) inferential statistics
b) descriptive statistics
c) sampling
d) none of the above

5. The process of making inferences about the characteristics of the population based on the sample information is termed

a) descriptive statistics
b) random sample
c) inferential statistics
d) sampling

6. In a random sample of 200 items, 5 items were defective. An estimate of the percentage of defective items in the population is

a) 5.0%
b) 2.5%
c) 200
d) none of the above

7. The entities on which data are collected are

a) variables
b) data sets
c) elements
d) none of the above

8. The numerical facts are called

a) qualitative measures
b) a population
c) a sample
d) statistics
e) none of the above

9. Labels or names used to identify attributes of elements are

a) quantitative data
b) qualitative data
c) simple data
d) none of the above

10. The labeling of parts as "defective" or "non-defective" is an example of

a) quantitative data
b) qualitative data
c) simple data
d) none of the above

11. A characteristic of interest for the elements is

a) a variable
b) an element
c) a data set
d) none of the above

12. In a questionnaire, respondents are asked to indicate whether their home is located in the city or the suburbs. The location is an example of

a) quantitative data
b) qualitative data
c) simple data
d) none of the above

13. Arithmetic operations are appropriate for

a) qualitative data
b) quantitative data
c) both quantitative and qualitative data
d) neither quantitative nor qualitative data
e) none of the above

14. Weight is an example of a variable with

a) qualitative data
b) quantitative data
c) both quantitative and qualitative data
d) neither quantitative nor qualitative data
e) none of the above

15. On a street, the houses are numbered from 300 to 450. The house numbers are examples of

a) qualitative data
b) quantitative data
c) both quantitative and qualitative data
d) neither quantitative nor qualitative data
e) none of the above

ANSWERS TO THE SELF-TESTING QUESTIONS

1. b
2. c
3. d
4. b
5. c
6. b
7. c
8. d
9. b
10. b
11. a
12. b
13. b
14. b
15. a

ANSWERS TO CHAPTER ONE EXERCISES

2. (a) 45
   (b) 6
   (c) 45
   (d) Sex, Race, and Teacher effectiveness are qualitative
   Age, Number of hours, and Grade point average are quantitative

3. (a) 8
   (b) 4
   (c) 8
   (d) Sales and Profits are quantitative
   Sales Rank, and Profits Rank are qualitative

4. (a) 11
   (b) 3
   (c) 11
   (d) Temperature is quantitative
   Weather Condition is qualitative

7. (a) Shrimp in the holding tank
   (b) Shrimp in the bucket
   (c) 50

8. (a) 27
   (b) 33.75%
   (c) 41.25%
   (d) 2
   (e) 80
   (f) both variables are quantitative

9. (a) 700 students
   (b) 20 students
   (c) 47
   (d) 42 to 52
   (e) 70

10. (a) All the automobiles on I-75
    (b) All the tenth cars
    (c) quantitative

11. No, the entire population is being studied.

12. Yes, testing is destructive.

13. See chapter outline and review.

14. See chapter outline and review.

15. (a) 45%
    (b) No, it is descriptive statistics.
    (c) statistical inference
    (d) 25%

CHAPTER TWO

DESCRIPTIVE STATISTICS I: TABULAR AND GRAPHICAL METHODS

CHAPTER OUTLINE AND REVIEW

In Chapter 1, you were introduced to the concept of statistics and were given a frequency distribution of the ages of 180 students at a local college, but you were not told how this frequency distribution was formulated. In Chapter 2 of your text, you were informed how such frequency distributions could be formulated and were introduced to several tabular and graphical procedures for summarizing data. Furthermore, you were shown how crosstabulations and scatter diagrams can be used to summarize data for two variables simultaneously. The terms which you should have learned from this chapter include:

A. **Qualitative Data:** Data which are measured by either nominal or ordinal scales of measurement. Each value serves as a name or label for identifying an item.

B. **Quantitative Data:** Data which are measured by interval or ratio scales of measurement. Quantitative data are numerical values on which mathematical operations can be performed.

C. **Bar Graph:** A graphical method of presenting qualitative data that have been summarized in a frequency distribution or a relative frequency distribution.

D. **Pie Chart:** A graphical device for presenting qualitative data by subdividing a circle into sectors which correspond to the relative frequency of each class.

E. **Frequency Distribution:** A tabular presentation of data, which shows the frequency of the appearance of data elements in several nonoverlapping classes. The purpose of the frequency distribution is to organize masses of data elements into smaller and more manageable groups. The frequency distribution can present both qualitative and quantitative data.

F. **Relative Frequency Distribution:** A tabular presentation of a set of data which shows the frequency of each class as a fraction of the total frequency. The relative frequency distribution can present both qualitative and quantitative data.

G. **Percent Frequency Distribution:** A tabular presentation of a set of data which shows the percentage of the total number of items in each class. The percent frequency of a class is simply the relative frequency multiplied by 100.

H. **Class:** A grouping of data elements in order to develop a frequency distribution.

I. **Class Width:** The length of the class interval. Each class has two limits. The lowest value is referred to as the lower class limit, and the highest value is the upper class limit. The difference between the upper and the lower class limits represents the class width.

J. **Class Midpoint:** The point in each class that is halfway between the lower and the upper class limits.

K. **Cumulative Frequency Distribution:** A tabular presentation of a set of quantitative data which shows for each class the total number of data elements with values less than the upper class limit.

L. **Cumulative Relative Frequency Distribution:** A tabular presentation of a set of quantitative data which shows for each class the fraction of the total frequency with values less than the upper class limit.

M.	**Cumulative Percent Frequency Distribution:**	A tabular presentation of a set of quantitative data which shows for each class the fraction of the total frequency with values less than the upper class limit.
N.	**Dot Plot:**	A graphical presentation of data, where the horizontal axis shows the range of data values and each observation is plotted as a dot above the axis.
O.	**Histogram:**	A graphical method of presenting a frequency or a relative frequency distribution.
P.	**Ogive:**	A graphical method of presenting a cumulative frequency distribution or a cumulative relative frequency distribution.
Q.	**Exploratory Data Analysis:**	The use of simple arithmetic and easy-to-draw pictures to look at data more effectively.
R.	**Stem-and-Leaf Display:**	An exploratory data analysis technique that simultaneously rank orders quantitative data and provides insight into the shape of the underlying distribution.
S.	**Crosstabulation:**	A tabular presentation of data for two variables. Rows and columns show the classes of categories for the two variables.
T.	**Scatter Diagram:**	A graphical method of presenting the relationship between two quantitative variables. One variable is shown on the horizontal and the other on the vertical axis.

In this chapter, you have also been informed about the role that computers play in statistical analysis. You have been shown how to use **Minitab** for your data analysis. At this point, I recommend that you ask your instructor or the staff at your computer center whether or not **Minitab** is available at your institution; and if it is available, how you can access it. If **Minitab** is not available, find out what other statistical software packages are available to you, and how you can sign on the computer and use the various statistical packages.

CHAPTER FORMULAS

$$\text{Relative Frequency of a Class} = \frac{\text{Frequency of the Class}}{n} \tag{2.1}$$

where n = total number of observations

$$\text{Approximate Class Width} = \frac{\text{Largest Data Value - Smallest Data Value}}{\text{Number of Classes}} \tag{2.2}$$

EXERCISES

***1.** A student has completed 20 courses in the School of Business Administration. Her grades in the 20 courses are shown below:

A	B	A	B	C
C	C	B	B	B
B	A	B	B	B
C	B	C	B	A

(a) Develop a frequency distribution for her grades.

Answer: To develop a frequency distribution we simply count her grades in each category. Thus, the frequency distribution of her grades can be presented as

Grade	Frequency
A	4
B	11
C	5
	20

(b) Develop a relative frequency distribution for her grades.

Answer: The relative frequency distribution is a distribution which shows the fraction or proportion of data items which fall in each category. The relative frequencies of each category can be computed by equation 2.1. Thus, the relative frequency distribution can be shown as follows:

Grade	Relative Frequency
A	$4/20 = 0.20$
B	$11/20 = 0.55$
C	$5/20 = 0.25$

(c) Develop a percent frequency distribution for her grades.

Answer: A percent frequency distribution is a tabular summary of a set of data showing the percent frequency for each class. The percent frequency of a class is simply the relative frequency multiplied by 100. Thus, we can multiply the relative frequencies which we found in Part b to arrive at the percent frequency distribution. Hence, the percent frequency distribution can be shown as follows:

Grade	Percent Frequency
A	20
B	55
C	25

(d) Develop a bar graph.

Answer: A bar graph is a graphical device for presenting the information of a frequency distribution for qualitative data. Bars of equal width are drawn to represent various classes (in this case, grades). The height of each bar represents the frequencies of various classes. Figure 2.1 shows the bar graph for the above data.

BAR GRAPH OF GRADES

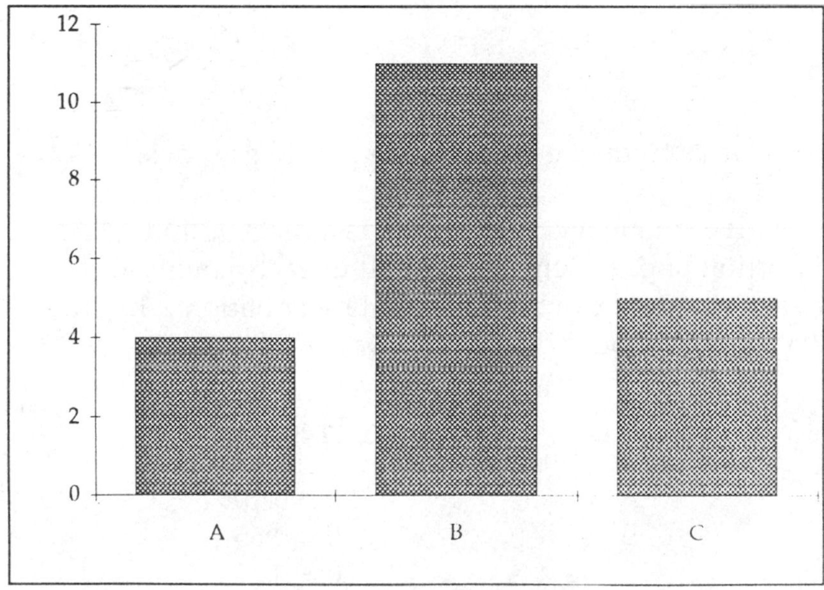

Figure 2.1

(e) Construct a pie chart

Answer: A pie chart is a pictorial device for presenting a relative frequency distribution of qualitative data. The relative frequency distribution is used to subdivide a circle into sections, where each section's size corresponds to the relative frequency of each class. Figure 2.2 shows the pie chart for the student's grades.

PIE CHART FOR GRADES

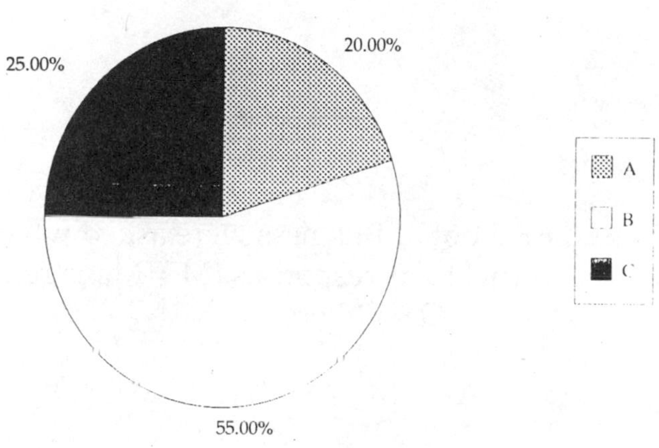

Figure 2.2

2. There are 800 students in the School of Business Administration at UTC. There are four majors in the school: Accounting, Finance, Management and Marketing. The following shows the number of students in each major:

Major	Number of Students
Accounting	240
Finance	160
Management	320
Marketing	80

(a) Develop a relative and a percent frequency distribution.

(b) Construct a bar chart.

(c) Construct a pie chart.

3. Thirty students in the School of Business were asked what their majors were. The following represents their responses (M = Management; A = Accounting; E = Economics; O = Others).

A	M	M	A	M	M	E	M	O	A
E	E	M	A	O	E	E	A	M	A
M	A	O	A	M	E	E	M	A	M

(a) Construct a frequency distribution.

(b) Construct a relative and a percent frequency distribution.

4. Twenty employees of ABC corporation were asked if they liked or disliked the new district manager. Below you are given their responses. Let L represent liked and D represent disliked.

L	L	D	L	D
D	D	L	L	D
D	L	D	D	L
D	D	D	D	L

(a) Construct a frequency distribution.

(b) Construct a relative and a percent frequency distribution.

5. Five hundred recent graduates indicated their majors as follows:

Major	Frequency
Accounting	60
Finance	100
Economics	40
Management	120
Marketing	80
Engineering	60
Computer Science	40
Total	500

(a) Construct a relative frequency distribution.

(b) Construct a percent frequency distribution.

*6. In a recent campaign, many airlines reduced their summer fares in order to gain a larger share of the market. The following data represent the prices of round-trip tickets from Atlanta to Boston for a sample of nine airlines:

120	140	140
160	160	160
160	180	180

Construct a dot plot for the above data.

Answer: The dot plot is one of the simplest graphical presentations of data. The horizontal axis shows the range of data values, and each observation is plotted as a dot above the axis. Figure 2.3 shows the dot plot for the above data. The four dots shown at the value of 160 indicate that four airlines were charging $160 for the round-trip ticket from Atlanta to Boston.

DOT PLOT FOR TICKET PRICES

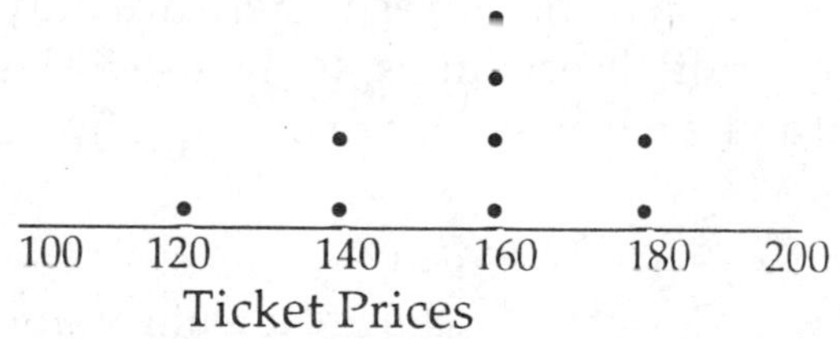

Figure 2.3

7. A sample of the ages of 10 employees of a company are shown below:

20	30	40	30	50
30	20	30	20	40

Construct a dot plot for the above data.

*8. The following data elements represent the amount of time (rounded to the nearest second) that 30 randomly selected customers spent in line before being served at a branch of First County Bank.

183	121	140	198	199
90	62	135	60	175
320	110	185	85	172
235	250	242	193	75
263	295	146	160	210
165	179	359	220	170

(a) Develop a frequency distribution for the above data.

Answer: The first step for developing a frequency distribution is to decide how many classes are needed. There are no "hard" rules for determining the number of classes; but generally, using anywhere from five to twenty classes is recommended, depending on the number of observations. Fewer classes are used when there are fewer observations, and more classes are used when there are numerous observations. In our case, there are only 30 observations. With such a limited number of observations, let us use 5 classes. The second step is to determine the width of each class. By using equation 2.2 which states

$$\text{Approximate Class Width} = \frac{\text{Largest Data Value - Smallest Data Value}}{\text{Number of Classes}}$$

we can determine the class width. In the above data set, the highest value is 359, and the lowest value is 60. Therefore,

$$\text{Approximate Class Width} = \frac{359 - 60}{5} = 59.8$$

We can adjust the above class width (59.8) and use a more convenient value of 60 for the development of the frequency distribution. Note that I decided to use five classes. If you had used 6 or 7 or any other reasonable number of classes, you would not have been wrong and would have had a frequency distribution with a different class width than the one shown above.

Now that we have decided on the number of classes and have determined the class width, we are ready to prepare a frequency distribution by simply counting the number of data items belonging to each class. For example, let us count the number of observations belonging to the 60 - 119 class. Six values of 60, 62, 75, 85, 980, and 110 belong to the class of 60 - 119. Thus, the frequency of this class is 6. Since we want to develop classes of equal width, the last class width is from 300 to 359.

THE FREQUENCY DISTRIBUTION OF WAITING TIMES AT FIRST COUNTY BANK

Waiting Times (Seconds)		Frequency
60 - 119		6
120 - 179		10
180 - 239		8
240 - 299		4
300 - 359		2
	Total	30

(b) What are the lower and the upper class limits for the first class of the above frequency distribution?

Answer: The lower class limit shows the smallest value which is included in a class. Therefore, the lower limit of the first class is 60. The upper class limit identifies the largest value included in a class. Thus, the upper limit of the first class is 119. (**Note**: The difference between the lower limits of adjacent classes provides the class width. Consider the lower class limits of the first two classes which are 60 and 120. We note that the class width is 120 - 60 = 60.)

(c) Develop a relative frequency distribution and a percent frequency distribution for the above.

Answer: The relative frequency for each class is determined by the use of equation 2.1.

$$\text{Relative Frequency of a Class} = \frac{\text{Frequency of the Class}}{n}$$

where n is the total number of observations. The percent frequency distribution is simply the relative frequencies multiplied by 100. Hence, the relative frequency distribution and the percent frequency distribution are developed as shown on the next page.

RELATIVE FREQUENCY AND PERCENT FREQUENCY DISTRIBUTIONS OF WAITING TIMES AT FIRST COUNTY BANK

Waiting Times (Seconds)	Frequency	Relative Frequency	Percent Frequency
60 - 119	6	6/30 = 0.2000	20.00
120 - 179	10	10/30 = 0.3333	33.33
180 - 239	8	8/30 = 0.2667	26.67
240 - 299	4	4/30 = 0.1333	13.33
300 - 359	2	2/30 = 0.0667	6.67
Total	30	1.0000	100.00

(d) Develop a cumulative frequency distribution.

Answer: The cumulative frequency distribution shows the number of data elements with values less than or equal to the upper limit of each class. For instance, the number of people who waited less than or equal to 179 seconds is 16 (6 + 10), and the number of people who waited less than or equal to 239 seconds is 24 (6 + 10 + 8). Therefore, the frequency and the cumulative frequency distributions for the above data will be as follows.

FREQUENCY AND CUMULATIVE FREQUENCY DISTRIBUTIONS FOR THE WAITING TIMES AT FIRST COUNTY BANK

Waiting Times (Seconds)	Frequency	Cumulative Frequency
60 - 119	6	6
120 - 179	10	16
180 - 239	8	24
240 - 299	4	28
300 - 359	2	30

(e) How many people waited less than or equal to 239 seconds?

Answer: The answer to this question is given in the table of the cumulative frequency. You can see that 24 people waited less than or equal to 239 seconds.

(f) Develop a cumulative relative frequency distribution and a cumulative percent frequency distribution.

Answer: The cumulative relative frequency distribution can be developed from the relative frequency distribution. It is a table which shows the fraction of data elements with values less than or equal to the upper limit of each class. Using the table of relative frequency, we can develop the cumulative relative and the cumulative percent frequency distributions as follows:

RELATIVE FREQUENCY AND CUMULATIVE RELATIVE FREQUENCY AND CUMULATIVE PERCENT FREQUENCY DISTRIBUTIONS OF WAITING TIMES AT FIRST COUNTY BANK

Waiting Times (Seconds)	Relative Frequency	Cumulative Relative Frequency	Cumulative Percent Frequency
60 - 119	0.2000	0.2000	20.00
120 - 179	0.3333	0.5333	53.33
180 - 239	0.2667	0.8000	80.00
240 - 299	0.1333	0.9333	93.33
300 - 359	0.0667	1.0000	100.00

NOTE: To develop the cumulative relative frequency distribution, we could have used the cumulative frequency distribution and divided all the cumulative frequencies by the total number of observations, that is, 30.

(g) Construct a histogram for the waiting times in the above example.

Answer: One of the most common graphical presentations of data sets is a histogram. We can construct a histogram by measuring the class intervals on the horizontal axis and the frequencies on the vertical axis. Then we can plot bars with the widths equal to the class intervals and the height equivalent to the frequency of the class which they represent. In Figure 2.4, the histogram of the waiting times is presented. As you note, the width of each bar is equal to the width of the various classes (60 seconds), and the height represents the frequency of the various classes. Note that the first class ends at 119; the next class begins at 120, and one unit exists between these two classes (and all other classes). To eliminate these spaces, the vertical lines are drawn halfway between the class limits. Thus, the vertical lines are drawn at 59.5, 119.5, 179.5, 239.5, 299.5, and 359.5.

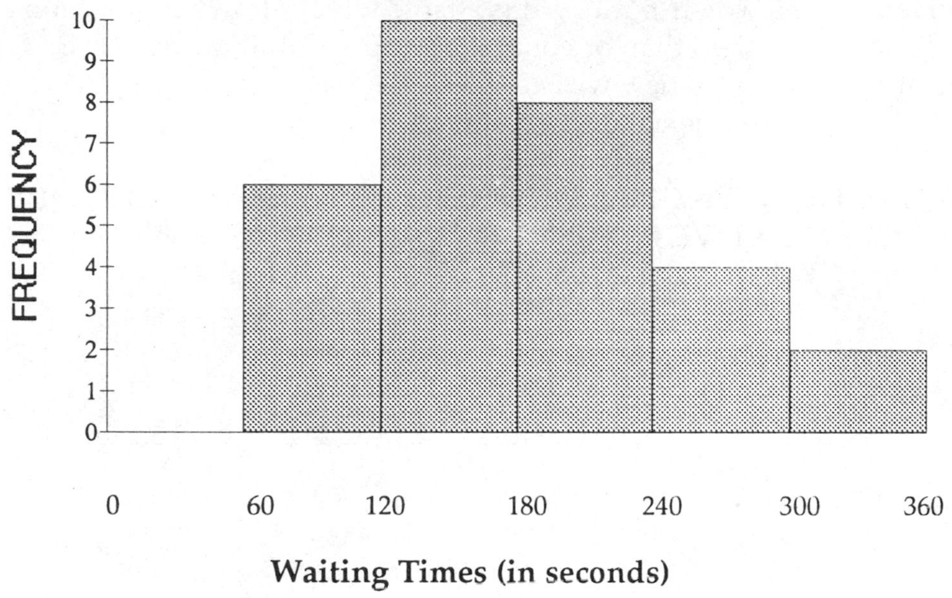

Figure 2.4

(h) Construct an ogive for the above example.

Answer: An ogive is a graphical representation of the cumulative frequency distribution or cumulative relative frequency distribution. It is constructed by measuring the class intervals on the horizontal axis and the cumulative frequencies (or cumulative relative frequencies) on the vertical axis. Then points are plotted halfway between class limits (i.e., 119.5, 179.5, 239.5, etc.) at a height equal to the cumulative frequency (or cumulative relative frequency). One additional point is plotted at 59.5 on the horizontal axis and 0 on the vertical axis. This point shows that there are no data values below the 60 - 119 class. Finally, these points are connected by straight lines. The result is an ogive which is shown in Figure 2.5.

OGIVE FOR THE CUMULATIVE FREQUENCY DISTRIBUTION OF THE WAITING TIMES AT FIRST COUNTY BANK
Waiting Times (in seconds)

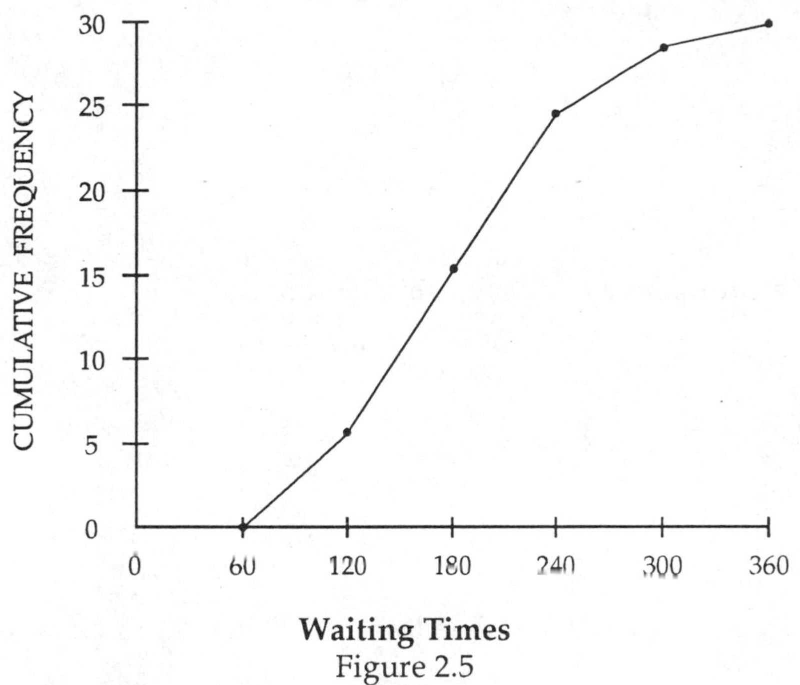

Figure 2.5

9. The following data set shows the number of hours of sick leave that some of the employees of Bastien's, Inc. have taken during the first quarter of the year (rounded to the nearest hour).

19	22	27	24	28	12
23	47	11	55	25	42
36	25	34	16	45	49
12	20	28	29	21	10
59	39	48	32	40	31

(a) Develop a frequency distribution for the above data. (Let the width of your classes be 10 units and start your first class as 10 - 19.)

(b) Develop a relative frequency distribution and a percent frequency distribution for the data.

(c) Develop a cumulative frequency distribution.

(d) How many employees have taken less than 40 hours of sick leave?

10. The grades of 20 students on their first statistics test are shown below:

71	52	66	76	78
71	68	55	77	91
72	75	78	62	93
82	85	87	98	65

(a) Develop a frequency distribution for the grades. (Let your first class be 50 - 59.)

(b) Develop a percent frequency distribution.

11. The sales record of a real estate company for the month of May shows the following house prices (rounded to the nearest $1,000). Values are in thousands of dollars.

105	55	45	85	75
30	60	75	79	95

(a) Develop a frequency distribution and a percent frequency distribution for the house prices. (Use 5 classes and have your first class be 20 - 39.)

(b) Develop a cumulative frequency and a cumulative percent frequency distribution for the above data.

(c) What percentage of the houses sold at a price below $80,000?

12. The hourly wages of 12 employees are shown below:

7	8	14	17
17	15	10	20
24	10	21	25

(a) Develop a frequency distribution. (Let your first class be 7 - 11.)

(b) Develop a cumulative frequency distribution.

13. A group of freshmen at an area university decided to sell magazine subscriptions to help pay for their Christmas party. Below is a list of the 18 students who participated and the number of subscriptions they sold.

Student	# of Subscriptions Sold
1	30
2	79
3	59
4	65
5	40
6	64
7	52
8	53
9	57
10	39
11	61
12	47
13	50
14	60
15	48
16	50
17	58
18	67

(a) Develop a frequency distribution and a percent frequency distribution. (Let the width of your classes be 10 units.)

(b) Develop a cumulative frequency and a cumulative percent frequency distribution.

(c) How many students sold less than 60 subscriptions?

14. The temperatures for the month of June in a midwestern city were recorded as follows: (Rounded to the nearest degree)

70	75	79	80	78	82
82	89	88	87	90	92
91	92	93	95	94	95
97	95	98	100	107	107
105	104	108	111	109	116

(a) Develop a frequency distribution and a relative frequency distribution for the above data. (Let the width of your classes be 10 units.)

(b) Develop a cumulative frequency and a cumulative relative frequency distribution.

(c) How many days was the temperature at least 90 degrees?

15. The frequency distribution below shows the yearly income distribution of a sample of 160 Kern County residents:

Yearly Income (in thousands of dollars)	Frequency
10 - 14	10
15 - 19	25
20 - 24	30
25 - 29	40
30 - 34	35
35 - 39	20
Total	160

(a) What percentage of the individuals in the sample had incomes of less than $25,000?

(b) How many individuals had incomes of at least $30,000?

16. The Alex Food Company bakes quiches and sells their products in the greater Los Angeles area. Their records over the past 60 days are shown below:

Sales Volume (Number of Quiches)	Number of Days
100 - 199	6
200 - 299	10
300 - 399	20
400 - 499	12
500 - 599	8
600 - 699	4
Total	60

(a) Develop a cumulative frequency distribution and a percent frequency distribution.

(b) What percentage of the days did the company sell at least 400 quiches?

17. The ages of 16 employees are shown below:

22	40	34	36
35	27	30	32
39	46	32	48
45	36	41	41

(a) Develop a frequency distribution. Let your first class be 20 - 25.

(b) Develop a cumulative frequency distribution.

18. For the following distribution, develop a percent frequency distribution.

Class	Frequency	Percent Frequency
40 - 59	48	
60 - 79	110	
80 - 99	82	
100 - 119	70	
120 - 139	58	
140 - 159	20	
160 - 179	12	

19. The frequency distribution below was constructed from data collected from a group of 250 students.

Height in Inches	Frequency
58 - 63	30
64 - 69	50
70 - 75	20
76 - 81	60
82 - 87	40
88 - 93	30
94 - 99	20
Total	250

(a) Construct a percent frequency distribution.

(b) Construct a cumulative frequency distribution.

(c) Construct a cumulative percent frequency distribution.

***20.** The test scores of 14 individuals on their first statistics examination are shown below:

95	87	52	43	77	84	78
75	63	92	81	83	91	88

a) Construct a stem-and-leaf display for these data.

Answer: To construct a stem-and-leaf display, the first digit of each data item is arranged in an ascending order and written to the left of a vertical line. Then, the second digit of each data item is written to the right of the vertical line next to its corresponding first digit as follows:

```
4 | 3
5 | 2
6 | 3
7 | 7 8 5
8 | 7 4 1 3 8
9 | 5 2 1
```

Now, the second digits are rank ordered horizontally, thus leading to the following stem-and-leaf display:

```
4  3
5  2
6  3
7  5 7 8
8  1 3 4 7 8
9  1 2 5
```

b) What does the above stem-and-leaf show?

Answer: Each line in the above display is called a stem, and each piece of information on a stem is a leaf. For instance, let us consider the fourth line:

7 5 7 8

The stem indicates that there are 3 scores in the seventies. These values are

75 77 78

Similarly, we can look at line five (where the first digit is 8) and see

8 1 3 4 7 8

This stem indicates that there are 5 scores in the eighties, and they are

81 83 84 87 88

At a glance, one can see the overall distribution for the grades. There is one score in the forties (43), one score in the fifties (52), one score in the sixties (63), three scores in the seventies (75, 77, 78), five scores in the eighties (81, 83, 84, 87, 88), and three scores in the nineties (91, 92, 95).

21. Construct a stem-and-leaf display for the following data:

22	44	36	45	49	57	38	47	51	12
18	48	32	19	43	31	26	40	37	52

***22.** The following is a crosstabulation of starting salaries (in $1,000's) of a sample of business school graduates by their gender.

Starting Salary

Gender	Less than 20	20 up to 25	25 and more	Total
Female	12	84	24	120
Male	20	48	12	80
Total	32	132	36	200

(a) What general comments can be made about the distribution of starting salaries and the gender of the individuals in the sample?

Answer: Using the frequency distribution at the bottom margin of the above table it is noted that majority of the individuals in the sample (132) have starting salaries in the range of $20,000 up to $25,000, followed by 36 individuals whose salaries are at least $25,000, and only 32 individuals had starting salaries of under $20,000. Now considering the right-hand margin it is noted that the majority of the individuals in the sample (120) are female while 80 are male.

(b) Compute row percentages and comment on the relationship between starting salaries and gender.

Answer: To compute the row percentages we divide the values of each cell by the row total and express the results as percentages. Let us consider the row representing females. The row percentages (across) are computed as
(12/120)(100)=10%; (84/120)(100)=70%; (24/120)(100)=20%
Continuing in the same manner and computing the row percentages for the other row we determine the following row percentages table:

Starting Salary

Gender	Less than 20	20 up to 25	25 and more	Total
Female	10%	70%	20%	100%
Male	25%	60%	15%	100%

From the above percentages it can be noted that the largest percentage of both genders' starting salaries are in the $20,000 to $25,000 range. However, 70% of females and only 60% of the males have starting salaries in this range. Also it can be noted that 10% of females' starting salaries are under $20,000, whereas, 25% of the males' starting salaries fall in this category.

(c) Compute column percentages and comment on the relationship between gender and starting salaries.

Column percentages are computed by dividing the values in each cell by column total and expressing the results as percentages. For instance for the category of "Less than 20" the column percentages are computed as (12/32)(100)=37.5 and (20/32)(100)=62.5 (rounded). Continuing in the same manner the column percentages will be as follows:

	Starting Salary		
Gender	Less than 20	20 up to 25	25 and more
Female	37.5%	63.6%	66.7%
Male	62.5%	36.4%	33.3%
Total	100%	100%	100%

Considering the "Less than 20" category it is noted that the majority (62.5%) are male. In the next category of "20 up to 25" the majority (63.6%) are female. Finally in the last category of "25 and more" the majority (66.7%) are female.

23. A survey of 400 college seniors resulted in the following crosstabulation regarding their undergraduate major and whether or not they plan to go to graduate school:

	Undergraduate Major			
Graduate School	Business	Engineering	Others	Total
Yes	35	42	63	140
No	91	104	65	260
Total	126	146	128	400

(a) Are majority of seniors in the survey planning to attend graduate school?

(b) Which discipline constitutes the majority of the individuals in the survey?

(c) Compute row percentages and comment on the relationship between the students' undergraduate major and their intention of attending graduate school.

(d) Compute the column percentages and comment on the relationship between the students' intention of going to graduate school and their undergraduate major.

***24.** The average grades of 8 students in statistics and the number of absences they had during the semester are shown below:

Student	Number of Absences (x)	Average Grade (y)
1	1	94
2	2	78
3	2	70
4	1	88
5	3	68
6	4	40
7	8	30
8	3	60

Develop a scatter diagram for the relationship between the number of absences (x) and their average grade (y).

Answer: A scatter diagram is a graphical method of presenting the relationship between two variables. The scatter diagram is shown in Figure 2.6. The number of absences (x) is shown in the horizontal axis and the average grade (y) on the vertical axis. The first student has one absence (x=1) and an average grade of 94 (y=94). Therefore, a point with coordinates of x=1 and y=94 is plotted on the scatter diagram. In a similar manner all other points for all 8 students are plotted.

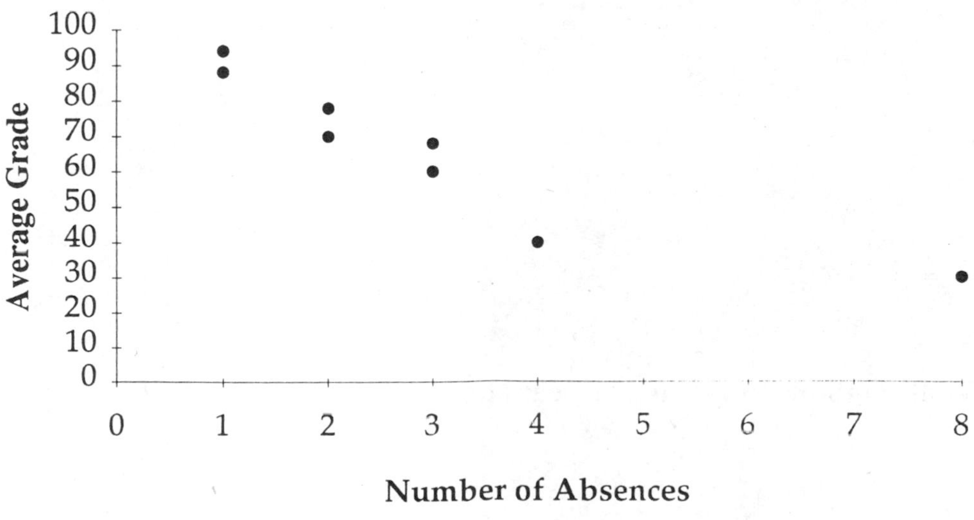

Figure 2.6

The scatter diagram shows that there is a negative relationship between the number of absences and the average grade. That is, the higher the number of absences, the lower the average grade appears to be.

25. You are given the following ten observations on two variables, x and y.

x	y
1	8
5	15
6	20
4	12
2	10
8	20
9	26
1	5
6	18
8	26

(a) Develop a scatter diagram for the relationship between x and y.

(b) What relationship, if any, appears to exist between x and y?

SELF-TESTING QUESTIONS

In the following multiple choice questions, circle the correct answer. An answer key is provided following the questions.

1. A tabular summary of a set of data, which shows the frequency of the appearance of data elements in several nonoverlapping classes is termed

a) the class width
b) a frequency polygon
c) a frequency distribution
d) a histogram
e) none of the above

2. A tabular summary of a set of data showing classes of the data and the fraction of the items belonging to each class is called

a) the class width
b) a relative frequency distribution
c) a cumulative relative frequency distribution
d) an ogive
e) none of the above

3. A histogram is

a) a graphical presentation of a frequency or relative frequency distribution
b) a graphical method of presenting a cumulative frequency or a cumulative relative frequency distribution
c) the history of data elements
d) none of the above

4. The length of the interval forming a class is called

a) the class midpoint
b) the lower class limit
c) the upper class limit
d) the class width
e) none of the above

5. A graphical method of presenting qualitative data by frequency distribution is termed

a) a frequency polygon
b) an ogive
c) a bar graph
d) none of the above

Answer questions 6 through 10 based on the following problem.

Michael's Rent-A-Car, a national car rental company, has kept a record of the number of cars they have rented for a period of 80 days. Their rental records are shown below:

Number of Cars Rented	Number of Days
0 - 19	5
20 - 39	15
40 - 59	30
60 - 79	20
80 - 99	10
Total	80

6. The class width of the above distribution is

a) 0 to 100
b) 20
c) 80
d) 5
e) none of the above

7. The lower limit of the first class is

a) 5
b) 80
c) 0
d) 20
e) none of the above

8. If one develops a cumulative frequency distribution for the above data, the last class will have a frequency of

a) 10
b) 100
c) 0 to 100
d) 80
e) none of the above

9. The percentage of days in which the company rented at least 40 cars is

a) 37.5%
b) 62.5%
c) 90.0%
d) 75.0%
e) 0.0%

10 The number of days in which the company rented less than 60 cars is

a) 30
b) 15
c) 60
d) 80
e) none of the above

11. The sum of frequencies for all classes will always equal

a) 1
b) the number of elements in a data set
c) the number of classes
d) a value between 0 to 1
e) none of the above

12. The relative frequency of a class is computed by

a) dividing the midpoint of the class by the sample size
b) dividing the frequency of the class by the midpoint
c) dividing the sample size by the frequency of the class
d) dividing the frequency of the class by the sample size
e) none of the above

13. If several frequency distributions are constructed from the same data set, the distribution with the narrowest class width will have the

a) fewest classes
b) most classes
c) same number of classes as the other distributions since all are constructed from the same data
d) none of the above

14. The sum of the relative frequencies for all classes will always equal

a) the sample size
b) the number of classes
c) one
d) any value larger than one
e) none of the above

15. In a cumulative relative frequency distribution, the last class will have a cumulative relative frequency equal to

a) one
b) zero
c) the total number of elements in the data set
d) none of the above

16. In a cumulative percent frequency distribution, the last class will have a cumulative percent frequency equal to

a) one
b) zero
c) the total number of elements in the data set
d) 100
e) none of the above

17. A tabular method which can be used to summarize the data on two variables simultaneously is called

a) simultaneous equations
b) an ogive
c) a histogram
d) crosstabulation
e) none of the above

ANSWERS TO THE SELF-TESTING QUESTIONS

1. c
2. b
3. a
4. d
5. c
6. b
7. c
8. d
9. d
10. e
11. b
12. d
13. b
14. c
15. a
16. d
17. d

ANSWERS TO CHAPTER TWO EXERCISES

2. (a)

Major	Relative Frequency	Percent Frequency
Accounting	0.3	30
Finance	0.2	20
Management	0.4	40
Marketing	0.1	10

(b)

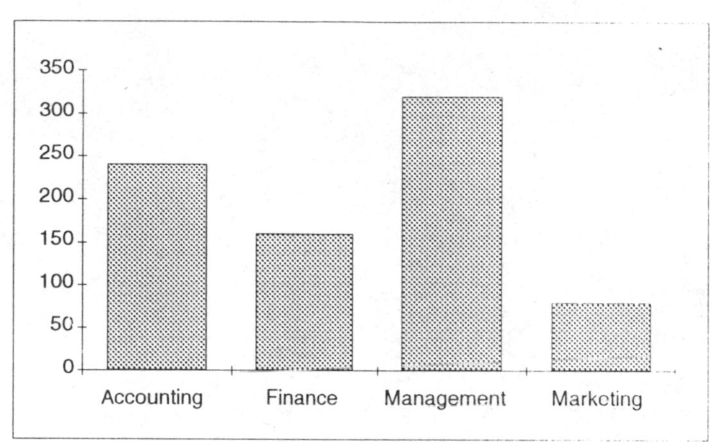

(c)

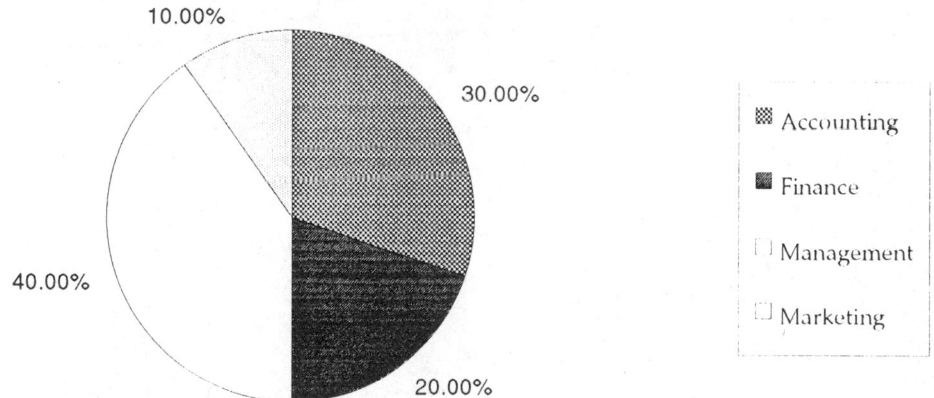

3. (a) and (b)

Major	Frequency	Relative Frequency	Percent Frequency
M	11	0.37	37
A	9	0.30	30
E	7	0.23	23
O	3	0.10	10

4. (a) and (b)

Preferences	Frequency	Relative Frequency	Percent Frequency
L	8	0.4	40
D	12	0.6	60
Total	20	1.0	100

5.

Major	Frequency	(a) Relative Frequency	(b) Percent Frequency
Accounting	60	0.12	12
Finance	100	0.20	20
Economics	40	0.08	8
Management	120	0.24	24
Marketing	80	0.16	16
Engineering	60	0.12	12
Computer Science	40	0.08	8
Total	500	1.00	100

7.

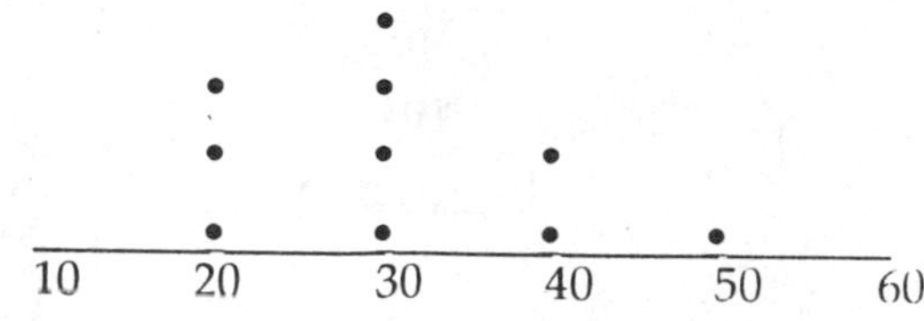

9.

Hours of Sick Leave Taken	(a) Freq.	(b) Relative Freq.	(b) Percent Freq.	(c) Cum. Freq.
10 - 19	6	0.20	20	6
20 - 29	11	0.37	37	17
30 - 39	5	0.16	16	22
40 - 49	6	0.20	20	28
50 - 59	2	0.07	7	30

(d) 22

10.

Class	(a) Frequency	(b) Percent Frequency
50 - 59	2	10
60 - 69	4	20
70 - 79	8	40
80 - 89	3	15
90 - 99	3	15
	20	100

11.

Sales Price (In Thousands of Dollars)	(a) Freq.	(a) Percent Freq.	(b) Cum. Freq.	(b) Cum. Percent Freq.
20 - 39	1	10	1	10
40 - 59	2	20	3	30
60 - 79	4	40	7	70
80 - 99	2	20	9	90
100 - 119	1	10	10	100

(c) 70%

12.

Class	(a) Frequency	(b) Cumulative Frequency
7 - 11	4	4
12 - 16	2	6
17 - 23	4	10
22 - 26	2	12
	12	

13.

No. of Subscriptions Sold	(a) Freq.	(a) Percent Freq.	(b) Cum. Freq.	(b) Cum. Percent Freq.
30 - 39	2	11	2	11
40 - 49	3	17	5	28
50 - 59	7	39	12	67
60 - 69	5	28	17	95
70 - 79	1	5	18	100

(c) 12

14.

Temperature	(a) Freq.	(a) Rel. Freq.	(b) Cum. Freq.	(b) Cum. Rel. Freq.
70 - 79	4	0.13	4	0.13
80 - 89	6	0.20	10	0.33
90 - 99	11	0.37	21	0.70
100 - 109	7	0.23	28	0.93
110 - 119	2	0.07	30	1.00

(c) 20

15. (a) 40.625%
(b) 55

16.

Sales Volume	(a) Cumulative Frequency	(a) Percent Frequency
100 - 199	6	10
200 - 299	16	17
300 - 399	36	33
400 - 499	48	20
500 - 599	56	13
600 - 699	60	7

(b) 40%

17.

Class	(a) Frequency	(b) Cumulative Frequency
20 - 25	1	1
26 - 31	2	3
32 - 37	6	9
38 - 43	4	13
44 - 49	3	16

18.

Class	Percent Frequency
40 - 59	0.120
60 - 79	0.275
80 - 99	0.205
100 - 119	0.175
120 - 139	0.145
140 - 159	0.050
160 - 179	0.030

19.

Height (In Inches)	Frequency	(a) Percent Frequency	(b) Cumulative Frequency	(c) Cumulative Percent Frequency
58 - 63	30	12	30	12
64 - 69	50	20	80	32
70 - 75	20	8	100	40
76 - 81	60	24	160	64
82 - 87	40	16	200	80
88 - 93	30	12	230	92
94 - 99	20	8	250	100
		100		

21.
```
1 2 8 9
2 2 6
3 1 2 6 7 8
4 0 3 4 5 7 8 9
5 1 2 7
```

23. (a) No, majority (260) will not attend graduate school
(b) Majority (146) are engineering major
(c)

	Undergraduate Major			
Graduate School	Business	Engineering	Others	Total
Yes	25%	30%	45%	100%
No	35%	40%	25%	100%

Majority who plan to go to graduate school are from "Other" majors. Majority of those who will not go to graduate school are engineering majors.

(d)

	Undergraduate Major		
Graduate School	Business	Engineering	Others
Yes	27.8%	28.8%	49.2%
No	72.2%	71.2%	50.8%
Total	100%	100%	100%

Approximately the same percentages of Business and engineering majors plan to attend graduate school (27.8% and 28.8% respectively). Of the "Other" majors approximately half (49.2%) plan to go to graduate school.

25. (a)

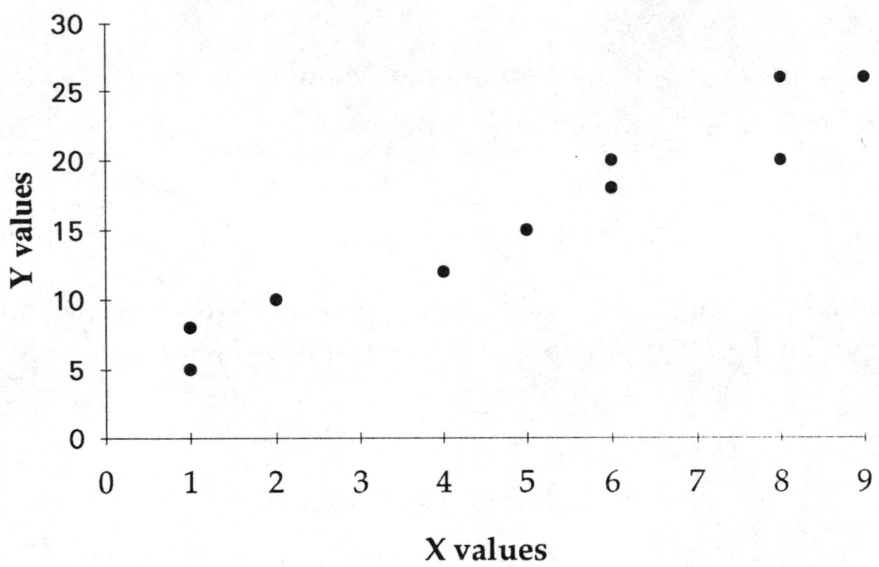

(b) A positive relationship between x and y appears to exist.

CHAPTER THREE

DESCRIPTIVE STATISTICS II: NUMERICAL METHODS

CHAPTER OUTLINE AND REVIEW

In this chapter, you have been introduced to various numerical measures of location and dispersion, both for ungrouped and grouped data. Furthermore, you have also been introduced to measures of association between two variables. More specifically, you have been introduced to the following measures and key concepts:

I. Measures of Location

A. Mean: The average value of a data set.

B. Trimmed Mean: The average value of a data set after a specified percentage of extreme values have been eliminated.

C. Median: The value of the middle item of a set of data, after the data set has been arranged in ascending (or descending) order. Therefore, the median is the value which divides the data set into two equal groups.

D. Mode: The most frequently occurring or the most common data value in a data set.

E. Percentiles: As the median divides the data set into two equal groups, the percentile divides the data set into one hundredths. Therefore, the *p*th percentile of a data set is a value such that at least *p* percent of the items take on this value or less (when the data set is arranged in ascending order). Hence, the median is the 50th percentile.

F. Quartiles: As the percentiles divide the data set into one hundredths, the quartiles divide the data set into four parts. The first quartile (Q_1), second quartile (Q_2), and third quartile (Q_3) are actually the 25th, 50th, and 75th percentiles.

G. Hinges: The lower and upper hinges are basically the 25th and 75th percentiles, respectively. The lower hinge is the median of the data elements in positions less than or equal to the median position of the entire data set. The upper hinge is the median of the data elements in positions more than or equal to the median position of the entire data set.

II. Measures of Dispersion

A. Range: The difference between the largest and the smallest values in a data set.

B. Interquartile Range: A measure of dispersion whose value is equal to the difference between the third and first quartiles ($IQR = Q_3 - Q_1$).

C. Variance: A measure of variability or scatterness of the data elements around their mean. Its value is equal to the average of the squared deviation of data elements about their mean (for the population).

D. Standard Deviation: A measure of absolute dispersion of the data elements around the mean whose value is equal to the square root of the variance.

E. Coefficient of Variation: A measure of relative dispersion whose value is the ratio of the standard deviation to the mean multiplied by 100. The coefficient of variation is a measure which is used for comparing two data sets and for determining which data set is more dispersed (relative to the average size of data values).

Besides the above measures, you should thoroughly understand the meaning of the following key concepts.

III. Other Key Concepts

A. Population Parameter: A descriptive measure of a population.

B. Sample Statistic: A descriptive measure of a sample.

C. Grouped Data: When the individual data items (observations) are summarized by means of a frequency distribution, the result is referred to as grouped data. It must be noted that when the data is grouped and is reported by means of a frequency distribution, the exact values of the individual observations are no longer known.

D. Five-Number Summary: An exploratory data analysis technique which uses the following five measures to summarize the data set: lowest value, first quartile, median, third quartile, and largest value.

E. Box Plot: A visual presentation of a five-number summary.

F. Z-Score: The number of standard deviations a data value is from the mean. It is referred to as either a standardized value or the Z - score.

G. **Chebyshev's Theorem:** A theorem which allows us to determine the percentage of data elements which fall within a specified number of standard deviations from the mean. More specifically, Chebyshev's Theorem states that at least $(1-1/k^2)$ of the data elements fall within plus or minus k standard deviations (for $k > 1$) of the mean.

H. **Outlier:** An unusually small or large data value.

I. **Fences:** Values which are used to identify outliers.

J. **Empirical Rule:** A rule applied to *bell-shaped* distributions which states what percentage of the items are within 1, 2, and 3 standard deviations of the mean. For data having a bell-shaped distribution, approximately:

- 68% of the data elements will be within plus or minus 1 standard deviation from the mean.
- 95% of the data elements will be within plus or minus 2 standard deviations from the mean.
- Almost all the items will be within 3 standard deviations from the mean.

K. **Covariance:** A numerical measure of linear association between two variables. Positive values indicate a positive relationship and negative values show a negative relationship.

L. **Correlation Coefficient:** A numerical measure of linear association between two variables. The coefficient of correlation ranges between -1 to +1. Values close to -1 indicate a strong negative linear relationship and values close to +1 indicate a strong positive correlation.

CHAPTER FORMULAS

Ungrouped Data

SAMPLE | *POPULATION*

Mean

$$\overline{X} = \frac{\Sigma X_i}{n} \qquad (3.1) \qquad\qquad \mu = \frac{\Sigma X_i}{N} \qquad (3.2)$$

where n = sample size | where N = size of population

Interquartile Range

$IQR = Q_3 - Q_1$ (3.3) (Same as for sample)

where: Q_3 = third quartile (i.e., 75th percentile)
Q_1 = first quartile (i.e., 25th percentile)

Variance

$$S^2 = \frac{\Sigma\left(X_i - \overline{X}\right)^2}{n-1} \qquad (3.5) \qquad\qquad \sigma^2 = \frac{\Sigma(X_i - \mu)^2}{N} \qquad (3.4)$$

or: | or:

$$S^2 = \frac{\Sigma X_i^2 - n\overline{X}^2}{n-1} \qquad\qquad \sigma^2 = \frac{\Sigma X_i^2 - N\mu^2}{N}$$

Standard Deviation

$$S = \sqrt{S^2} \qquad (3.6) \qquad\qquad \sigma = \sqrt{\sigma^2} \qquad (3.7)$$

Coefficient of Variation (C.V.)

$$C.V. = \left(\frac{S}{\overline{X}}\right)(100) \qquad\qquad C.V. = \left(\frac{\sigma}{\mu}\right)(100) \qquad (3.8)$$

CHAPTER FORMULAS
(continued)

SAMPLE *POPULATION*

Z-Score

$$Z_i = \frac{X_i - \overline{X}}{S} \qquad Z_i = \frac{X_i - \mu}{\sigma} \qquad (3.9)$$

where Z_i = number of standard deviations X_i is from the mean

Covariance

$$s_{xy} = \frac{\Sigma(X_i - \overline{X})(Y_i - \overline{Y})}{n-1} \qquad (3.10) \qquad \sigma_{XY} = \frac{\Sigma(X_i - \mu_X)(Y_i - \mu_Y)}{N} \qquad (3.11)$$

Pearson Product Moment Correlation Coefficient: Sample Data

$$r_{XY} = \frac{S_{XY}}{S_X S_Y} \qquad (3.12) \qquad \rho_{XY} = \frac{\sigma_{XY}}{\sigma_X \sigma_Y} \qquad (3.14)$$

where

r_{XY} = Sample correlation coefficient

S_{XY} = Sample covariance

S_X = Sample standard deviation of X

S_Y = Sample standard deviation of Y

where

ρ_{XY} = Population correlation coefficient

σ_{XY} = Population covariance

σ_X = Population standard deviation of X

σ_Y = Population standard deviation of Y

r_{XY} can also be computed by the alternate formula as:

$$r_{XY} = \frac{\Sigma X_i Y_i - (\Sigma X_i \, \Sigma Y_i)/n}{\sqrt{\Sigma X_i^2 - (\Sigma X_i)^2/n}\sqrt{\Sigma Y_i^2 - (\Sigma Y_i)^2/n}} \qquad (3.13)$$

CHAPTER FORMULAS
(continued)

Grouped Data

SAMPLE *POPULATION*

Mean

$$\overline{X} = \frac{\Sigma f_i M_i}{n} \quad (3.15) \qquad \mu = \frac{\Sigma f_i M_i}{N} \quad (3.17)$$

where

f_i = frequency of class i

M_i = midpoint of class i

Variance

$$S^2 = \frac{\Sigma f_i (M_i - \overline{X})^2}{n-1} \quad (3.16) \qquad \sigma^2 = \frac{\Sigma f_i (M_i - \mu)^2}{N} \quad (3.18)$$

or or

$$S^2 = \frac{\Sigma f_i M_i^2 - n\overline{X}^2}{n-1} \qquad \sigma^2 = \frac{\Sigma f_i M_i^2 - N\mu^2}{N}$$

EXERCISES

***1.** A sample of 9 gasoline stations in Chattanooga had the following prices for a gallon of gas:

Gas Station #	Price Per Gallon (X)
1	$ 1.14
2	1.19
3	1.25
4	1.21
5	1.17
6	1.19
7	1.22
8	1.24
9	1.19
	$\Sigma X_i = 10.80$

a) What is the mean price per gallon among the 9 stations?

Answer: The mean refers to the average price, and the average is determined by adding the prices of all nine stations and dividing the sum by 9 (the number of gas stations).

$$\overline{X} = \frac{\Sigma X_i}{n}$$

The $\Sigma X_i = \$10.80$, as shown above on the bottom of column two. Therefore, the mean will be

$$\overline{X} = \frac{\Sigma X_i}{n} = \frac{10.80}{9} = \$1.20$$

Had the above data represented a population, the calculations would have been the same only we would have used the following notations:

$$\mu = \frac{\Sigma X_i}{N}$$

(b) If gas station number one's price was $2.50 per gallon, could a different mean be computed so that it would be a better indication of the central location of data?

Answer: When there are unusually large or unusually small data values in a data set, we can trim the data by dropping the extreme values and then computing the *trimmed mean*. In this example, we would drop the gas station with the unusually high price of $2.50 per gallon and compute the *trimmed mean* for the remaining eight gas stations. Thus the trimmed mean would be the average price for the last eight gas stations which is $1.19 (rounded).

(c) What is the mode for the above data?

Answer: The mode is the most frequently occurring item in a data set. In the above example, the mode is $1.19 which represents the most common price. Three gas stations had a price of $1.19 per gallon.

(d) What is the median of the above data?

Answer: The median is the middle value of a data set, after the data has been arranged in ascending order. Hence, to determine the median, the data is first arranged in ascending order as follows:

X_i	
$1.14	
1.17	
1.19	
1.19	
1.19	median
1.21	
1.22	
1.24	
1.25	

The data element which is located in the middle of the data set is the median. As shown above, the median for this example is $1.19. It must be noted that in the above example the number of data points was odd. (There were 9 gas stations.) If the data set had an even number of items, then the median would have been the average of the middle two items. Assume we had a sample of 10 gas stations with the following prices (after the prices have been arranged in ascending order):

X_i	
$1.14	
1.17	
1.19	
1.19	
1.19	middle 2 values
1.21	
1.21	
1.22	
1.24	
1.25	

Then the median would be the average of the middle 2 values or

$$\text{median} = \frac{1.19 + 1.21}{2} = \$1.20$$

The median is also referred to as the second quartile (Q_2) or the 50th percentile.

(e) Assume in the above example that gas station number one's price was $1.95 per gallon. Then would the mean or the median be a better measure of location?

Answer: Since the mean is influenced by extreme values and under the above assumption one gas station has an extreme price of $1.95, the median would be a better measure of location. Note that with the extreme price of $1.95 the mean changes from $1.20 to $1.368, while the median does not change and is still $1.20.

(f) Compute the 25th percentile (i.e., the first quartile).

Answer: The 25th percentile for the above example is that price where at least 25% of the gas stations charge equal to or less than that price. To determine that price, we do the following step-by-step procedure.

STEP 1: Arrange the prices of the 9 gas stations in ascending order:

1.14, 1.17, 1.19, 1.19, 1.19, 1.21, 1.22, 1.24, 1.25

STEP 2: Determine the position of the 25th percentile by computing an index "i" as follows:

i = (p/100) (n)

In the above formula, p represents the percentile of interest (in this case 25) an n is the number of data values (in this case 9). Thus, the index i is computed as follows:

i = (25/100) (9) = 2.25

STEP 3: If i is not an integer value, the next integer value greater than i will represent the position of the percentile of interest. In this exercise, i is not an integer value (2.25). Therefore, the position of the 25th percentile is the next integer value greater than 2.25 or the 3rd position. Now considering the data elements after they were arranged in ascending order (see Step 1), we note that the 3rd position has a value of $1.19 (indicated by an asterisk * on the previous page), which is the 25th percentile. This indicates that at least 25% of gas stations have prices of $1.19 or less.

Note: In Step 3, above, if i was an integer, the pth percentile would have been the average of the data values in positions i and i+1.

(g) Compute the 75th percentile (i.e., the third quartile).

Answer: The 75th percentile or the third quartile (Q_3) for the above example is that price where at least 75% of the gas stations charge equal to or less than that price. The procedure for computing the 75th percentile is similar to that of computing the 25th percentile (as shown in Part f).

STEP 1: Arrange the prices of the 9 gas stations in ascending order:

                                                  *
1.14, 1.17, 1.19, 1.19, 1.19, 1.21, 1.22, 1.24, 1.25

STEP 2: Determine the position of the 75th percentile by computing an index "i" as follows:

i = (75/100) (9) = 6.75

STEP 3: In this exercise, i is not an integer value (6.75). Therefore, the position of the 75th percentile is the next integer value greater than 6.75 or the 7th position. Now considering the data elements after they were arranged in ascending order (see Step 1), we note that the 7th position has a value of $1.22, which is the 75th percentile. This indicates that at least 75% of gas stations have prices of $1.22 or less.

(h) Compute the lower and upper hinges. Compare your answers to Parts f and g.

Answer: The lower and upper hinges are basically the first and the third quartiles, but the computational procedures are slightly different. To find the lower hinge, first we need to consider the data in the positions less than or equal to the median position. In Part d of this problem, we determined the median to be $1.19. Therefore, the data in the positions less than or equal to the median position are as follows.

1.14, 1.17, 1.19*, 1.19, 1.19

The median of the above data points is the lower hinge. Thus, $1.19 (in the third position as marked by " * ") is the lower hinge. Comparing this value with that of the first quartile, which was computed in Part e of this problem, we note they are the same.

To determine the upper hinge, first consider the data elements in the positions greater than or equal to the median position (refer to Part d):

1.19, 1.21, 1.22*, 1.24, 1.25

The median, or upper hinge, of the above data set is $1.22, which is the same value we computed for the third quartile (see Part g).

(i) Compute the interquartile range.

Answer: The interquartile range is the difference between the third and the first quartiles. In Parts f and g of this problem, we computed the first and the third quartiles as $Q_1 = 1.19$ and $Q_3 = 1.22$. Therefore, the interquartile range is:

$$IQR = Q_3 - Q_1 = 1.22 - 1.19 = 0.03$$

2. The price of a selected stock over a five day period is shown below.

17, 11, 13, 17, 16

Using the above data, compute the mean, the median, and the mode.

3. Briefly explain the meanings of the following terms:

(a) Parameter

(b) Statistic

(c) Mean

(d) Median

(e) Mode

(f) Percentile

4. The price of gold at the end of each month of 1995 is shown below.

Month	Price Per Ounce
January	$225.00
February	230.00
March	225.00
April	236.00
May	270.00
June	382.00
July	322.00
August	324.00
September	320.00
October	310.00
November	368.00
December	388.00

(a) Determine the average price of gold for the year 1995.

(b) Determine the mode for the above data.

(c) Fully explain the meaning of the value which you determined in Part b.

(d) Determine the median.

(e) Fully explain the meaning of the value which you found in Part d.

(f) Compute the 30th and the 80th percentiles.

5. The prices of a sample of eight selected men's colognes are shown below:

Product	Size of Container (Fl. Oz.)	Price
A	2	$4.00
B	4	12.00
C	3	12.00
D	4	20.00
E	2	10.00
F	5	30.00
G	6	42.00
H	2	16.00

(a) Determine the median and the mode for the price *per ounce.*
Hint: The prices shown above are not the price *per ounce.* They represent the prices of each container.

(b) Determine the average of the per ounce prices for the eight colognes.

(c) Compute the 85th percentile and explain its meaning.

6. In order to determine the average length of local calls in an office, fifteen calls were selected randomly. The duration of each call is shown below:

Call Number	Duration (in minutes)
1	2
2	12
3	10
4	3
5	5
6	6
7	3
8	5
9	8
10	4
11	5
12	4
13	5
14	4
15	9

(a) Determine the average duration of the calls.

(b) Determine the median and the mode.

(c) Compute the 80th percentile.

7. A sample of 10 employees in the graphics department of Design, Inc. is selected. The employees' ages are given below:

Employee Number	Age
1	34
2	35
3	39
4	24
5	62
6	40
7	18
8	35
9	28
10	35

(a) What is the average age of the employees in the graphics department of Design, Inc.?

(b) Determine the mode.

(c) Determine the median.

(d) Compute the first and the third quartiles.

(e) Compute the lower and upper hinges. Compare your answers to Part d.

(f) Compute the interquartile range.

8. In exercise 9 of Chapter 2, you were given the following information regarding the number of hours of sick leave that a sample of the employees of Bastien's, Inc. have taken during the first quarter of the year (rounded to the nearest hour).

19	22	27	24	28	12
23	47	11	55	25	42
36	25	34	16	45	49
12	20	28	29	21	10
59	39	48	32	40	31

(a) Determine the average hours of sick leave.

(b) Determine the mode and the median.

9. In exercise 13 of Chapter 2, you were given the following information regarding the number of magazine subscriptions which were sold by a group of students.

Student	# of Subscriptions Sold
1	30
2	79
3	59
4	65
5	40
6	64
7	52
8	53
9	57
10	39
11	61
12	47
13	50
14	60
15	48
16	50
17	58
18	67

Determine the mean, the mode and the median for the above data.

10. In exercise 14 of Chapter 2, the temperatures for the month of June in a midwestern city were given as follows.

70	75	79	80	78	82
82	89	88	87	90	92
91	92	93	95	94	95
97	95	98	100	107	107
105	104	108	111	109	116

Determine the mean, the mode, and the median for the above data.

***11.** In exercise 1 of this chapter, the gasoline prices for 9 selected gas stations were given. Refer to exercise 1 and answer the following questions:

(a) What is the range of the gasoline prices?

Answer: The range is a measure of dispersion whose value is equal to the difference between the largest and the smallest values of a data set. Hence, the range of gasoline prices is

Range = 1.25 - 1.14 = $0.11

(b) Calculate the sample variance.

Answer: The variance is a measure of dispersion which is defined as

$$S^2 = \frac{\Sigma(X_i - \overline{X})^2}{n-1}$$

To calculate the variance, the first step will be to determine $\Sigma(X_i - \overline{X})^2$. (Remember $\overline{X}$ was calculated previously, and its value was $1.20.)

Station Number	X_i	$(X_i - \overline{X})$	$(X_i - \overline{X})^2$
1	$1.14	1.14 - 1.20 = - 0.06	0.0036
2	1.19	1.19 - 1.20 = - 0.01	0.0001
3	1.25	1.25 - 1.20 = 0.05	0.0025
4	1.21	1.21 - 1.20 = 0.01	0.0001
5	1.17	1.17 - 1.20 = - 0.03	0.0009
6	1.19	1.19 - 1.20 = - 0.01	0.0001
7	1.22	1.22 - 1.20 = - 0.02	0.0004
8	1.24	1.24 - 1.20 = 0.04	0.0016
9	1.19	1.19 - 1.20 = 0.01	0.0001

$$\Sigma(X_i - \overline{X})^2 = 0.0094$$

Therefore, the variance is calculated as

$$S^2 = \frac{\Sigma(X_i - \overline{X})^2}{n-1} = \frac{0.0094}{9 - 1} = 0.001175$$

(c) If the above data represented a population, how would the variance be computed?

Answer: The variance of a population would be computed as follows.

$$\sigma^2 = \frac{\Sigma(X_i - \mu)^2}{N} = \frac{0.0094}{9} = 0.00104$$

(d) Determine the sample's standard deviation.

Answer: The standard deviation is the square root of the variance. Therefore, it is

$$S = \sqrt{S^2} - \sqrt{0.001175} - 0.03428$$

(e) Is there another approach for computing the sample variance?

Answer: Yes, the following formula may be easier to use and reduces rounding errors.

$$S^2 = \frac{\Sigma X_i^2 - n\overline{X}^2}{n-1}$$

We have already calculated ΣX_i to be 10.80 and $\overline{X}$ to be 1.20 (see problem 1). The only other value in the above formula which needs to be calculated is ΣX_i^2.

X_i	X_i^2
1.14	1.2996
1.19	1.4161
1.25	1.5625
1.21	1.4641
1.17	1.3689
1.19	1.4161
1.22	1.4884
1.24	1.5376
1.19	1.4161

$$\Sigma X_i^2 = 12.9694$$

Then

$$S^2 = \frac{\Sigma X_i^2 - n\overline{X}^2}{n-1} = \frac{12.9694 - (9)(1.20)^2}{9-1} = \frac{0.0094}{8} = 0.001175$$

(f) If a sample of gas stations in Atlanta shows a mean price of $1.31 with a standard deviation of 0.036, then which city's sample shows a more dispersed price distribution?

Answer: To compare the relative dispersion of two groups, we need to calculate the coefficient of variation, which is a measure of relative dispersion.

$$\text{Coefficient of Variation} = \left(\frac{S}{\overline{X}}\right)(100)$$

$$\text{Coefficient of Variation for Chattanooga} = \left(\frac{.03428}{1.2}\right)(100) = 2.85\%$$

$$\text{Coefficient of Variation for Atlanta} = \left(\frac{.036}{1.31}\right)(100) = 2.74\%$$

Therefore, the sample in Atlanta shows a lesser degree of relative dispersion than Chattanooga even though Atlanta had greater absolute dispersion (i.e., the standard deviation of Atlanta was larger than the standard deviation of Chattanooga).

(g) Compute the Z-score for the gas station which had a price of \$1.25 per gallon.

Answer: The Z-score represents the number of standard deviations a particular data element is from the mean. The mean and the standard deviation for this problem were computed previously, and their values were \$1.20 and \$0.03428, respectively. Thus, the Z-score for the gas station with the price of \$1.25 is

$$Z = \frac{X_i - \overline{X}}{S} = \frac{1.25 - 1.20}{.03428} = 1.459 \text{ (Rounded)}$$

The above Z-score indicates that the gas station with the price of \$1.25 is approximately 1.459 standard deviations above the mean.

12. In exercise 4 of this chapter, you were presented with the following information regarding the end-of-the-month gold prices for the year 1995.

Month	Price Per Ounce
January	\$225.00
February	230.00
March	225.00
April	236.00
May	270.00
June	382.00
July	322.00
August	324.00
September	320.00
October	310.00
November	368.00
December	388.00

(a) Calculate the standard deviation, assuming the data represent a sample.

(b) Calculate the standard deviation, assuming the data represent a population.

(c) If in 1994 the average price of gold was $400 an ounce with a standard deviation of $70, which year (1994 or 1995) shows a more dispersed price distribution?

(d) Compute the Z-score for December's price.

13. In exercise 5 of this chapter, the prices of a sample of eight selected men's colognes were given as shown below.

Product	Size of Container (Fl. Oz.)	Price
A	2	$4.00
B	4	12.00
C	3	12.00
D	4	20.00
E	2	10.00
F	5	30.00
G	6	42.00
H	2	16.00

(a) Compute the range for the price *per ounce.*

(b) Compute the variance and the standard deviation for the price *per ounce.*

(c) Determine the coefficient of variation and explain its meaning.

14. Global Engineers hired the following number of Class 1 engineers during the first six months of the past year. (Assume the data represent a sample.)

Month	No. of Class 1 Engineers Hired
January	3
February	2
March	4
April	2
May	6
June	0

(a) Determine the mean, the median, the mode and the range for the above data.

(b) Compute the variance and the standard deviation.

(c) Compute the first and the third quartiles.

(d) Compute the lower and the upper hinges.

(e) Compute the Z-scores for the months of May and June.

15. A consumer product testing group tested the gas water heaters produced by 11 leading producers (all heaters were of the same size and the same specifications) and determined the following annual operating costs:

Manufacturer	Annual Operating Costs (in dollars)
A	\$121
B	124
C	134
D	128
E	119
F	115
G	131
H	124
I	125
J	133
K	132

(a) Find the average operating costs among the 11 producers.

(b) Find the mode and the median.

(c) Determine the range.

(d) Compute the variance and the standard deviation. (Assume the data represent a sample.)

***16.** The flashlight batteries produced by one of the northern manufacturers are known to have an average life of 60 hours with a standard deviation of 4 hours.

(a) At least what percentage of flashlights will have a life of 54 to 66 hours?

Answer: Chebyshev's theorem states that for any data set, at least $(1 - 1/k^2)$ of the data elements fall within plus or minus k (for $k > 1$) standard deviations of the mean. In the above example, the distance from 54 to the mean of 60 is 6 hours. Since the standard deviation is 4 hours, the 6 hours represents 1.5 standard deviations (i.e., $k = 1.5$). Applying Chebyshev's theorem, we note that

$$(1 - 1/k^2) = (1 - 1/(1.5)^2) = 0.56$$

which indicates that at least 56% of the batteries will have a life of 54 to 66 hours.

(b) At least what percentage of the batteries will have a life of 52 to 68 hours?

Answer: Again applying Chebyshev's theorem, with $k = 2$ standard deviations (note that there are 8 hours from the lower point of 52 to the mean of 60, which represents 2 standard deviations)

$$(1 - 1/k^2) = (1 - 1/(2)^2) = 0.75$$

indicating that at least 75% of the batteries have a life of 52 to 68 hours.

(c) Determine an interval for the batteries' lives that will be true for at least 80% of the batteries.

Answer: Once again, by using Chebyshev's theorem,

$$(1 - 1/k^2) = 0.80$$

we can solve for k in the above equation by first subtracting 1 from both sides of the equation:

$$- 1/k^2 = - 0.20$$

then multiplying both sides by $- k^2$

$$1 = 0.2\ k^2$$

and now solving for k^2

$k^2 = 1/0.20 = 5$

Hence, $k = \sqrt{5}$ or 2.236 standard deviations. Therefore, 80% of the values will be in the interval of the mean plus or minus 2.236 standard deviations or, $60 \pm (2.236)(4)$. This indicates that at least 80% of the batteries will have lives of 51.056 to 68.944 hours.

17. In a statistics class, the average grade on the final examination was 75 with a standard deviation of 5.

(a) At least what percentage of the students received grades between 50 and 100?

(b) Determine an interval for the grades that will be true for at least 70% of the students.

***18.** The weights of 12 individuals who enrolled in a fitness program are shown below:

Individual	Weight (Pounds)
1	100
2	105
3	110
4	130
5	135
6	138
7	142
8	145
9	150
10	170
11	240
12	300

(a) Provide a five-number summary for the data.

Answer: A five-number summary is a method of summarizing data by means of the following five values.

1. Smallest value
2. First quartile (Q_1)
3. Median
4. Third quartile (Q_3)
5. Largest value

For the above data, the smallest weight is 100 pounds; the first quartile $Q_1 = 120$ pounds (i.e., the average weight of individuals 3 and 4); the median weight is 140 pounds; the third quartile $Q_3 = 160$ pounds (i.e., the average weight of individuals 8 and 9); and the largest value is 300 pounds. Therefore, the five-number summary for the above data is

Smallest	Q_1	Q_2 (Median)	Q_3	Largest
100	120	140	160	300

This five-number summary indicates that approximately 25% of the data elements are located between adjacent values.

(b) Show the box plot for the weight data.

Answer: A box plot is a visual presentation of the five-number summary. Figure 3.1 shows the box plot for the weight data.

Box Plot for the Weight Data

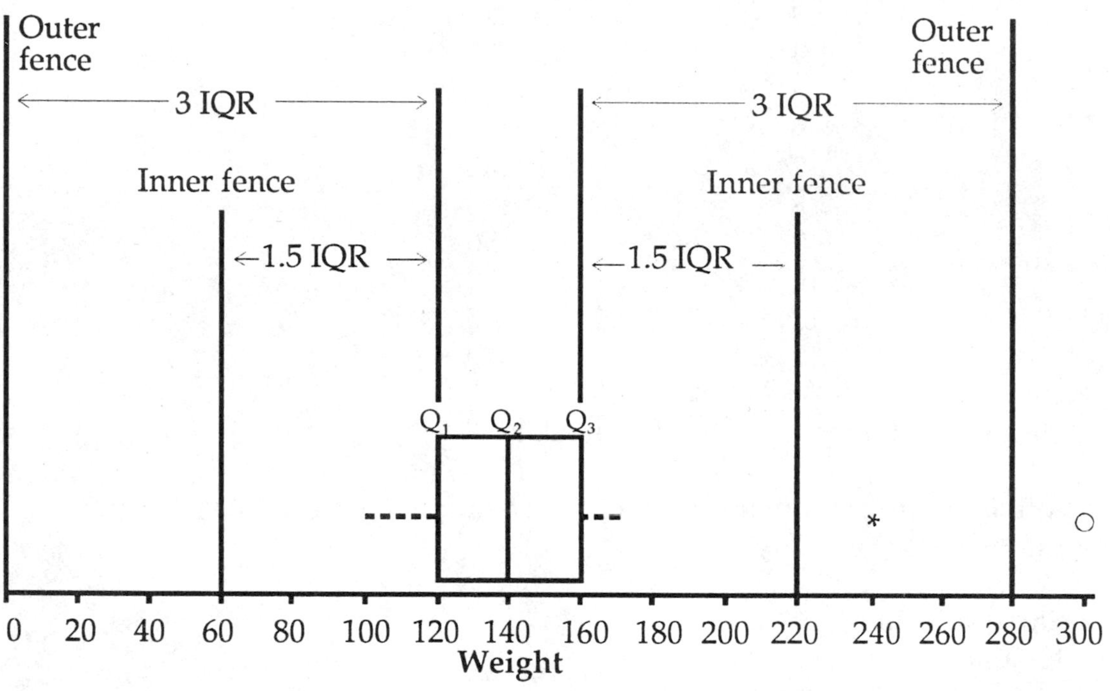

Figure 3.1

The following procedure is used for constructing the box plot.

1. A box is drawn to contain the middle 50% of the data elements. Thus the ends of this box are located at the first and the third quartiles. In this example, the box begins at $Q_1 = 120$ and ends at $Q_3 = 160$. The width of the box shows the interquartile range.

2. At the location of the median (in our case at 140) a vertical line is drawn in the box to show at what point data is divided into two equal parts.

3. The *inner fences* are drawn at 1.5 IQR (remember IQR = Q_3 - Q_1 = 160 - 120 = 40) below and above Q_1 and Q_3. Therefore, the inner fences are drawn at 120 - (1.5)(40) = 60 and 160 + (1.5)(40) = 220. The *outer fences* are drawn at 3 IQR below and above Q_1 and Q_3. Thus, the outer fences are drawn at 120 - (3)(40) = 0 and 160 + (3)(40) = 280. The fences are used for identifying *outliers.* Data elements which fall between the inner and outer fences are considered *mild outliers,* and data falling outside the outer fences are considered *extreme outliers.*

4. Dashed lines, known as *whiskers,* are drawn from the ends of the box to the smallest and largest data values *inside the inner fences.* In our example, the whiskers end at weights of 100 and 170.

5. The location of mild outliers are marked by "*" and the location of extreme outliers are marked by "o." Thus in our example, we note there is one mild outlier (240) and one extreme outlier (300). In Figure 3.1, for the purpose of demonstration, the inner and outer fences and various distances and labels have been shown. Ordinarily, the box plot does not show the above, but simply will appear as shown in Figure 3.2.

The Usual Appearance of the Box Plot for the Weight Data

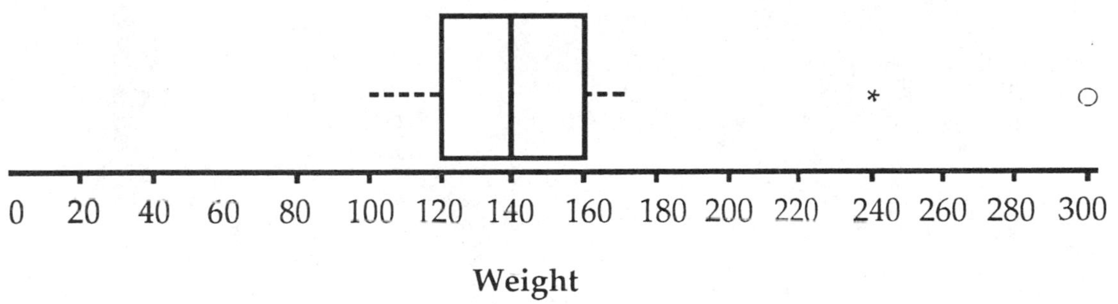

Figure 3.2

19. The annual salaries of the employees of MBS accounting firm are shown below.

Employee	Salary ($1,000s)
1	75
2	18
3	26
4	29
5	24
6	35
7	40
8	60
9	45
10	38
11	24
12	27
13	37
14	32
15	20

Provide a five-number summary for the above data.

***20.** In exercise 24 of Chapter 2 the average grades of a sample of 8 students in statistics and the number of absences they had during the semester were given as follows.

Student	Number of Absences (X_i)	Average Grade (Y_i)
1	1	94
2	2	78
3	2	70
4	1	88
5	3	68
6	4	40
7	8	30
8	3	60

(a) Compute the sample covariance.

Answer: The sample covariance, which is a measure of association between two variables, is defined as

$$s_{xy} = \frac{\Sigma(X_i - \overline{X})(Y_i - \overline{Y})}{n-1}$$

To calculate the covariance, we need to compute $\Sigma(X_i - \overline{X})(Y_i - \overline{Y})$, where $\overline{X}$ is the average of X values (24/8 = 3) and $\overline{Y}$ is the average of Y values (528/8 = 66). Thus, the numerator for the above formula can be computed by means of the following table:

	(X_i)	(Y_i)	$(X_i - \overline{X})$	$(Y_i - \overline{Y})$	$(X_i - \overline{X})(Y_i - \overline{Y})$
	1	94	-2	28	-56
	2	78	-1	12	-12
	2	70	-1	4	-4
	1	88	-2	22	-44
	3	68	0	2	0
	4	40	1	-26	-26
	8	30	5	-36	-180
	3	60	0	-6	0
Totals	24	528	0	0	-322

Now the sample covariance can be computed as

$$S_{xy} = \frac{\Sigma(X_i - \overline{X})(Y_i - \overline{Y})}{n-1} = \frac{-322}{8-1} = -46$$

(b) If the above data represented a population, how would the covariance be computed?

Answer: The covariance of the population is computed as follows:

$$\sigma_{XY} = \frac{\Sigma(X_i - \mu_X)(Y_i - \mu_Y)}{N} = \frac{-322}{8} = -40.25$$

(c) Interpret the meaning of the sample covariance which was found in Part a.

Answer: Covariance is a measure of linear relationship between two variables. If the covariance is negative it indicates a negative, or if it is positive, a positive linear relationship between variables. In our case the sample covariance was computed to be -46; thus, it indicates that there is a negative linear relationship between the number of absences and the average grade. This indicates that as the number of absences increase, the average grade tends to decrease.

(d) Is there any problem in using sample covariance as a measure of strength of linear relationship between variables?

Answer: The problem with using the sample covariance as a measure of strength of association between variables is the fact that the magnitude of the covariance depends on the units of measurement. Assume we are interested in determining the association between the monthly income of individuals and their age. If the monthly income figures are given in dollars the magnitude of the covariance would be much smaller than if the figures were given in cents, where in fact the degree of association is the same.

(e) Is there a measure of association which avoids the problem of units of measurement?

Answer: The Pearson Product Moment Correlation Coefficient (or simply sample correlation coefficient) is a measure which overcomes the problem of units of measurement. The sample correlation coefficient is computed by dividing the sample covariance by the product of the standard deviations of the two variables. The sample coefficient of correlation is computed as

$$r_{XY} = \frac{S_{XY}}{S_X S_Y}$$

where

r_{XY} = Sample correlation coefficient
S_{XY} = Sample covariance
S_X = Sample standard deviation of X
S_Y = Sample standard deviation of Y

The sample covariance was computed in Part a. Its value was $S_{XY} = -46$. Now we need to compute the standard deviation of the X and Y values. The standard deviation of X values is computed as

$$S_X = \sqrt{\frac{\Sigma\left(X_i - \overline{X}\right)^2}{n-1}} = \sqrt{\frac{36}{8-1}} = 2.268$$

Similarly the standard deviation of Y values is computed as

$$S_Y = \sqrt{\frac{\Sigma\left(Y_i - \overline{Y}\right)^2}{n-1}} = \sqrt{\frac{3440}{8-1}} = 22.168$$

Now the coefficient of correlation can be computed as

$$r_{XY} = \frac{S_{XY}}{S_X S_Y} = \frac{-46}{(2.268)(22.168)} = -0.915$$

The coefficient of correlation ranges between -1 to +1. A value of -1 indicates a perfect negative, and a value of +1 a perfect positive correlation. In our case the coefficient of correlation is -0.915 which shows a very strong negative relationship between the number of absences and the average grade, indicating that as the number of absences increases the average grade decreases.

(f) Is there another approach for computing the sample coefficient of correlation?

Answer: Yes, the following equation (alternate formula) can be used to compute the sample coefficient of correlation.

$$r_{XY} = \frac{\Sigma X_i Y_i - (\Sigma X_i \Sigma Y_i)/n}{\sqrt{\Sigma X_i^2 - (\Sigma X_i)^2/n}\sqrt{\Sigma Y_i^2 - (\Sigma Y_i)^2/n}}$$

The following computations are needed in order to use the alternate formula:

	X_i	Y_i	$X_i Y_i$	X_i^2	Y_i^2
	1	94	94	1	8836
	2	78	156	4	6084
	2	70	140	4	4900
	1	88	88	1	7744
	3	68	204	9	4624
	4	40	160	16	1600
	8	30	240	64	900
	3	60	180	9	3600
Totals	24	528	1262	108	38288

Now the sample coefficient of correlation is computed:

$$r_{XY} = \frac{\Sigma X_i Y_i - (\Sigma X_i \Sigma Y_i)/n}{\sqrt{\Sigma X_i^2 - (\Sigma X_i)^2/n}\sqrt{\Sigma Y_i^2 - (\Sigma Y_i)^2/n}}$$

$$= \frac{1262 - (24)(528)/8}{\sqrt{108 - (24)^2/8}\sqrt{38288 - (528)^2/8}} = -0.915$$

As you note, this is the same value which was determined in Part e, but the alternate formula is much easier to use.

21. In exercise 25 of Chapter 2 you were given the following ten observations on two variables, x and y. Assume the data represent a sample.

x	y
1	8
5	15
6	20
4	12
2	10
8	20
9	26
1	5
6	18
8	26

(a) Compute the sample covariance.

(b) Interpret the meaning of the sample covariance which was found in Part a.

(c) Compute the coefficient of correlation.

***22.** The income distribution for a sample of 155 Walker County residents is shown below.

Yearly Income (in thousands of dollars)	Frequency f_i	Class Midpoint M_i	f_iM_i
10 - 14	12	12	144
15 - 19	23	17	391
20 - 24	35	22	770
25 - 29	40	27	1080
30 - 34	35	32	1120
35 - 39	10	37	370
	155		$\Sigma f_iM_i = 3875$

(a) Compute the mean income for the above sample.

Answer: The mean of the grouped data is

$$\bar{X} = \frac{\Sigma f_i M_i}{n}$$

To calculate Σf_iM_i, first the class midpoints (M_i) are determined. The class midpoints are located halfway between the class limits. For instance, the midpoint of the first class is (10 + 14)/2 = 12. The class midpoints for all classes are shown above. Next, the product of the class midpoints and their respective frequencies are determined (shown above in the last column). The sum of the values of the column of Σf_iM_i yields 3875. Therefore, the mean income will be

$$\bar{X} = \frac{\Sigma f_i M_i}{n} = \frac{3875}{155} = 25$$

Since the incomes were in thousands of dollars, the mean income is $25,000.00.

(b) Compute the sample variance.

Answer: The variance of a sample for grouped data is

$$S^2 = \frac{\Sigma f_i (M_i - \overline{X})^2}{n-1}$$

To compute the numerator, first the difference between each midpoint and the sample mean (25) is determined. These values are shown below in the third column designated "deviation." Next, each deviation is squared, as shown in the fourth column. Then, each squared deviation is multiplied by its respective frequency, as shown in the last column. The sum of the values in the last column (7130) is the numerator for the computation of the variance.

		Deviation	Squared Deviation	Frequency	
Income	M_i	$(M_i - \overline{X})$	$(M_i - \overline{X})^2$	f_i	$f_i (M_i - \overline{X})^2$
10 - 14	12	- 13	169	12	2028
15 - 19	17	- 8	64	23	1472
20 - 24	22	- 3	9	35	315
25 - 29	27	2	4	40	160
30 - 34	32	7	49	35	1715
35 - 39	37	12	144	10	1440

$$\Sigma f_i (M_i - \overline{X})^2 = 7130$$

Therefore, the sample variance is

$$S^2 = \frac{\Sigma f_i (M_i - \overline{X})^2}{n-1} = \frac{7130}{155 - 1} = 46.2987 \text{ (Rounded)}$$

The standard deviation is $\sqrt{46.2987} = 6.8$ (i.e., \$6,800).

23. In exercise 11 of Chapter 2, the sales records of a real estate company were given. The following frequency distribution shows the sale prices and their respective frequencies: Assume the data represent a sample.

Sale Price (in thousands of dollars)	Number of Houses Sold
20 - 39	1
40 - 59	2
60 - 79	4
80 - 99	2
100 - 119	1

(a) Determine the mean.

(b) Determine the standard deviation.

24. In exercise 9 of Chapter 2, the hours of sick leave which the employees of Bastien, Inc. had taken during the first quarter of the year were given. From that data, the following frequency distribution has been developed.

Hours of Sick Leave Taken	Number of Employees
10 - 19	6
20 - 29	11
30 - 39	5
40 - 49	6
50 - 59	2
Total	30

(a) Determine the mean.

(b) Determine the variance and the standard deviation. Assume the data represent a population.

25. The following frequency distribution has been formulated from the data of exercise 13 of Chapter 2. Assume the data represent a sample.

No. of Magazine Subscriptions Sold	No. of Students Who Sold in this Range
30 - 39	2
40 - 49	3
50 - 59	7
60 - 69	5
70 - 79	1

(a) Determine the mean.

(b) Compute the variance and the standard deviation.

26. The following frequency distribution has been developed from the data given in exercise 14 of Chapter 2. Assume the data represent a population.

Temperature	Number of Days
70 - 79	4
80 - 89	6
90 - 99	11
100 - 109	7
110 - 119	2

(a) Determine the mean.

(b) Compute the standard deviation.

27. Compute the mean and the standard deviation for the following frequency distribution. Assume the data represent a sample.

Class	Frequency
45 - 47	3
48 - 50	6
51 - 53	8
54 - 56	2
57 - 59	1

28. The following data show the yearly salaries of a sample of football coaches at state supported universities. The figures do not include outside income.

University	Salary (in $1,000s)
A	55
B	46
C	70
D	49
E	64
F	61
G	55
H	96

(a) Determine the mean yearly salary.

(b) Compute the standard deviation.

(c) Determine the mode.

(d) Determine the median.

(e) Determine the 70th percentile.

29. In 1994, the average donation to the United Help Organization was $500 with a standard deviation of $60. In 1995, the average donation was $650 with a standard deviation of $65. The donations in which year show a more dispersed distribution?

30. The following frequency distribution shows the ACT scores of a **sample** of students at a local university.

Score	Frequency	MIDPOINT	$f_i m_i$	$(m_i - \bar{x})$	$(m_i - \bar{x})^2$	$f(m_i - \bar{x})^2$
16 - 18	1	17	17	-9	81	81
19 - 21	2	20	40	-6	36	72
22 - 24	4	23	92	-3	9	36
25 - 27	12	26	312	0	0	0
28 - 30	11	29	319	3	9	99
	30		780			288

(a) Determine the mean. $\frac{780}{30} = 26$

$\bar{x} = \frac{\Sigma(f_i m_i)}{n}$

(b) Determine the standard deviation.

$s^2 = \frac{f(m_i - \bar{x})^2}{n-1} = \frac{288}{29} = 9.931$

$s = \sqrt{9.931} = 3.15$

$\text{Coeff} = \frac{3.15}{26} \times 100 = 12.1\%$

31. The following data represent the daily demand (y in thousands of units) and the unit price (x) for a product.

x	y
4	40
7	32
8	34
3	32
6	68
7	18
12	14
9	28

(a) Compute the sample covariance for the above data.

(b) Compute the sample correlation coefficient.

SELF-TESTING QUESTIONS

In the following multiple choice questions, circle the correct answer. An answer key is provided following the questions.

1. A numerical value used as a summary measure for a population of data is called a

a) sample
b) statistic
c) range
d) parameter
e) none of the above

2. The average value of a data set is called the

a) mean
b) median
c) mode
d) range
e) percentile

3. The most frequently occurring data value in a data set is the

a) median
b) arithmetic mean
c) population parameter
d) range
e) mode

4. A statistic is

a) a descriptive measure of a population
b) a descriptive measure of a sample
c) the average value of the data set
d) none of the above

5. A measure of central location which splits the data set into two equal groups is called the

a) mean
b) mode
c) median
d) standard deviation
e) none of the above

6. The coefficient of variation is

a) the same as the variance
b) a measure of central tendency
c) a measure of absolute dispersion
d) a measure of relative dispersion
e) none of the above

Use the following data set to answer questions 7 through 11. Assume the data set represents a sample.

6, 8, 3, 7, 6, 0

7. The median of the above data set is

a) 5
b) 6
c) 6.5
d) 3
e) 0

8. The mode of the above data set is

a) 8
b) 0
c) 3
d) 6
e) none of the above

9. The mean of the above data set is

a) 5
b) 6
c) 7
d) 8
e) none of the above

10. The standard deviation of the above data set is

a) 2.7
b) 8.8
c) 2.9
d) 0
e) 1.6

11. The range of the above data set is

a) 6
b) 8
c) 3
d) 5
e) 0

The following frequency distribution shows the time (in minutes per week) that a sample of business students used the computer terminals. Answer questions 12 and 13 based on the following frequency distribution.

Time	f
20 - 39	2
40 - 59	4
60 - 79	6
80 - 99	4
100 - 119	2

12. The mean of the above distribution is

a) 3.6
b) 5.2
c) 69.5
d) 80
e) 9

13. The standard deviation of the above distribution is

a) 564.70
b) 312.72
c) 70
d) 18.92
e) 23.76

14. An exploratory data analysis which uses the lowest value, 25th percentile, median, 75th percentile and largest value is known as a

a) median analysis
b) box analysis
c) 5 - number summary
d) none of the above

15. A box plot is

a) the same as an ogive
b) a frequency distribution
c) the same as a frequency polygon
d) a visual presentation of a 5 - number summary
e) none of the above

16. The symbol μ is used to represent

a) the mean of the sample
b) the mean of the population
c) the standard deviation of the sample
d) the standard deviation of the population
e) the size of the population

17. The symbol σ is used to represent

a) the mean of the sample
b) the mean of the population
c) the standard deviation of the sample
d) the standard deviation of the population
e) the size of the population

18. The symbol $\overline{X}$ is used to represent

a) the mean of the sample
b) the mean of the population
c) the standard deviation of the sample
d) the standard deviation of the population
e) the size of the population

19. The symbol S is used to represent

a) the mean of the sample
b) the mean of the population
c) the standard deviation of the sample
d) the standard deviation of the population
e) the size of the population

20. The symbol N is used to represent

a) the mean of the sample
b) the mean of the population
c) the standard deviation of the sample
d) the standard deviation of the population
e) the size of the population

21. The sum of the deviation of the individual data elements from their mean is always

a) equal to zero
b) equal to one
c) negative
d) positive
e) none of the above

22. The median is always the same as

a) the first quartile
b) the third quartile
c) the mode
d) the mean
e) none of the above

23. The interquartile range

a) is a measure of location
b) is a measure of dispersion
c) is always equal to the median
d) is always equal to the mode
e) none of the above

24. The ratio of the standard deviation to the mean is

a) the variance
b) the range
c) the coefficient of variation
d) always greater than 1
e) none of the above

25. The value of the variance can never be

a) positive
b) zero
c) larger than the standard deviation
d) negative
e) none of the above

26. A numerical measure of linear association between two variables is

a) the standard deviation
b) the coefficient of variation
c) the mean
d) the covariance
e) none of the above

27. The range of the correlation coefficient is

a) 0 to +1
b) -1 to +1
c) -1 to infinity
d) -1 to 0
e) none of the above

ANSWERS TO THE SELF-TESTING QUESTIONS

1. d
2. a
3. e
4. b
5. c
6. d
7. b
8. d
9. a
10. c
11. b
12. c
13. e
14. c
15. d
16. b
17. d
18. a
19. c
20. e
21. a
22. e
23. b
24. c
25. d
26. d
27. b

ANSWERS TO CHAPTER THREE EXERCISES

2. Mean = 14.8 Median = 16 Mode = 17

3. See the chapter outline

4. (a) $300.00
   (b) $225.00
   (c) Most common price
   (d) $315.00
   (e) The value of the middle item
   (f) 30th is 236 80th is 368

5. Hint: first find the price per ounce.
   (a) Median = $5.00 Mode = $5.00
   (b) $5.00
   (c) $7.00

6. (a) 5.667
   (b) Median = Mode = 5
   (c) 8.5

7. (a) 35
   (b) Mode = 35
   (c) Median = 35
   (d) First quartile = 28
   Third quartile = 39
   (e) Lower hinge = 28
   Upper hinge = 39
   (f) IQR = 11

8. (a) 30.3
   (b) Median = 28 There are 3 modes: 12, 28, 25

9. Mean = 54.39 Median = 55 Mode = 50

10. Mean = 93.63 Mode = 95 Median = 93.5

12. (a) 61.64
    (b) 59.02
    (c) Coefficient of variation:
        1994 17.5%
        1995 20.54% (more dispersed)
    (d) 1.428 (rounded)

13. (a) $6.00
    (b) $S^2 = 4$ $S = 2$
    (c) 40%

14. (a) Mean = 2.833 Median = 2.5 Mode = 2.0 Range = 6
    (b) $S^2 = 4.166$ $S = 2.041$
    (c) First quartile = 2
        Third quartile = 4
    (d) Lower hinge = 2
        Upper hinge = 4
    (e) Z-score for May = 1.55 (rounded)
        Z-score for June = -1.39 (rounded)

15. (a) Mean = 126
    (b) Mode = 124 Median = 125
    (c) 19
    (d) $S^2 = 38.203$ $S = 6.181$

17. (a) 96%
    (b) $75 \pm (1.826)(5) = 65.87$ to 84.13

19. Lowest = 18
    $Q_1 = 24$ (i.e., the 25th percentile)
    Median (Q_2) = 32
    $Q_3 = 40$ (i.e., the 75th percentile)
    Highest = 75

21. (a) 20.67 (rounded)
    (b) Since the covariance is positive it indicates that there is a positive correlation between x and y. That means as x increases so does y.
    (c) 0.9673 (rounded)

23. (a) Mean = 69.5 (in thousands of dollars)
    (b) S = 23.094 (in thousands of dollars)

24. (a) 30.17 (rounded)
    (b) Variance = 144.556 (rounded) Standard deviation = 12.023 (rounded)

25. (a) Mean = 54.5
    (b) Variance = 117.65 (rounded) Standard deviation = 10.85 (rounded)

26. (a) Mean = 93.5
    (b) Standard deviation = 11.06 (rounded)

27. Mean = 50.8 Standard deviation = 3.14 (rounded)

28. (a) $62,000
    (b) $15,802.35
    (c) $55,000
    (d) 58,000
    (e) 64,000

29. Coefficient of variation in 1994 = 12% (more dispersed)
    Coefficient of variation in 1995 = 10%

30. (a) 26
    (b) 3.15 (rounded)

31. (a) -187.428 (rounded)
    (b) -0.8248 (rounded)

CHAPTER FOUR

INTRODUCTION TO PROBABILITY

CHAPTER OUTLINE AND REVIEW

There are many situations where a decision maker must make decisions in an environment of future uncertainties. Therefore, the decision maker is faced with the "chance" or likelihood that a particular situation may prevail.

In this chapter, you have studied the concept which deals with such uncertainties. The topic of study has been the concept of probability, the terminology used in the study of probability, and the various laws governing probability. The following is a brief description of the main points involved in the study of probability.

A. **Probability:** A numerical measure of the likelihood of the occurrence of an event. Probability ranges between 0 and 1.0, where 0 indicates the event will not occur and 1.0 indicates that the event will definitely occur. Furthermore, if there are k possible outcomes, then the sum of the probabilities of all k outcomes must be equal to 1.0.

B. **Experiment:** Any process which results in well defined outcomes. In any experiment, only one outcome is possible.

C. **Sample Point:** Each individual outcome of an experiment.

D. **Sample Space:** The collection of all possible sample points in an experiment.

E. **Counting Rules:** Methods of identifying the number of sample points in an experiment. You have been introduced to two counting rule methods: first, the counting rule for multiple step experiments and, second, the counting rule which is used for counting the number of ways n objects can be selected from N objects (combination). For more detail on these counting rules, refer to the section of chapter formulas.

F. **Tree Diagram:** A graphical method of presenting the sample points of an experiment.

G. **Classical Method:** A method of assigning probabilities which assumes that if there are n possible outcomes to an experiment, each of the experimental outcomes has the same chance of occurring. Hence, the probability of the occurrence of each outcome is 1/n. In other words, the classical method assumes equal probabilities for various experimental outcomes.

H. **Relative Frequency Method:** The assignment of probabilities based on historical data.

I. **Subjective Method:** The assignment of probabilities based on the judgment of the experimenter.

J. **Event:** Any specific collection of sample points in an experiment.

K. **Venn Diagram:** A graphical method of showing the sample space and the operations involving events.

L. Union of Two Events:

That event which contains all the sample points which are in one event or the other event or in both events. Hence, the union of events A and B is that event which contains all the sample points that are in both. The symbol " $\cup$ " is used to denote the union of two events. In the following Venn diagram the area which is shaded shows the union of events A and B.

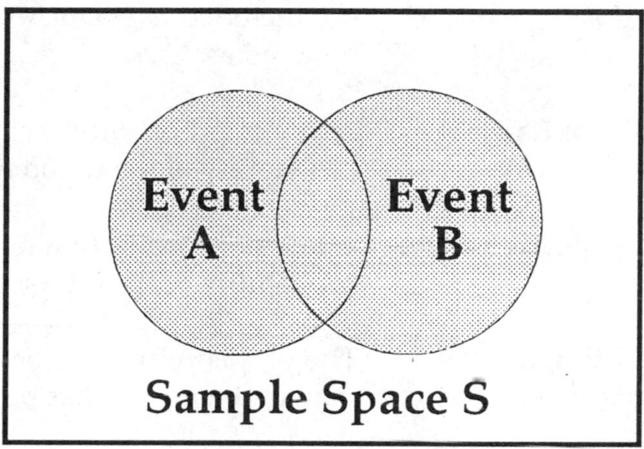

M. Intersection of Two Events:

That event which contains all the sample points which are in both events. The symbol " $\cap$ " denotes intersection of events. The shaded area in the following Venn diagram shows the intersection of events A and B.

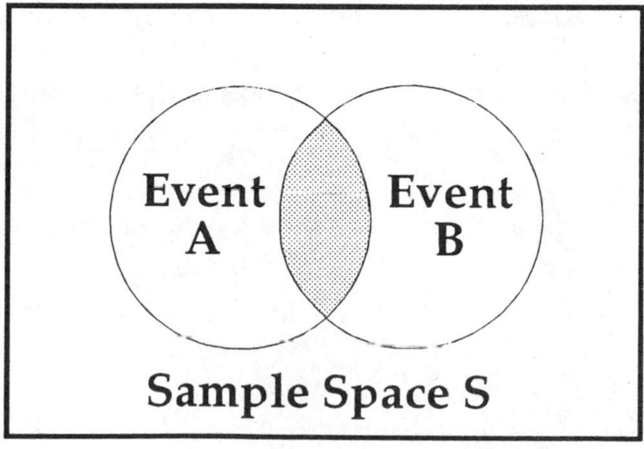

N.	**Mutually Exclusive Events:**	Two events are said to be mutually exclusive if they have no sample points in common (which means their intersection is not possible). Hence, if one event occurs, the other cannot occur.
O.	**Collectively Exhaustive Events:**	A group of events which contain all the sample points in the sample space.
P.	**Complement of Event A:**	The collection of all the sample points which are not included in event A. The complement of A is denoted as A^c.
Q.	**Addition Law:**	A law of probability which is used to determine the probability of the union of events.
R.	**Multiplication Law:**	A law of probability which is used to determine the probability of the intersection of events.
S.	**Conditional Probability:**	The probability of the occurrence of an event, given that another event has occurred.
T.	**Independent Events:**	Two events are said to be independent if the occurrence of one does not affect the occurrence of the other.
U.	**Prior Probabilities:**	The initial estimates of the probabilities of the occurrences of various events.
V.	**Posterior Probabilities:**	The resulting probabilities after the probabilities are revised in the light of new information.
W.	**Bayes' Theorem:**	A method of determining posterior probabilities.

CHAPTER FORMULAS

A Counting Rule for Multiple-step Experiments

Total number of outcomes = $(n_1)(n_2)\ldots(n_k)$

where (n_1) = number of possible outcomes on the first step

(n_2) = number of possible outcomes on the second step
etc.

and k = number of steps

The number of Combinations of N objects taken n at a time

$$\binom{N}{n} = \frac{N!}{n!(N-n)!} \qquad (4.1)$$

where $N! = N(N-1)(N-2)\ldots(2)(1)$
$n! = n(n-1)(n-2)\ldots(2)(1)$

and $0! = 1$

Range of the probability of each outcome

$$0 \leq P(E_i) \leq 1.0 \quad \text{for all i's} \qquad (4.2)$$

Sum of all the experimental outcome probabilities

$$\Sigma\, P(E_i) = 1.0 \qquad (4.3)$$

Sum of the probability of Event A and its Complement

$$P(A) + P(A^c) = 1.0$$

Computing probability using the Complement

$$P(A) = 1 - P(A^c) \qquad (4.4)$$

CHAPTER FORMULAS
(continued)

Addition Law (the probability of the union of two events)

$$P(A \cup B) = P(A) + P(B) - P(A \cap B) \tag{4.5}$$

Conditional Probability

$$P(A \mid B) = \frac{P(A \cap B)}{P(B)} \tag{4.6}$$

or

$$P(B \mid A) = \frac{P(A \cap B)}{P(A)} \tag{4.7}$$

Two Events A and B are Independent if

$$P(A \mid B) = P(A) \tag{4.8}$$

or

$$P(B \mid A) = P(B) \tag{4.9}$$

Multiplication Law (the probability of the intersection of two events)

$$P(A \cap B) = P(A)\, P(B \mid A) \tag{4.10}$$

or

$$P(A \cap B) = P(B)\, P(A \mid B) \tag{4.11}$$

Multiplication Law for Independent Events

$$P(A \cap B) = P(A)\, P(B) \tag{4.12}$$

CHAPTER FORMULAS
(continued)

Bayes' Theorem (Two - Event Case)

$$P(A_1 \mid B) = \frac{P(A_1)\,P(B \mid A_1)}{P(A_1)\,P(B \mid A_1) + P(A_2)\,P(B \mid A_2)} \tag{4.16}$$

and

$$P(A_2 \mid B) = \frac{P(A_2)\,P(B \mid A_2)}{P(A_1)\,P(B \mid A_1) + P(A_2)\,P(B \mid A_2)} \tag{4.17}$$

Bayes' Theorem in General

$$P(A_i \mid B) = \frac{P(A_i)\,P(B \mid A_i)}{P(A_1)\,P(B \mid A_1) + P(A_2)\,P(B \mid A_2) + \ldots + P(A_n)\,P(B \mid A_n)} \tag{4.18}$$

for $A_i = A_1, A_2, \ldots, A_n$

where $P(A_i)$ = prior probability of event i
$P(B \mid A_i)$ = conditional probability of event B given A_i
$P(A_i \mid B)$ = posterior probability of A_i given B

EXERCISES

***1.** Some of the disks produced by a computer disk manufacturer are defective. From the production line, 2 disks are selected and inspected.

(a) What constitutes an experiment for the above situation?

Answer: An experiment is any process which results in well defined outcomes. In the above situation, the selection and classification of disks (as to being defective or non-defective) represent an experiment.

(b) How many sample points exist in the above experiment? List the sample points.

Answer: There are 2 possible outcomes in the selection of the first disk, that is, defective or non-defective ($n_1 = 2$) and 2 possible outcomes on the selection of the second disk ($n_2 = 2$). Hence, there are $(n_1)(n_2) = (2)(2) = 4$ distinct sample points or experimental outcomes. Letting E_i represent the experimental outcomes, then the sample points are as follows:

Sample Points	First Disk	Second Disk
E_1	defective	defective
E_2	defective	non-defective
E_3	non-defective	defective
E_4	non-defective	non-defective

(c) What constitutes the sample space? Use set notation to define the sample space.

Answer: The sample space refers to the collection of all possible sample points. Therefore, for the above experiment, the sample space is shown by the following set notation:

$$S = \{E_1, E_2, E_3, E_4\}$$

(d) Construct a tree diagram for the above experiment.

Answer: Figure 4.1 is a tree diagram for the previous experiment. The first step corresponds to the selection of the first disk, and the second step corresponds to the selection of the second disk. Note that the sample points are given on the right side of the diagram.

TREE DIAGRAM

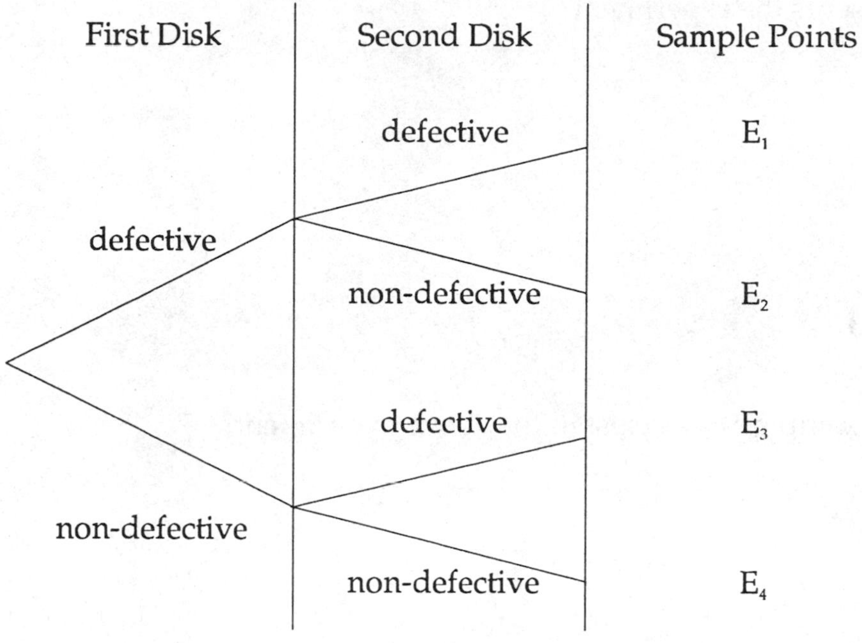

Figure 4.1

2. An experiment consists of selecting a student body president and vice president. All undergraduate students, freshmen through seniors, are eligible for the offices.

(a) How many sample points (possible outcomes as to the classifications) exist?

(b) Enumerate all the sample points.

3. Three applications for admission to a local university are checked, and it is determined whether each applicant is male or female.

(a) What represents the experiment?

(b) How many sample points exist in the above experiment?

(c) List the sample points and use set notation to show the sample space.

4. Assume your favorite football team has 2 games left to finish the season. The outcome of each game can be win, lose, or tie.

(a) How many possible outcomes exist?

(b) List the sample points.

5. Each customer entering a department store will either buy or not buy some merchandise. An experiment consists of following 3 customers and determining whether or not they purchase any merchandise. How many sample points exist in the above experiment? (Note that each customer is either a purchaser or non-purchaser.) List the sample points.

6. An experiment consists of tossing 4 coins successively.

(a) How many sample points exist?

(b) Enumerate the sample points and use set notation to define the sample space.

7. From nine cards numbered 1 through 9, two cards are drawn. Consider the selection and classification of the cards as odd or even as an experiment.

(a) How many sample points are there for this experiment?

(b) Use set notation to define the sample space.

***8.** A company plans to interview 10 recent graduates for possible employment. The company has three positions available. How many groups of three can the company select?

Answer: In this situation, we are interested in determining the number of combinations of 10 jobs (i.e., N = 10) taken three at a time (i.e., n = 3). Therefore, we use the combinations formula as follows:

$$\binom{N}{n} = \frac{N!}{n!(N-n)!} = \frac{10!}{3!(10-3)!} = 120$$

9. A student has to take 7 more courses before she can graduate. If none of the courses are prerequisite to others, how many groups of three courses can she select for the next semester?

$$\binom{7}{3} = \frac{7!}{3!(7-3)} = \frac{7\cdot6\cdot5\cdot4\cdot3\cdot2\cdot1}{3\cdot2\cdot1\,(4\cdot3\cdot2\cdot1)} = \frac{210}{6} = 35$$

10. How many committees consisting of 3 female and 5 male students can be selected from a group of 5 female and 8 male students? (Hint: Find the possible groups for each gender and then multiply the results.)

***11.** The sales records of an automobile dealer in Dallas indicate the following weekly sales volume over the last 200 weeks.

Number of Cars Sold	Number of Weeks
0	8
1	14
2	25
3	60
4	50
5	23
6	12
7	8
Total	200

(a) Determine the probabilities associated with the various sales volumes.

Answer: Since historical data are available, the probabilities can be assigned based on the relative frequency method. For example, out of the 200 weeks, there were 8 weeks where the sales level was zero. Hence, the probability of zero sales is 8/200 = 0.04. Similarly, the probability of selling 1 automobile in any given week is 14/200 = 0.07. Then the probabilities associated with the various sales levels will be as follows:

Sales Level	Probability of the Sales Level
0	0.040
1	0.070
2	0.125
3	0.300
4	0.250
5	0.115
6	0.060
7	0.040
Total	1.000

(b) Have the two basic requirements of the probability assignment been met?

Answer: The first requirement is that the range of probability for any experimental outcome be between zero and 1. As can be seen from the above probabilities, this requirement has been met. The second requirement is that the sum of the probabilities of all experimental outcomes be equal to 1. We note that the sum of the probabilities of the various sales levels is 1. Hence, the second requirement has also been met.

12. A recent survey conducted among 800 married couples showed the following number of children per couple.

Number of Children	Number of Couples
None	40
1	100
2	480
3	120
4	40
≥ 5	20
Total	800

(a) Which approach would you recommend for assigning probabilities to the sample points?

(b) What is the probability for a couple having 5 or more children?

(c) Assign probabilities to the sample points and verify that your assignment satisfies the two basic requirements of probability assignment.

***13.** A market analyst assigns the following subjective probabilities for the Dow Jones Industrial averages at the close of tomorrow's market.

Dow Jones Compared to Today		Probability
Higher	(E_1)	0.6
The Same	(E_2)	0.2
Lower	(E_3)	0.1

(a) Are the above probability assignments valid?

Answer: One of the requirements of probability assignment is that the sum of the probabilities of all experimental outcomes equal to 1. In the above example, all possible outcomes are given; but the sum of their probabilities does not add to 1. Hence, the assignment is not valid.

(b) Assume that in the previous example an analyst predicts that the chance of the market going up tomorrow is 10 times as large as the chance of the market going down. Furthermore, she feels that the chance of the market remaining the same is 9 times as large as the market going down. Assign valid probabilities to the market's outcome.

Answer: Assigning a weight of 1 to E_3, we note that E_1 is 10 times as likely as E_3, and E_2 is 9 times as likely as E_3. Hence, the probabilities can be calculated as follows:

Market Condition		Weights	Probability
Higher	(E_1)	10	10/20 = 0.50
The Same	(E_2)	9	9/20 = 0.45
Lower	(E_3)	1	1/20 = 0.05
Total		20	1.00

14. A sportscaster assigned the following subjective probabilities to the outcome of the forthcoming game for the home team.

Experimental Outcome	Probability
E_1 (Win)	0.5
E_2 (Tie)	0.3
E_3 (Lose)	0.3

(a) Are the above probability assignments valid? Why or why not? Explain fully.

(b) Some local coaches feel that the home team's chance of winning the forthcoming game is 6 times as large as losing the game and 2 times as large as having a tie game. Assign valid probabilities for the win, loss, and tie game.

***15.** Your favorite professional football team (we shall refer to them as the "Favorites") has 2 games left to finish the season, which they will play in the next 2 weeks. If we define the experiment as playing the 2 games,

(a) How many sample points are there in this experiment?

Answer: There are 3 possible outcomes (win, lose, and tie) for each game. Therefore, there is a total of 3 x 3 = 9 possible outcomes.

(b) List all the experimental outcomes for the Favorites.

Answer: All the experimental outcomes (sample points) are

Experimental Outcome	First Game	Second Game
E_1	Win	Win
E_2	Win	Tie
E_3	Win	Lose
E_4	Tie	Win
E_5	Tie	Tie
E_6	Tie	Lose
E_7	Lose	Win
E_8	Lose	Tie
E_9	Lose	Lose

(c) Use set notation to define the sample space for the Favorites.

Answer: $S = \{E_1, E_2, E_3, E_4, E_5, E_6, E_7, E_8, E_9\}$

(d) Assume 200 football experts have given their opinions on the possible outcomes of the two forthcoming games. The following table shows the results of the experts' predictions.

		First Game			
		Win	Tie	Lose	Total
Second Game	Win	80	15	5	100
	Tie	30	20	10	60
	Lose	10	25	5	40
	Total	120	60	20	200

Based on the predictions of the 200 experts, assign probabilities for all the possible outcomes of the two future games.

Answer: Based on the experts' predictions, we note that 80 experts felt that the Favorites will win both games. Therefore, we can assign a probability of winning both games as $P(E_1) = 80/200 = 0.4$ In a similar fashion, we can determine the probability of winning the first game and tying the second as $P(E_2) = 30/200 = 0.15$. Continuing in a similar manner, the following probabilities can be assigned.

Experimental Outcome	First Game	Second Game	Probabilities $P(E_i)$
E_1	Win	Win	0.400
E_2	Win	Tie	0.150
E_3	Win	Lose	0.050
E_4	Tie	Win	0.075
E_5	Tie	Tie	0.100
E_6	Tie	Lose	0.125
E_7	Lose	Win	0.025
E_8	Lose	Tie	0.050
E_9	Lose	Lose	0.025
		Total	1.000

(e) Assume the Favorites need to win at least one game to go to the play-offs. What is the probability of their going to the play-offs?

Answer: Let us define the following events.

W_1 = the event of winning the first game

W_2 = the event of winning the second game

W_p = the event of going to the play-offs

Referring to the answer to Part d of the exercise, we note that sample points E_1, E_2 and E_3 correspond to event W_1; sample points E_1, E_4 and E_7 correspond to event W_2; and sample points E_1, E_2, E_3, E_4 and E_7 correspond to event W_p. In set notation, the events are

$$W_1 = \{E_1, E_2, E_3\}$$

$$W_2 = \{E_1, E_4, E_7\}$$

$$W_p = \{E_1, E_2, E_3, E_4, E_7\}$$

Since the probability of any event is equal to the sum of the probabilities of the sample points in the event, the probabilities of the above events will be

$$P(W_1) = P(E_1) + P(E_2) + P(E_3)$$

$$= 0.40 + 0.15 + 0.05$$

$$= 0.60$$

Similarly,

$$P(W_2) = P(E_1) + P(E_4) + P(E_7)$$

$$= 0.4 + 0.075 + 0.025 = 0.50$$

and

$$P(W_p) = P(E_1) + P(E_2) + P(E_3) + P(E_4) + P(E_7)$$

$$= 0.400 + 0.150 + 0.050 + 0.075 + 0.0250$$

$$= 0.70$$

Therefore, the probability of winning the first game is 0.6; the probability of winning the second game is 0.5; and the probability of going to the play-offs is 0.70.

(f) Find the probability of the union of events W_1 and W_2 (i.e., $P(W_1 \cup W_2)$) and explain what the union shows.

Answer: The union of two events refers to the sample points belonging to one or the other or both. In this case, the union of W_1 and W_2 refers to all the sample points belonging to W_1 or W_2 or both. From Part e of the exercise, we have

$$W_1 = \{E_1, E_2, E_3\}$$

and

$$W_2 = \{E_1, E_4, E_7\}$$

Therefore, the union of W_1 and W_2 is

$$W_1 \cup W_2 = \{E_1, E_2, E_3, E_4, E_7\}$$

Note that the sample points in $W_1 \cup W_2$ are situations in which at least one of the games is won and the Favorites go to the play-offs. Therefore,

$$P(W_1 \cup W_2) = P(E_1, E_2, E_3, E_4, E_7) = 0.70$$

This probability was actually calculated in Part e of this exercise and was denoted as $P(W_p)$.

(g) Find the probability of winning both games.

Answer: In this question, we are interested in the intersection of events W_1 and W_2. The intersection of W_1 and W_2 is that event which contains all the sample points belonging to both W_1 and W_2. Recall that

$$W_1 = \{E_1, E_2, E_3\}$$

and

$$W_2 = \{E_1, E_4, E_7\}$$

The only sample point which is in both W_1 and W_2 is E_1. Denoting the intersection of the two events as $W_1 \cap W_2$, we have

$$W_1 \cap W_2 = \{E_1\}$$

Therefore, the probability of winning both games is

$$P(W_1 \cap W_2) = P(E_1) = 0.4$$

(h) Are the events W_1 and W_2 mutually exclusive? If they are not mutually exclusive, can you find the union of the two events?

Answer: Two or more events are said to be mutually exclusive if the events have no sample points in common. Recall

$$W_1 = \{E_1, E_2, E_3\} \qquad \text{and} \qquad W_2 = \{E_1, E_4, E_7\}$$

In this case, we note that W_1 and W_2 have the sample point E_1 in common. Therefore, they are not mutually exclusive. When events are not mutually exclusive, we can find their union as follows.

$$P(W_1 \cup W_2) = P(W_1) + P(W_2) - P(W_1 \cap W_2)$$

In Part e, we determined

$$P(W_1) = 0.6$$

$$P(W_2) = 0.5$$

and in Part f, we calculated $P(W_1 \cap W_2) = 0.4$.

Therefore,

$$P(W_1 \cup W_2) = 0.6 + 0.5 - 0.4 = 0.70$$

Note that this is the same answer which we had found in Parts e and f.

(i) Let L_1 represent the event that the Favorites lose the first game. Are the events W_1 and L_1 mutually exclusive? Find $P(W_1 \cup L_1)$.

Answer: The sample points which correspond to the loss of the first game are

$L_1 = \{E_7, E_8, E_9\}$

and we recall

$W_1 = \{E_1, E_2, E_3\}$

Since events L_1 and W_1 do not have any sample points in common, they are mutually exclusive. The probability of the union of mutually exclusive events is the sum of the probabilities of all the sample points in each event. Therefore,

$$P(W_1 \cup L_1) = P(W_1) + P(L_1)$$

$$= P(E_1) + P(E_2) + P(E_3) + P(E_7) + P(E_8) + P(E_9)$$

$$= 0.400 + 0.150 + 0.050 + 0.025 + 0.050 + 0.025 = 0.7$$

(j) Define T_1 as the event that the Favorites tie the first game. Are the events W_1, T_1, and L_1 mutually exclusive and collectively exhaustive?

Answer: The sample points which correspond to each of the three events are

$W_1 = \{E_1, E_2, E_3\}$

$T_1 = \{E_4, E_5, E_6\}$

$L_1 = \{E_7, E_8, E_9\}$

Since the three events do not have any sample points in common, they are mutually exclusive. Furthermore, the union of the three events are mutually exclusive and collectively exhaustive. This simply means that the Favorites will either win, tie, or lose the first game. Therefore,

$$P(W_1 \cup T_1 \cup L_1) = 1$$

(k) If $P(W_1)$ is the probability of winning the first game, what is the probability of not winning the first game?

Answer: Since W_1 is the event of winning the first game, then its complement, or W_1^c is the event of not winning the first game. Recall

$$W_1 = \{E_1, E_2, E_3\}$$

$$P(W_1) = 0.6$$

The complement of W_1 (denoted as W_1^c) must contain all the sample points which are not in W_1. Hence,

$$W_1^c = \{E_4, E_5, E_6, E_7, E_8, E_9\}$$

Therefore, the probability of W_1^c is

$$P(W_1^c) = P(E_4) + P(E_5) + P(E_6) + P(E_7) + P(E_8) + P(E_9)$$

$$= 0.075 + 0.100 + 0.125 + 0.025 + 0.050 + 0.025$$

$$= 0.4$$

A more direct approach to the above is to note that

$$P(W_1) + P(W_1^c) = 1$$

Since we had previously calculated the $P(W_1) = 0.6$, we can simply calculate $P(W_1^c)$ as

$$P(W_1^c) = 1 - P(W_1)$$

$$= 1 - 0.60$$

$$= 0.40$$

16. Assume a sample space is given as

$S = \{E_1, E_2, E_3, E_4, E_5\}$

and the following probabilities are assigned to the sample points:

$P(E_1) = 0.01$ A C
$P(E_2) = 0.19$ B C
$P(E_3) = 0.40$ A B
$P(E_4) = 0.30$ C
$P(E_5) = \underline{0.10}$ A
Total 1.00

Let

$A = \{E_1, E_3, E_5\}$
$B = \{E_2, E_3\}$
$C = \{E_1, E_2, E_4\}$

(a) Find P(A), P(B) and P(C).

P(A) = .01 + .40 + .10 = .51
P(B) = .19 + .40 = .59
P(C) = .01 + .19 + .30 = .50

(b) Find $(A \cap C)$ and $P(A \cap C)$.

$A \cap C = E_1$
$P(A \cap C) = .01$

(c) Are events A and C mutually exclusive? Why or why not?

No Because they both have E_1

(d) Find $P(C^c)$.

$P(C^c) = 1.00 - .50 = .50$

(e) Find $(A \cup B)$ and $P(A \cup B)$.

$A \cup B = E_1, E_2, E_3, E_5$
$P(A \cup B) = .01 + .19 + .40 + .10 = .70$

(f) Let

$X = \{E_1, E_3, E_5\}$
$Y = \{E_2, E_4\}$

Are events X and Y mutually exclusive? Are they collectively exhaustive? Explain.

They are mutually exclusive & exhaustive because they have no points in common & because the sum of their probabilities is equal to 1.00

17. Assume you have applied to two different universities (let's refer to them as universities A and B) for your graduate work. In the past, 25% of students (with similar credentials as yours) who applied to University A were accepted; while University B accepted 35% of the applicants. Assume that acceptance at the two universities are independent events.

.75 .65 .130 .5125

(a) What is the probability that you will be accepted in both universities?

$P(A) = .25 \quad P(B) = .35 \quad P(A|B) = .25$ .0875

Mult. Law (Probability of the intersection of 2 pts)

$P(A \cap B) = P(A)\,P(B) = .25 \cdot .35 = .0875$

(b) What is the probability that you will be accepted to at least one graduate program?

union $P(A \cup B) = P(A) + P(B) - P(A \cap B)$

$P(A \cup B) = .25 + .35 - .0875 = .5125$

(c) What is the probability that one and only one of the universities will accept you?

	A	B
Accept	.25	.35
Not Acc.	.75	.65
	1.00	1.00

P(A) + P(B) − P(A|B) − P(B|A) = .25 + .35 − .0875 − .0875 = .425

or P(A) − P(A|B) = .1625
P(B) − P(B|A) = .2625 = .425

(d) What is the probability that neither university will accept you? .4875

= 1 − P(A∪B) = 1 − .5125 = .4875

18. The disk manufacturer (stated in exercise 1) has determined that 3% of his disks are defective. He has just sold 2 disks to a customer.

(a) What is the probability that both of the disks purchased by the customer are defective? 0.0009

A − Disk 1 .03 P = P(.03) P(.03) = .0009
B − Disk 2 .03
P(A∩B) = P(A) P(B) = .03 · .03 = .0009

	Disk 1	Disk 2
Def	.03	.03
ND	.97	.97
	1.00	1.00

(b) What is the probability that one of the disks is defective? .0582

P(A) − P(A∩B) = .03 − .0009 = .0291
P(B) − P(A∩B) = .03 − .0009 = .0291
.0582

(c) What is the probability that neither one of the disks is defective? .9409

P(A) + P(B) − P(A∩B) = 1 − (.03 + .03 − .0009) = .9409
1 − .06 + .0009

(d) Are the events described in a, b, and c of this exercise mutually exclusive and collectively exhaustive? Explain.

19. In exercise 12 of this chapter, the results of a survey of 800 married couples and the number of children they had were given. Based on that data, the following probabilities are determined.

Number of Children	Probability
0	0.050
1	0.125
2	0.600
3	0.150
4	0.050
≥ 5	0.025
Total	1.000

If a couple is selected at random, what is the probability that the couple will have

(a) Less than 4 children?

.050+.125+.600+.150= .925

(b) More than 2 children? .225

(c) Either 2 or 3 children? .75

20. Assume that in your hand you hold an ordinary six sided die and a dime. You toss both the die and the dime on a table.

(a) What is the probability that a head appears on the dime and a six on the die?

(b) What is the probability that a tail appears on the dime and any number more than 3 appears on the die?

(c) What is the probability that a number larger than 2 appears on the die?

21. Refer to exercise 15 of this chapter. Let

W_2 = the event of winning the second game
T_2 = the event of a tie on the second game
L_2 = the event of losing the second game

(a) Find $P(W_2)$, $P(T_2)$ and $P(L_2)$.

$P(W_2) = (E_1 E_4 E_7) = .400 + .075 + .025 = .50$
$P(T_2) = (E_2 E_5 E_8) = .15 + .10 + .05 = .30$
$P(L_2) = (E_3 E_6 E_9) = .05 + .125 + .025 = .20$

(b) Find $P(L_1 \cap L_2)$. .025

$P(L_1) = E_7 E_8 E_9$
$P(L_2) = E_3 E_6 E_9$ = .025

(c) Find $P(T_1 \cap T_2)$. .1

$P(T_1) = E_4 E_5 E_6$
$P(T_2) = E_2 E_5 E_8$ } $= E_5 = .10$

(d) Find $P(L_2 \cup T_2)$. .5

$P(L_2)$ $E_3 E_6 E_9$ = .20
$P(T_2)$ $E_2 E_5 E_8$ = .30
.50

(e) Are L_2 and T_2 mutually exclusive?

yes

22. On a very short quiz, there are one multiple choice question with 5 possible choices (a, b, c, d, e) and one true or false question. Assume you are taking the quiz but do not have any idea what the correct answer is to either question. You mark an answer anyway. Assume that the events are independent of each other.

(a) What is the probability that you have given the correct answer to both questions?

#1-5
±2-2 = 5·2=10
1/10

(b) What is the probability that only one of the two answers is correct?

1of2 or 5/10

(c) What is the probability that neither answer is correct?

5/10 - 1/10 = 4/10

(d) What is the probability that only your answer to the multiple choice question is correct?

(e) What is the probability that you have answered only the true or false question correctly?

5/10 - 1/10 = 4/10

23. Each year the IRS randomly audits 10% of the tax returns. Assume a married couple has filed separate returns and the events of their being audited are independent of each other.

(a) What is the probability that both the husband and the wife will be audited?

(b) What is the probability that only one of them will be audited?

(c) What is the probability that neither one of them will be audited?

(d) What is the probability that at least one of them will be audited?

***24.** Assume $P(A) = 0.40$, $P(B) = 0.50$, and $P(A \mid B) = 0.3$

(a) Are events A and B independent? Explain.

Answer: Two events, A and B, are independent if $P(A \mid B) = P(A)$. Since in this situation $P(A \mid B)$ and $P(A)$ are not equal, the two events are not independent.

(b) Find $P(A \cap B)$ Hint: $P(A \cap B) = P(B \cap A)$

Answer: The multiplication law of probability states that

$$P(B \cap A) = P(B)P(A \mid B)$$

Therefore,

$$P(B \cap A) = (0.5)(0.3) = 0.15$$

(c) Find $P(B \mid A)$.

Answer: Once again using the Multiplication Law, we note that

$$P(A \cap B) = P(A)P(B \mid A)$$

Solving for $P(B \mid A)$, we have

$$P(B \mid A) = \frac{P(A \cap B)}{P(B)}$$

Substituting the values, we conclude

$$P(B \mid A) = \frac{0.15}{0.4} = 0.375$$

25. Assume that P(A) = 0.7, P(B) = 0.8, and P(B ∩ A) = 0.56.

(a) Find P(A | B) and P(B | A).

$P(B|A) = \frac{P(A \cap B)}{P(A)} = \frac{.56}{.7} = .8$

$P(A|B) = \frac{P(B \cap A)}{P(B)} = \frac{.56}{.8} = .7$

(b) Are events A and B independent? Explain.

Yes. Because $P(A|B) = P(A)$

26. Assume that two events A and B are mutually exclusive; and furthermore, P(A) = 0.3 and P(B) = 0.4.

(a) Explain what is meant by "mutually exclusive" events.

no points in common

(b) Find P(A ∩ B).

$P(A \cup B) = P(A) + P(B) - P(A \cap B)$

$.3 + .4$

0

(c) Find P(A ∪ B).

$P(A \cup B) = P(A) + P(B) - P(A \cap B)$

$P(A \cup B) = .3 + .4 - 0$

$P(A \cup B) = .7$

***27.** Tammy is a general contractor and has submitted two bids for two projects (A and B). The probability of getting project A is 0.65. The probability of getting project B is 0.77. The probability of getting at least one of the two projects is 0.90.

(a) What is the probability that she will get both projects?

Answer: We are interested in finding $P(A \cap B)$. The Addition Law states

$$P(A \cup B) = P(A) + P(B) - P(A \cap B)$$

In this problem $P(A \cup B) = 0.90$, $P(A) = 0.65$, and $P(B) = 0.77$. Substituting these values in the above equation, we will have

$$0.90 = 0.65 + 0.77 - P(A \cap B)$$

Now solving for $P(A \cap B)$, we have

$$P(A \cap B) = 0.65 + 0.77 - 0.90 = 0.52$$

(b) Are the events of getting the two projects mutually exclusive?

Answer: If the events are mutually exclusive, their intersection must be zero. In Part a, we found that the intersection of these two events was $P(A \cap B) = 0.52$. Thus, the events are not mutually exclusive.

(c) Are the two events independent? Explain, using probabilities.

Answer: The two events A and B are independent if $P(A \mid B) = P(A)$. For this problem, we can find $P(A \mid B)$ as

$$P(A \mid B) = \frac{P(A \cap B)}{P(B)} = \frac{0.52}{0.77} = 0.675$$

Since this value is not equal to the probability of A we conclude that the events are not independent.

28. Assume you are taking two courses this semester (A and B). The probability that you will pass course A is 0.835, the probability that you will pass both courses is 0.276. The probability that you will pass at least one of the courses is 0.981.

(a) What is the probability that you will pass course B?

$P(A) = .835$
$P(B) =$
$P(A \cap B) = .276$
$P(A|B) = .981$
$P(A \cup B) = .981$

$P(A \cup B) = P(A) + P(B) - P(A \cap B)$
$.981 = .835 + \quad - .276$
$P(B) = P(A \cap B) - P(A) + P(A \cup B)$
$P(B) = .276 - .835 + .981$
$P(B) = .422$

(b) Are the passing of the two courses independent events? Use probability information to justify your answer.

$P(A|B) = \frac{P(A \cap B)}{P(B)}$

$P(A|B) = \frac{.276}{.422} = .654$

No $P(A|B) = .654$ & Does not equal $P(A) = .835$

(c) Are the events of passing the courses mutually exclusive? Fully explain.

No - the intersection is not $\emptyset$

***29.** An automobile dealer has kept records on the customers who visited his store. 40% of the people who visited his dealership were female. Furthermore, his records show that 35% of the females who visited his store purchased an automobile, while 20% of the males who visited his store purchased an automobile. Let

A_1 = the event that the customer is female

A_2 = the event that the customer is male

(a) What is the probability that a customer entering the store will buy an automobile?

Answer: In this case, we want to determine the probability that a customer, regardless of the sex of the customer, will buy a car. Let B = the event that the customer will buy a car. From the statement of the problem, we know the conditional probabilities. That is, the probability that a customer will buy a car under the condition that the customer is a female, is 0.35; and the probability that the customer will buy a car under the condition that the customer is male is 0.20. What was just said can be expressed as

$$P(B \mid A_1) = 0.35$$

$$P(B \mid A_2) = 0.20$$

Therefore, the probability that a customer will buy a car, that is P(B), can be computed as

$$P(B) = P(A_1)\,P(B \mid A_1) + P(A_2)\,P(B \mid A_2) = (0.4)\,(0.35) + (0.6)\,(0.2) = 0.26$$

This indicates that the probability of a customer purchasing a car (regardless of the customer's sex) is 0.26. For details see Figure 4.2.

(b) A car salesperson has just informed us that he sold a car to a customer. What is the probability that the customer was female?

Answer: Based on the information that a purchase was made, we can revise the probability of a customer being a female. Bayes' theorem states

$$P(A_1 \mid B) = \frac{P(A_1)\,P(B \mid A_1)}{P(A_1)\,P(B \mid A_1) + P(A_2)\,P(B \mid A_2)}$$

Substituting the values in the previous equation, we have

$$P(A_1 \mid B) = \frac{(0.4)\,(0.35)}{(0.4)\,(0.35)\ +\ (0.6)\,(0.2)} = \frac{0.14}{0.26} = 0.538$$

Hence, the probability that the customer was a female has been revised from 0.4 to 0.538.

(c) Prepare a table in order to calculate $P(A_1 \mid B)$,$P(A_2 \mid B)$ and $P(B)$.

Answer: The following table summarizes the Bayes' theorem calculations. The calculations are the same as those shown in parts a and b. However, the tabular approach is clearer.

Event	Prior Probabilities $P(A_i)$	Conditional Probabilities $P(B \mid A_i)$	Joint Probabilites $P(A_i \cap B)$	Posterior Probabilities $P(A_i \mid B)$
A_1	0.4	0.35	0.14	$\frac{.14}{.26} = .538$
A_2	0.6	0.20	0.12	$\frac{.12}{.26} = .462$
			$P(B) = 0.26$	

You must note that the joint probability in the above table is the product of the prior and the conditional probabilities. Furthermore, you can see that the probability that a customer is female, given that she purchased a car is 0.538; while the probability that a customer is male, given that he purchased a car is 0.462.

(d) Draw a complete probability tree for the above problem.

PROBABILITY TREE

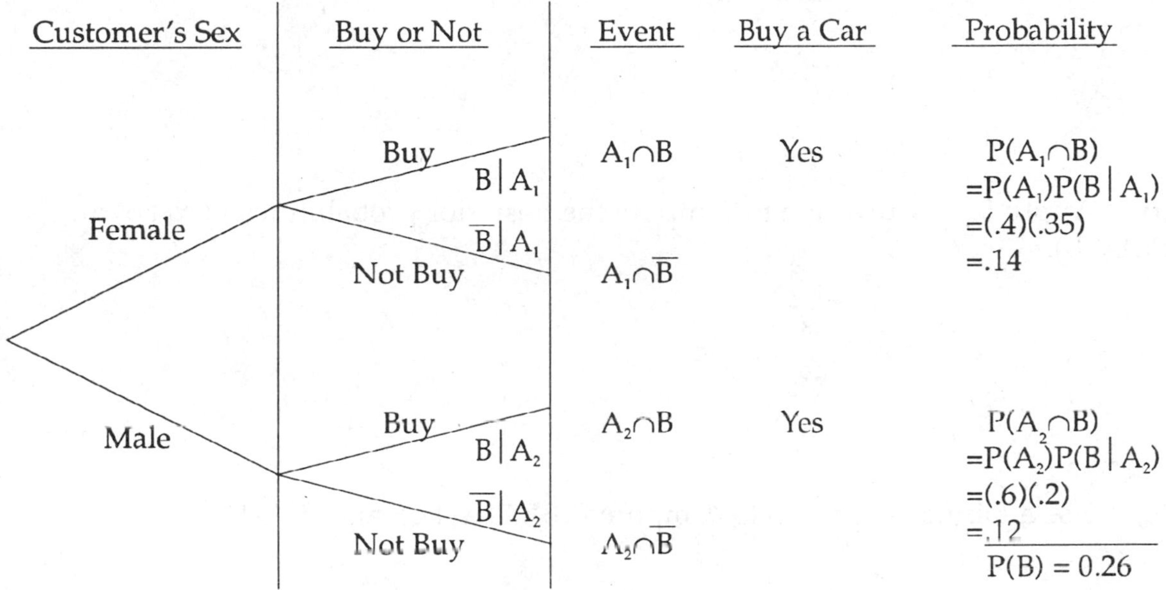

Figure 4.2

30. Refer to exercise 29 and let B = the event that the customer will not buy a car.

(a) A customer visited the store but did not purchase a car. What is the probability that the customer was male? Hint: You want to find $P(A_2 | B^C)$.

(b) Prepare a table in order to calculate $P(A_1 | B^C)$, $P(A_2 | B^C)$ and $P(B^C)$.

31. The prior probabilities for events A_1 and A_2 are $P(A_1) = 0.1$ and $P(A_2) = 0.9$, and the conditional probabilities of event B given A_1 and A_2 are $P(B \mid A_1) = 0.8$ and $P(B \mid A_2) = 0.7$.

(a) Compute $P(B \cap A_1)$, $P(B \cap A_2)$, and $P(B)$.

(b) Apply Bayes' theorem to compute the posterior probabilities $P(A_1 \mid B)$ and $P(A_2 \mid B)$.

(c) Use a tabular approach to compute $P(B)$, $P(A_1 \mid B)$, and $P(A_2 \mid B)$.

(d) Use a tabular approach to compute $P(B^C)$, $P(A_1 \mid B^C)$, and $P(A_2 \mid B^C)$.

32. A local university offers a review course for those who are planning to take the CPA exam. Thirty percent of the people who took the exam had attended the university's CPA review course. The results of the test showed that 60% of those who attended the CPA review course passed the exam; while among those who did not attend the course, only 20% passed the exam.

(a) An individual has passed the exam. What is the probability that he attended the CPA review course?

(b) Let

A_1 = the event that the individual attended the CPA review course
A_2 = the event that the individual did not attend the CPA review course
B = the event that the individual passed the exam

Use a tabular approach to compute $P(B)$, $P(A_1 \mid B)$, and $P(A_2 \mid B)$.

(c) Use a tabular approach to compute $P(B^C)$, $P(A_1 \mid B^C)$, and $P(A_2 \mid B^C)$.

33. The table below shows part of the results of applying Bayes' theorem to a situation:

Events	$P(A_i)$	$P(B \mid A_i)$	$P(A_i \cap B)$	$P(A_i \mid B)$
A_1	.3	.8		
A_2	.2	.4		
A_3	.5	.6		

(a) Identify the prior, conditional, joint, and posterior probabilities.

(b) Complete the above table and find $P(A_1 \cap B)$, $P(A_2 \cap B)$, $P(B)$, $P(A_1 \mid B)$, and $P(A_2 \mid B)$.

34. The prior probabilities for events A_1, A_2, A_3 and A_4 are $P(A_1) = 0.1$, $P(A_2) = 0.6$, $P(A_3) = 0.25$, and $P(A_4) = .05$. The conditional probabilities of event B are $P(B \mid A_1) = .7$, $P(B \mid A_2) = 0.5$, $P(B \mid A_3) = 0.4$, and $P(B \mid A_4) = 0.9$.

(a) Use the tabular approach to find $P(B)$, $P(A_1 \mid B)$, $P(A_2 \mid B)$, $P(A_3 \mid B)$, and $P(A_4 \mid B)$.

(b) Use the tabular approach to find $P(B^C)$, $P(A_1 \mid B^C)$, $P(A_2 \mid B^C)$, $P(A_3 \mid B^C)$, and $P(A_4 \mid B^C)$.

35. A survey of business students who had taken the Graduate Management Admission Test (GMAT) indicated that students who have spent at least five hours studying GMAT review guides have a probability of 0.85 of scoring above 400. Students who do not review have a probability of 0.65 of scoring above 400. It has been determined that 70% of the business students review for the test.

(a) Find the probability of scoring above 400.

(b) Given that a student scored above 400, what is the probability that he/she reviewed for the test?

36. Michael O. Ahmadi has applied for scholarships to two universities. The probability that he will receive a scholarship from University A is 0.55; the probability that he will receive a scholarship from University B is 0.65. The probability that he will receive scholarships from both universities is 0.4.

(a) What is the probability that he will receive at least one scholarship? (Hint: find the probability of A or B or both.)

.8

$P(A) + P(B) - P(A \cap B)$
$.55 + .65 - .40$

(b) What is the probability that he will not receive a scholarship from either university?

.2

$1 - .80 = .20$

***37.** What is the probability that in a group of three people, at least 2 of them will have the same birth dates (i.e., the same date and not necessarily the same year)?

Answer: First let us compute the probability that none of them have the same birth dates and then subtract the results from one, thus arriving at the probability that at least two of them have the same birth dates. The probability that the first individual has a birth date is 365/365; the probability that the second person was not born on the same day of the year is 364/365; and the probability that the third person was not born on any of the other two person's birth date is 363/365. Now we can compute the probability that none of them have the same birth date as

$$\frac{365}{365} \cdot \frac{364}{365} \cdot \frac{363}{365} = 0.9917958$$

Finally, the probability that at least 2 of them were born in the same day is

$$1 - 0.9917958 = 0.0082042$$

38. What is the probability that in a group of four people, at least 2 of them will have the same birth dates (i.e., the same date and not necessarily the same year)?

SELF-TESTING QUESTIONS

In the following multiple choice questions, circle the correct answer. An answer key is provided following the questions.

1. The set of all possible sample points (experimental outcomes) is called

a) a sample
b) an event
c) the sample space
d) none of the above

2. A method of assigning probabilities which assumes that the experimental outcomes are equally likely is referred to as the

a) objective method
b) classical method
c) subjective method
d) none of the above

3. A method of assigning probabilities based on historical data is called the

a) classical method
b) subjective method
c) relative frequency method
d) none of the above

4. The probability assigned to each experimental outcome must be

a) any value larger than zero
b) smaller than zero
c) at least one
d) between zero and one

5. If two events are mutually exclusive, then their intersection

a) will always be equal to zero
b) can have any value larger than zero
c) must be larger than zero, but less than one
d) none of the above

6. The union of events A and B is

a) the same as the intersection of events A and B
b) always equal to zero
c) that event which contains all sample points belonging to A and B
d) that event which contains all sample points belonging to A or B or both

7. Given that an event E has a probability of 0.25, then the probability of the complement of event E

a) cannot be determined with the above information
b) can have any value between zero and one
c) must be 0.75
d) none of the above

8. Two events X and Y are independent if

a) $P(Y \mid X) = P(X)$
b) $P(Y \mid X) = P(Y)$
c) both a and b are satisfied
d) none of the above

9. If $P(A) = 0.2$, $P(B) = 0.6$, and $P(A \mid B) = 0.4$, then $P(A \cap B)$ is

a) 0.80
b) 0.08
c) 0.12
d) 0.24

10. If $P(A) = 0.5$, $P(B) = 0.3$, and $P(A \cap B) = 0.1$, then $P(B \mid A)$ is

a) 0.33
b) 0.20
c) 0.15
d) 0.05

11. Of five letters (A, B, C, D, and E), two letters are to be selected at random. How many possible selections are there?

a) 20
b) 7
c) 5!
d) 10
e) none of the above

12. An experiment consists of three steps. There are three possible outcomes on the first step, four possible outcomes on the second step, and five possible outcomes on the third step. The total number of experimental outcomes is

a) (3!)(4!)(5!)
b) 60
c) 20
d) 10
e) none of the above

13. The range of probability is

a) any value larger than zero
b) any value between minus infinity to plus infinity
c) zero to one
d) any value between -1 to 1
e) none of the above

14. If a penny is tossed six times and comes up heads all six times, the probability of heads on the seventh trial is

a) less than the probability of tails
b) 1/64
c) 0.5
d) larger than the probability of tails
e) none of the above

15. If X and Y are independent events with P(X) = 0.3 and P(Y) = 0.5, then $P(X \cup Y)$ is

$P(x \cup y) = P(x) + P(y) - P(A \cap B)$

a) 0.80
b) 0.15
c) 0.20
d) 0.65
e) none of the above

ANSWERS TO THE SELF-TESTING QUESTIONS

1. c
2. b
3. c
4. d
5. a
6. d
7. c
8. b
9. d
10. b
11. d
12. b
13. c
14. c
15. d

ANSWERS TO CHAPTER FOUR EXERCISES

2. (a) 16
   (b)

F,F	S,F	J,F	Sr,F
F,S	S,S	J,S	Sr,S
F,J	S,J	J,J	Sr,J
F,Sr	S,Sr	J,Sr	Sr,Sr

3. (a) The selection and classification by sex
   (b) (2)(2)(2) = 8
   (c)

MMM	FMM	FFM	FFF
	MFM	FMF	
	MMF	MFF	

   Let E_i represent each of the above sample points.
   Then:
   $S = \{E_1, \ldots, E_8\}$

4. (a) (3)(3) = 9
   (b) Sample points are:

WW	LW	TW
WL	LL	TL
WT	LT	TT

5. (2)(2)(2) = 8
   Let P = purchaser
   N = non-purchaser

PPP	PPN	PNN	NNN
	PNP	NPN	
	NPP	NNP	

6. (a) (2)(2)(2)(2) = 16
   (b)

HHHH	HHHT	TTTH	HHTT	TTTT
	HHTH	TTHT	HTHT	
	HTHH	THTT	HTTH	
	THHH	HTTT	THHT	
			THTH	
			TTHH	

Defining each of the above outcomes as E_i, the set notation will be as follows: $S = \{E_1, \ldots, E_{16}\}$.

7. (a) (2)(2) = 4
   (b) Let 0 = odd
   E = even
   S = {00, 0E, E0, EE}

9. 35

10. 560

12. (a) Relative frequency
    (b) 0.025
    (c)

Number of Children	P
0	0.050
1	0.125
2	0.600
3	0.150
4	0.050
5	0.025
	1.000

14. (a) No. Sum of probabilities is greater than 1.
    (b) 0.6, 0.3, 0.1

16. (a) $P(A) = P(E_1) + P(E_3) + P(E_5) = 0.01 + 0.4 + 0.1 = 0.51$
    $P(B) = 0.59$
    $P(C) = 0.50$
    (b) $(A \cap C) = \{E_1\}$
    $P(A \cap C) = P(E_1) = 0.01$
    (c) A and C are not mutually exclusive because they have E_1 in common.
    (d) $P(C^c) = 1 - P(C) = 1 - 0.5 = 0.5$
    (e) $(A \cup B) = \{E_1, E_2, E_3, E_5\}$
    $P(A \cup B) = 0.7$
    (f) X and Y are mutually exclusive and collectively exhaustive because they do not have any sample points in common and the sum of their probabilities is equal to 1.

17. (a) 0.0875
    (b) 0.5125
    (c) 0.425
    (d) 0.4875

18. (a) 0.0009
    (b) 0.0582
    (c) 0.9409
    (d) Yes, they do not have any sample points in common, and the sum of their probabilities is equal to 1.

19. (a) 0.925
    (b) 0.225
    (c) 0.750

20. (a) 1/12
    (b) 3/12
    (c) 8/12

21. (a) $P(W_2) = 0.5$ $P(T_2) = 0.3$ $P(L_2) = 0.2$
    (b) 0.025
    (c) .1
    (d) 0.5
    (e) Yes, they do not have any sample points in common, and the sum of their probabilities is equal to 1.

22. (a) 1/10
    (b) 5/10
    (c) 4/10
    (d) 1/10
    (e) 4/10

23. (a) 0.01
    (b) 0.18
    (c) 0.81
    (d) 0.19

25. (a) 0.7, 0.8
    (b) Yes, $P(A \mid B) = P(A)$

26. (a) Two events are said to be mutually exclusive if they have no sample points in common (which means their intersection is not possible). Hence, if one event occurs, the other cannot occur.

    (b) 0
    (c) 0.7

28. (a) 0.422
(b) No, $P(A \mid B) = 0.276/0.422 = 0.654$ which is not equal to $P(A) = 0.835$
(c) No, the intersection is not zero.

30. (a) 0.6486

(b)

Event	$P(A_i)$	$P(B^c \mid A_i)$	$P(A_i \cap B^c)$	$P(A_i \mid B^c)$
A_1	.4	.65	.26	.26/.74 = .3514
A_2	.6	.80	.48	.48/.74 = .6486
			$P(B^c) = .74$	

31. (a) $P(B \cap A_1) = 0.08$
$P(B \cap A_2) = 0.63$
$P(B) = 0.71$

(b) $P(A_1 \mid B) = 0.08/.71 = 0.1127$
$P(A_2 \mid B) .= 0.63/0.71 = 0.8873$

(c)

Event	$P(A_i)$	$P(B \mid A_i)$	$P(B \cap A_i)$	$P(A_i \mid B)$
A_1	.1	.8	.08	.08/.71 = .1127
A_2	.9	.7	.63	.63/.71 = .8873
			$P(B) = .71$	

(d)

Event	$P(A_i)$	$P(B^C \mid A_i)$	$P(B^C \cap A_i)$	$P(A_i \mid B^C)$
A_1	.1	.2	.02	.02/.29 = .069
A_2	.9	.3	.27	.27/.29 = .931
			$P(B^C) = .29$	

32. (a) .18/.32 = .5625

(b)

Event	$P(A_i)$	$P(B \mid A_i)$	$P(B \cap A_i)$	$P(A_i \mid B)$
A_1	.3	.60	.18	.18/.32 = .5625
A_2	.7	.20	.14	.14/.32 = .4375
			$P(B) = .32$	

(c)

Event	$P(A_i)$	$P(B^c \mid A_i)$	$P(B^c \cap A_i)$	$P(A_i \mid B^c)$
A_1	.3	.40	.12	.12/.68 = .1765
A_2	.7	.80	.56	.56/.68 = .8235
			$P(B^c)$ = .68	

33. (a) and (b)

Event	Prior $P(A_i)$	Conditional $P(B \mid A_i)$	Joint $P(A_i \cap B)$	Posterior $P(A_i \mid B)$
A_1	.3	.8	.24	.24/.62 = .3871
A_2	.2	.4	.08	.08/.62 = .1290
A_3	.5	.6	.30	.30/.62 = .4839
			$P(B)$ = .62	

34. (a)

Event	$P(A_i)$	$P(B \mid A_i)$	$P(A_i \cap B)$	$P(A_i \mid B)$
A_1	0.10	.7	.070	.07/.515 = .1359
A_2	0.60	.5	.300	.30/.515 = .5825
A_3	0.25	.4	.100	.10/.515 = .1942
A_4	0.05	.9	.045	.045/.515 = .0874
			$P(B)$ = .515	

(b)

Event	$P(A_i)$	$P(B^c \mid A_i)$	$P(A_i \cap B^c)$	$P(A_i \mid B^c)$
A_1	0.10	.3	.030	.03/.485 = .0618
A_2	0.60	.5	.300	.31/.485 = .6186
A_3	0.25	.6	.150	.15/.485 = 3093
A_4	0.05	.1	.005	.005/.485 = .0103
			$P(B^c)$ = .485	

35.

Event	$P(A_i)$	$P(B \mid A_i)$	$P(A_i \cap B)$	$P(A_i \mid B)$
A_1	.70	.85	.595	.595/.79 = .7532
A_2	.30	.65	.195	.195/.79 = .2468
			$P(B)$ = .790	

(a) .79
(b) .595/.790 = 0.7532

36. (a) 0.8
    (b) 0.2

38. $1 - \left(\frac{365}{365} \cdot \frac{364}{365} \cdot \frac{363}{365} \cdot \frac{362}{365}\right) = 0.01635$

CHAPTER FIVE

DISCRETE PROBABILITY DISTRIBUTIONS

CHAPTER OUTLINE AND REVIEW

In Chapter 4, you learned what is meant by an experiment, how one determines the probability of the outcome of an experiment, and various laws of probability. In Chapter 5, you have been introduced to the concept of a random variable and the probability distribution of a discrete random variable. You were also introduced to three major discrete probability distributions: *Binomial*, *Poisson*, and *Hypergeometric* distributions. The main concepts which you should have learned in Chapter 5 are as follows.

A. Random Variable: A numerical description of the outcome of an experiment. For instance, if an experiment consists of inspecting 10 tires produced by a manufacturer, then the random variable is the number of defective tires whose value could be any number from zero to 10.

B. Discrete Random Variable: A random variable that can assume only a countable number of values. For example, if the random variable is the number of defective tires in a group of 10 tires, the variable under consideration is a discrete random variable because it can take only values of 0 or 1 or 2 or 3 or . . . 10 (i.e., we cannot have 1.7 or 2.2 defective tires).

C. **Continuous Random Variable:** A variable that can assume any value in an interval. For example, if a random variable is the weight of an item which may weigh from 5 to 7 ounces, the random variable weight is a continuous random variable because it can take any value in the interval of 5 to 7 ounces (such as, 5.1, 6.22, etc.).

D. **Probability Distribution:** A description of how the probabilities are distributed over the values the random variable can assume.

E. **Discrete Probability Distribution:** A description of how the probabilities of a discrete random variable are distributed over various values of the random variable by means of a probability function denoted by f(x).

F. **Probability Function:** A function such as f(x) which gives the probability that x can assume a particular value.

G. **Expected Value:** When the values of the random variable are multiplied by their respective probabilities and the results are summed, this summation is known as the expected value. Hence, the expected value is actually a weighted average of the values of the random variable, where the probabilities are the weights. The expected value can be viewed as a long run average.

H. **Variance:** A measure of dispersion or variability of a random variable.

I. **Standard Deviation:** The positive square root of the variance.

J. **Binomial Probability Distribution:** A distribution which is applied to discrete random variables and is used for determining the probability of x successes in n trials. For the binomial distribution to be applicable, the following assumptions must be satisfied.

1. The experiment consists of a sequence of n identical trials.
2. For each trial, there are two possible outcomes (one called success, the other failure).
3. The probabilities of success (or failure) remain the same as successive trials are made.
4. The trials are independent of each other.

K. **Binomial Probability Function:** A function used to determine probabilities in a binomial experiment.

L. **Poisson Probability Distribution:** A distribution (like the binomial distribution) which is applied to discrete random variables. It is used to determine the probability of x occurrences of an event over a designated interval. For the Poisson distribution to be applicable, the following assumptions must be satisfied.

1. The probability of the occurrence of an event is the same for any two intervals of the same length.
2. The occurrence of an event in any interval is independent of the occurrence in any other interval.

M. **Poisson Probability Function:** The function used to compute Poisson probabilities.

N. **Hypergeometric Probability Function:** The function used to compute the probability of x successes in n trials when the trials are not independent.

CHAPTER FORMULAS

Required Conditions for a Discrete Probability Function

$$f(x) \geq 0 \tag{5.1}$$

$$\Sigma f(x) = 1 \tag{5.2}$$

Expected Value of a Discrete Random Variable

$$E(x) = \mu = \Sigma(x\, f(x)) \tag{5.3}$$

Variance of a Discrete Random Variable

$$\text{Variance }(x) = \sigma^2 = \Sigma(x - \mu)^2 f(x) \tag{5.4}$$

Number of Experimental Outcomes Providing Exactly x Successes in n Trials

$$\binom{n}{x} = \frac{n!}{x!(n - x)!} \tag{5.5}$$

where $n! = n(n - 1)(n - 2)\ldots(2)(1)$ (Remember: $0! = 1$) (5.6)

Probability of a Particular Sequence of Outcomes (in a Binomial Experiment) Resulting in Exactly x Successes in n Trials

$$p^x (1 - p)^{n - x} \tag{5.7}$$

where p = probability of success
x = number of successes
n = number of trials

Chapter Formulas
(Continued)

Binomial Probability Function

$$f(x) = \frac{n!}{x!(n-x)!} p^{x} (1-p)^{n-x} \qquad (5.8)$$

where x = 0 ,1, 2, ..., n

The Mean of a Binomial Distribution

$$\mu = n p \qquad (5.9)$$

The Variance of a Binomial Distribution

$$\sigma^{2} = n p (1-p) \qquad (5.10)$$

Poisson Probability Function

$$f(x) = \frac{\mu^{x} e^{-\mu}}{x!} \qquad \text{for } x = 0, 1, 2, \ldots \qquad (5.11)$$

where μ = average number of occurrences in an interval
e = 2.71828
x = 0, 1, 2, ..., n

Hypergeometric Probability Function

$$f(x) = \frac{\binom{r}{x}\binom{N-r}{n-x}}{\binom{N}{n}} \qquad \text{for } 0 \le x \le r \qquad (5.12)$$

where f(x) = probability of x successes
N = number of elements in the population
r = number of elements in the population labeled success
n = number of elements in the sample

EXERCISES

***1.** An experiment consists of counting the number of bad checks received by a grocery store in a given day.

(a) Identify the random variable.

Answer: A random variable is a numerical description of the outcome of an experiment. In this case, the random variable is the number of bad checks received in a given day.

(b) What is the value that the random variable can take?

Answer: The value that the random variable can take ranges between zero and the number of checks written in a given day.

(c) Is the random variable discrete or continuous?

Answer: Our random variable (the number of bad checks) can take only a countable number of values. That is, the number of bad checks could be 1, 2, 3, etc. Therefore, the random variable is a discrete random variable.

2. An experiment consists of making 80 telephone calls in order to sell a particular insurance policy.

(a) What is the random variable?

(b) What values can the random variable take?

(c) Is the random variable discrete or continuous?

3. An experiment consists of determining the speed of automobiles on a highway by the use of radar equipment.

(a) What is the random variable?

(b) What values can the random variable take?

(c) Is the random variable discrete or continuous?

4. Write at least 4 experiments in which the random variable is discrete and 4 experiments in which the random variable is continuous. Identify the random variable in each, and give the values that the random variable can take.

***5.** G.R.I.P.E. Corporation has kept a record of the number of grievances filed per week for the last 50 weeks. The results are shown below.

Number of Grievances	Number of Weeks
0	2
1	18
2	25
3	4
4	1
Total	50

(a) Develop a probability distribution for the above data.

Answer: A probability distribution describes the distribution of random variables and their corresponding probabilities. In the above situation, we note that the probability of no grievance is 2/50 or symbolically, $f(0) = 2/50 = 0.04$. Similarly, the probability of 1 grievance or $f(1) = 18/50 = 0.36$. Continuing in the same manner, the probability distribution can be shown as follows, where x represents the number of grievances and f(x) is the probability of that number of grievances.

x	f(x)
0	0.04
1	0.36
2	0.50
3	0.08
4	0.02
Total	1.00

(b) Is the above a proper probability distribution?

Answer: We note that both required conditions for a discrete probability distribution have been met. First, $f(x) \geq 0$, which means the probability for each event is greater than or equal to zero; and in this case, each probability is greater than zero. The second required condition is that $\Sigma f(x) = 1$, which means the sum of all the probabilities must be equal to one. In the above situation, this second condition is also satisfied. Hence, the distribution which was determined in Part a is a proper probability distribution.

(c) Determine the cumulative probability distribution for the above problem.

Answer: The cumulative probability distribution F(x), gives the probability that a random variable can take on a certain value or less. In this case, we note that the probability of zero grievances is 0.04; and the probability of one or less grievance is 0.04 + 0.036 = 0.40. Symbolically, we can write the sum of the probability of zero grievances and one grievance as
F(1) = f(0) + f(1) = 0.04 + 0.36 = 0.40. Similarly, we can compute the cumulative probability for the entire distribution:

x	F(x)
0	0.04
1	0.40
2	0.90
3	0.98
4	1.00

6. The management of the grocery store mentioned in exercise 1, has kept a record of bad checks received per day for a period of 200 days. The data is shown below.

Number of Bad Checks Received	Number of Days
0	8
1	12
2	20
3	60
4	40
5	30
6	20
7	10

(a) Develop a probability distribution for the above data.

(b) Is the probability distribution, which you found in Part a, a proper probability distribution? Explain.

(c) Determine the cumulative probability distribution F(x).

(d) What is the probability that in a given day the store receives four or less bad checks?

(e) What is the probability that in a given day the store receives more than 3 bad checks?

7. The police records of a metropolitan area kept over the past 300 days show the following number of fatal accidents.

Number of Fatal Accidents x	Number of Days f(x)
0	45
1	75
2	120
3	45
4	15

(a) Develop a probability distribution for the daily fatal accidents.

(b) Determine the cumulative probability F(x).

(c) What is the probability that in a given day there will be less than 3 accidents?

(d) What is the probability that in a given day there will be at least 1 accident?

***8.** Referring to exercise 5 of this chapter, we note the following probability distribution existed for the number of grievances filed with the G.R.I.P.E. Corporation.

x	f(x)
0	0.04
1	0.36
2	0.50
3	0.08
4	0.02

(a) Determine the expected value of the number of grievances and explain its meaning.

Answer: The expected value of a discrete random variable is a weighted average of all possible values of the random variable, where the probabilities are the weights. In other words, the expected value of a discrete random variable x is as follows.

$$E(x) = \mu = \Sigma(x\ f(x))$$

Hence, the expected value can be calculated as follows.

x	f(x)	x f(x)
0	0.04	(0)(0.04) = 0.00
1	0.36	(1)(0.36) = 0.36
2	0.50	(2)(0.50) = 1.00
3	0.08	(3)(0.08) = 0.24
4	0.02	(4)(0.02) = 0.08

$$E(x) = \mu = \Sigma(x\ f(x)) = 1.68$$

The expected value of the number of grievances is 1.68. This figure represents the mean or the average value of the random variable. Obviously, the G.R.I.P.E. Corporation will never have 1.68 grievances, but this figure shows the "long run" average value of the number of grievances.

(b) Determine the variance and the standard deviation.

Answer: The variance of a discrete random variable is as follows.

$$\text{Variance }(x) = \sigma^2 = \Sigma((x - \mu)^2 f(x))$$

We have already calculated the mean to be $\mu = 1.68$. Hence, the variance can be calculated as shown below.

x	x - μ	$(x - \mu)^2$	f(x)	$(x - \mu)^2 f(x)$
0	-1.68	2.8224	0.04	0.1129
1	-0.68	0.4624	0.36	0.1665
2	0.32	0.1024	0.50	0.0512
3	1.32	1.7424	0.08	0.1394
4	2.32	5.3824	0.02	0.1076

$$\sigma^2 = \Sigma(x - \mu)^2 f(x) = 0.5776$$

Since the standard deviation is the square root of the variance, the standard deviation can be determined as

$$\sigma = \sqrt{\sigma^2} = \sqrt{0.5776} = 0.76$$

9. Referring to exercise 6, we note that the number of bad checks received per day and their respective probabilities are as follows:

Number of Bad Checks Received Per Day	Probability
0	0.04
1	0.06
2	0.10
3	0.30
4	0.20
5	0.15
6	0.10
7	0.05

(a) What is the expected number of bad checks received per day?

(b) Determine the variance in the number of bad checks received per day.

(c) What is the standard deviation?

10. Compute All is a computer consulting firm. The number of new clients that they have obtained each month has ranged from 0 to 6. The number of new clients has the probability distribution shown below.

Number of New Clients	Probability
0	0.05
1	0.10
2	0.15
3	0.35
4	0.20
5	0.10
6	0.05

(a) What is the expected number of new clients per month?

(b) Determine the variance and the standard deviation.

11. The number of electrical outages in a city varies from day to day. Assume that the number of electrical outages in the city (x) has the following probability distribution.

x	f(x)
0	0.80
1	0.15
2	0.04
3	0.01

(a) Determine the mean and the standard deviation for the number of electrical outages.

(b) If each outage costs the power company $1500, what is the expected daily cost?

12. The demand for wood heaters has been increasing in recent years. A local store which sells wood heaters has determined the following probability distribution for the demand for wood heaters (x) per week.

x	f(x)
0	0.01
1	0.02
2	0.10
3	0.35
4	0.20
5	0.11
6	0.08
7	0.05
8	0.04
9	0.03
10	0.01

(a) Determine the expected demand per week.

(b) Determine the variance and the standard deviation.

13. Oriental Reproductions, Inc. is a company which produces handmade carpets with oriental designs. The production records show that the monthly production has ranged from 1 to 5 carpets. The production levels and their respective probabilities are shown below.

Production Per Month x	Probability f(x)
1	0.01
2	0.04
3	0.10
4	0.80
5	0.05

(a) Determine the expected monthly production level.

(b) Determine the standard deviation for the production.

***14.** A production process has been producing 10% defective items. A random sample of four items is selected from the production line.

(a) What is the probability that the first 3 selected items are non-defective and the last item is defective?

Answer: If we let D represent the outcome where the selected item is defective and N represent the selection of a non-defective item, then the outcome whose probability we want to find is

(N, N, N, D)

Since these events are independent, the probability of their joint occurrence is equal to the product of their individual probabilities. Therefore, the probability of 3 non-defectives followed by a defective item is (0.9)(0.9)(0.9)(0.1) = 0.0729. Now let us view the above situation in a more generalized form. Recall that when an experiment involves n trials, the probability of obtaining any one sequence of outcomes which result in exactly x successes, as given by equation 5.7, is

$$p^{x}(1-p)^{n-x}$$

where p = probability of success (in the above situation $p = 0.9$)
x = number of successes (in our case, the number of non-defective items which is equal to 3)
n = number of trials (in our situation, $n = 4$)

Hence, the probability of the above sequence of outcomes is as follows.

$$(0.9)^3(1-0.9)^{4-3} = (0.9)^3(.1)^1 = 0.0729$$

(b) If a sample of 4 items is selected, how many outcomes contain exactly 3 non-defective items?

Answer: In Part a of this exercise, we looked at one possible outcome which contained exactly 3 non-defective items. We can enumerate all the outcomes which contain exactly 3 non-defective items:

(N, N, N, D)
(N, N, D, N)
(N, D, N, N)
(D, N, N, N)

As you can see, there are 4 possible outcomes which contain exactly 3 non-defective items. Now let us look at the situation posed in Part b in a more generalized form. Using equation 5.5, we can determine the number of experimental outcomes providing exactly x successes in n trials:

$$\binom{n}{x} = \frac{n!}{x!(n-x)!} = \frac{4!}{3!(4-3)!} = 4$$

(c) What is the probability that a random sample of 4 contains 3 non-defective items?

Answer: In Part a, we determined the probability of 3 non-defective items (in a specific sequence) to be 0.0729; and in Part b, we determined that there were 4 possible sequences. Hence, the probability of 3 non-defective items (in any sequence) is

(4) (0.0729) = 0.2916

In general, we can use the binomial probability function to determine the probability of x successes in n trials. This equation states

$$f(x) = \frac{n!}{x!(n-x)!} p^{x} (1-p)^{n-x}$$

where $x = 0, 1, 2, \ldots n$

Therefore, for our example,

$$f(3) = \frac{4!}{3!(4-3)!} (0.9)^{3} (.1)^{1} = 0.2916$$

(d) Determine the probability distribution for the number of non-defective items in a sample of four.

Answer: In a sample of four, the number of non-defectives can be 0, 1, 2, 3, and 4. We can determine the probability distribution as

x	f(x)
0	$\frac{4!}{0!\,4!}(0.9)^0(0.1)^4 = 0.0001$
1	$\frac{4!}{1!\,3!}(0.9)^1(0.1)^3 = 0.0036$
2	$\frac{4!}{2!\,2!}(0.9)^2(0.1)^2 = 0.0486$
3	$\frac{4!}{3!\,1!}(0.9)^3(0.1)^1 = 0.2916$
4	$\frac{4!}{4!\,0!}(0.9)^4(0.1)^0 = \underline{0.6561}$
	1.0000

(e) Are there tables available where one can readily read the probabilities?

Answer: Yes, Table 5 of Appendix B (in your textbook) provides us with probabilities for various values of p, n, and x. In Part c of this problem, we computed the probability that a sample of four (n = 4) contains three non defective (x = 3) elements, where the probability of a non-defective element was 0.9. the probability was determined to be 0.2916. Now let us determine this probability by using the table of binomial probabilities. Refer to Table 5 of Appendix B and turn to page b-16. On the left, you are given the sample size. Locate a sample size of 4. At the top of the page, you are given the probability of success. Locate the p of 0.9. Now you can read the binomial probability for x = 3, n = 4, and p = 0.9 as f(3) = 0.2916. As you note, this is the same value which was determined in Part c.

15. Forty percent of all registered voters in a national election are female. If a random sample of 5 voters is selected,

(a) What is the probability that the sample contains 2 female voters?

(b) What is the probability that there are no females in the sample?

(c) What is the probability that every member of the selected sample is female?

16. The records of a department store show that 20% of their customers who make a purchase return the merchandise in order to exchange it. What is the probability that in the next 6 purchases

(a) Three customers return the merchandise for exchange?

(b) Four customers return the merchandise for exchange?

(c) None of the customers return the merchandise for exchange?

17. A cosmetics salesperson who calls potential customers to sell her products has determined that 30% of her telephone calls result in a sale. Determine the probability distribution for her next three calls. Note that the next three calls could result in 0, 1, 2, or 3 sales.

***18.** Refer to exercise 14 of this chapter.

(a) Determine the expected number of non-defective items in a sample of four.

Answer: We know that the expected number of non-defectives can be calculated as shown below.

$$\mu = \Sigma(x\ f(x))$$

In Part d of exercise 14, we determined the probability distribution for the number of non-defectives in a sample of four. Hence, the expected number of non-defective items can be calculated as

x	f(x)	x f(x)
0	0.0001	0.0000
1	0.0036	0.0036
2	0.0486	0.0972
3	0.2916	0.8748
4	0.6561	2.6244
		E(x) = 3.6000

(b) Is there another approach for determining the expected number of non-defectives? If yes, explain.

Answer: In the case of a binomial distribution, the expected value of a random variable can be determined by

$$E(x) = n\,p$$

Therefore, in our example where n = 4 and p = 0.9, we can determine the expected number of non-defective items as

$$E(x) = (4)\,(0.9) = 3.6$$

(c) Find the standard deviation for the number of non-defectives.

Answer: In the case of a binomial distribution, the variance of a random variable is given by the following.

$$\sigma^2 = n\,p\,(1 - p)$$

Therefore, the variance is

$$\sigma^2 = (4)(.9)(1 - .9) = 0.36$$

and the standard deviation is

$$\sigma = \sqrt{\sigma^2} = \sqrt{0.36} = 0.6$$

19. In exercise 17, you were asked to determine the probability distribution for the number of telephone calls resulting in a sale (from a sample of three), where past data indicated that 30% of the calls resulted in a sale. The following table gives this probability distribution.

x	f(x)	x f(x)
0	0.343	
1	0.441	
2	0.189	
3	0.027	

(a) Using the above data, find the expected number of sales and show that $np = \Sigma x\, f(x)$.

(b) Find the standard deviation for the number of sales.

20. In Part d of exercise 14, the probability distribution for the number of non-defective items was determined.

(a) Use the data in problem 14 and find the probability distribution for the number of defective items.

(b) Determine the expected number of defective items.

(c) Determine the standard deviation for the number of defectives item.

IN EXERCISES 21 THROUGH 23, USE THE BINOMIAL PROBABILITY TABLES.

21. In a large western university, 15% of the students are graduate students. If a random sample of 20 students is selected, what is the probability that the sample contains

(a) Exactly four graduate students?

(b) No graduate students?

(c) Exactly twenty graduate students?

(d) More than nine graduate students?

(e) Less than five graduate students?

(f) What is the expected number of graduate students?

22. In a southern state, it was revealed that 5% of all automobiles in the state did not pass inspection. What is the probability that of the next ten automobiles entering the inspection station

(a) None will pass inspection?

(b) All will pass inspection?

(c) Exactly two will not pass inspection?

(d) More than three will not pass inspection?

(e) Less than two will not pass inspection?

(f) Find the expected number of automobiles not passing inspection.

(g) Determine the standard deviation for the number of cars not passing inspection.

23. Twenty percent of the applications received for a particular credit card are rejected. What is the probability that among the next fifteen applications

(a) None will be rejected?

(b) All will be rejected?

(c) Less than 2 will be rejected?

(d) More than four will be rejected?

(e) Determine the expected number of rejected applications and its variance.

***24.** During the registration period (in a local college), students consult their advisors about course selection. A particular advisor noted that during each half hour an average of eight students came to see her for advising.

(a) What is the probability that during a half hour period exactly four students will consult her?

Answer: If we assume that the probability of a student consulting his/her advisor is the same for any two time periods of equal length and, furthermore, if we assume that consulting or not consulting the advisor in any time period is independent of consulting or not consulting the advisor in any other time period, we can conclude that the Poisson probability distribution is applicable. The Poisson probability function is

$$f(x) = \frac{\mu^{x} e^{-\mu}}{x!} \qquad \text{for } x = 0, 1, 2, \ldots$$

where μ is the average number of occurrences in an interval and e is a constant whose value is 2.7182818. . . Therefore, we can find the probability of four students consulting their advisor as

$$f(4) = \frac{(8)^4 \ (e)^{8}}{4!} = \frac{(4096)\ (0.0003355)}{24} = 0.0573$$

(b) Are there tables available where one can readily read probabilities for specific values of x and μ?

Answer: Yes, there are tables which give probabilities for specific values of x and μ. Refer to the Poisson probability tables (in your textbook), and you will note that with a $\mu = 8$ and $x = 4$ the probability can be read as $f(4) = 0.0573$, which is the same value which we found in Part a.

(c) What is the probability that less than three students will consult their advisor?

Answer: In this question, we are interested in determining the probability that 0, 1, or 2 students will consult their advisor. We can read the individual probabilities from the Poisson probability tables and sum the individual probabilities in order to find the probability of $x \leq 2$. In other words,

$$f(x \leq 2) = f(2) + f(1) + f(0)$$

$$= 0.0107 + 0.0027 + 0.0003$$

$$= 0.0137$$

25. An insurance company has determined that each week an average of nine claims are filed in their Atlanta branch. What is the probability that during the next week

(a) Exactly seven claims will be filed?

(b) No claims will be filed?

(c) Less than four claims will be filed?

(d) At least eighteen claims will be filed?

26. A local university reports that 3% of their students take their general education courses on a pass/fail basis. Assume that fifty students are registered for a general education course.

(a) What is the expected number of students who have registered on a pass/fail basis?

(b) What is the probability that exactly five are registered on a pass/fail basis?

(c) What is the probability that more than three are registered on a pass/fail basis?

(d) What is the probability that less than four are registered on a pass/fail basis?

27. Only 0.02% of credit card holders of a company report the loss or theft of their credit cards each month. The company has 15,000 credit cards in the city of Memphis. What is the probability that during the next month in the city of Memphis

(a) No one reports the loss or theft of his/her credit cards?

(b) Every credit card is lost or stolen?

(c) Six people report the loss or theft of their cards?

(d) At least nine people report the loss or theft of their cards?

(e) Determine the expected number of reported lost or stolen credit cards.

(f) Determine the standard deviation for the number of reported lost or stolen cards.

***28.** A retailer of electronic equipment received six VCRs from the manufacturer. Three of the VCRs were damaged in the shipment. The retailer sold two VCRs to two customers.

(a) What is the probability that both customers received damaged VCRs?

Answer: Let us denote "D" as damaged and "G" as good in the population of the six VCRs as

D D D G G G

The probability that the first customer received a damaged VCR is 3/6. Once the first customer has received a damaged one, then the population will consist of five:

D D G G G

Now, the probability that the second customer received a damaged one is 2/5. Therefore, the probability that both customers received damaged VCRs is the product of the two probabilities or (3/6) (2/5) = 6/30 = 0.2.

(b) Can a binomial formula be used for the solution of the above problem?

Answer: No, in a binomial experiment, trials are independent of each other. Hence, the probability remains the same as successive trials are made. In the above problem, the probability of having purchased a defective VCR changes from customer to customer.

(c) What kind of probability distribution does the above satisfy, and is there a function for solving such problems?

Answer: This problem represents a hypergeometric probability distribution, which is closely related to the binomial distribution. The only difference between the two distributions is that in a hypergeometric distribution the probability of outcomes changes from trial to trial. The hypergeometric probability function, as shown below, yields the probability of x successes in n trials when the trials are not independent.

$$f(x) = \frac{\binom{r}{x}\binom{N-r}{n-x}}{\binom{N}{n}}$$

where $f(x)$ = probability of x successes
N = number of elements in the population
r = number of elements in the population labeled success
n = number of elements in the sample

In the case of the above problem we have

N = 6 (total number of elements)
r = 3 (number of defective parts)
n = 2 (number of elements in the sample)

The hypergeometric function for x = 2 is shown below.
(Remember N - r = 6 - 3 = 3 and n - x = 2 - 2 = 0.)

$$f(2) = \frac{\binom{3}{2}\binom{3}{0}}{\binom{6}{2}} = \frac{\left(\frac{3!}{2!\ 1!}\right)\left(\frac{3!}{0!\ 3!}\right)}{\left(\frac{6!}{2!\ 4!}\right)} = \frac{3}{15} = 0.2$$

Note, this is the same answer which was obtained in Part a of this problem.

(d) What is the probability that one of the two customers received a defective VCR?

Answer: The hypergeometric function can be solved for the following values.

N = 6
r = 3
n = 2
x = 1
N - r = 6 - 3 = 3
n - x = 2 - 1 = 1

as

$$f(1) = \frac{\binom{3}{1}\binom{3}{1}}{\binom{6}{2}} = \frac{\left(\frac{3!}{1!\ 2!}\right)\left(\frac{3!}{1!\ 2!}\right)}{\left(\frac{6!}{2!\ 4!}\right)} = \frac{9}{15} = 0.6$$

29. Compute the hypergeometric probabilities for the following values of n and x. Assume $N = 8$ and $r = 5$.

(a) $n = 5,\ x = 2$

(b) $n = 6,\ x = 4$

(c) $n = 3,\ x = 0$

(d) $n = 3,\ x = 3$

30. Seven students have applied for merit scholarships. This year 3 merit scholarships were awarded. If a random sample of 3 applications (from the population of seven) is selected, what is the probability that

(a) Two students received scholarships?

(b) Not any of the students received scholarships?

31. Determine the probability of being dealt 4 aces in a 5-card poker hand.

SELF-TESTING QUESTIONS

In the following multiple choice questions, circle the correct answer. An answer key is provided following the questions.

1. The variance is a measure of dispersion or variability in the random variable. It is a weighted average of the

a) square root of the deviations from the mean
b) square root of the deviations from the median
c) squared deviations from the median
d) squared deviations from the mean

2. A numerical description of the outcome of an experiment is a

a) random description
b) random outcome
c) random number
d) random variable

3. A random variable that can take on only a finite or countable number of values is known as

a) a continuous random variable
b) a discrete random variable
c) a discrete probability function
d) none of the above

4. A random variable that may take on any value in an interval or collection of intervals is known as

a) a continuous random variable
b) a discrete random variable
c) a continuous probability function
d) none of the above

5. A weighted average of the value of a random variable where the probability function provides weights is known as

a) a probability function
b) a random variable
c) the expected value
d) none of the above

6. The following represents the probability distribution for the daily demand of microcomputers at a local store.

Demand	Probability
0	0.1
1	0.2
2	0.3
3	0.2
4	0.2

The expected daily demand is

a) 1.0
b) 2.2
c) 2, since it has the highest probability
d) of course 4, since it is the largest demand level
e) 0

7. Refer to question 6 above. The probability of having a demand for at least two microcomputers is

a) 0.7
b) 0.3
c) 0.4
d) 1.0

8. The probability distribution for a discrete random variable which is used to compute the probability of x successes in n trials is known as the

a) normal probability distribution
b) standard normal distribution
c) binomial probability distribution
d) none of the above

9. The probability distribution for a discrete random variable which is used to compute the probability of x occurrences of an event over a specified interval is known as

a) the normal probability distribution
b) the standard normal distribution
c) a discrete random variable
d) a linear function
e) the Poisson probability distribution

Answer questions 10 through 12 based on the following information.

The student body of a large university consists of 60% female students. A random sample of 8 students is selected.

10. What is the probability that among the students in the sample exactly two are female?

a) 0.0896
b) 0.2936
c) 0.0413
d) 0.0007
e) none of the above

11. What is the probability that among the students in the sample at least 7 are female?

a) 0.1064
b) 0.0896
c) 0.0168
d) 0.8936
e) none of the above

12. What is the probability that among the students in the sample at least 6 are male?

a) 0.0413
b) 0.0079
c) 0.0007
d) 0.0499
e) none of the above

13. Which of the following is a required condition for a discrete probability function?

a) $\sum f(x) = 0$
b) $\sum f(x) = 1$
c) $f(x) < 0$
d) $f(x) > 0$
e) none of the above

14. Which of the following is not a characteristic of a binomial experiment?

a) the experiment consists of a sequence of n identical trials
b) at least 2 outcomes are possible
c) probabilities remain the same as successive trials are made
d) the trials are independent of each other

15. The expected value of a binomial probability distribution is

a) $E(x) = pn(1 - n)$
b) $E(x) = np(1 - p)$
c) $E(x) = np$
d) $E(x) = p(1 - n)$

ANSWERS TO THE SELF-TESTING QUESTIONS

1. d
2. d
3. b
4. a
5. c
6. b
7. a
8. c
9. e
10. c
11. a
12. d
13. b
14. b
15. c

ANSWERS TO CHAPTER FIVE EXERCISES

2. (a) Number of sales
   (b) 0 to 80
   (c) Discrete

3. (a) Speed of automobiles
   (b) 0 to maximum possible speed
   (c) Continuous

6.

Number Bad Checks	(a) Probability	(c) F(x)
0	.04	.04
1	.06	.10
2	.10	.20
3	.30	.50
4	.20	.70
5	.15	.85
6	.10	.95
7	.05	1.00

   (b) Yes, the sum of the probabilities is equal to 1.
   (d) 0.7
   (e) 0.5

7.

x	(a) p(x)	(b) F(x)
0	0.15	0.15
1	0.25	0.40
2	0.40	0.80
3	0.15	0.95
4	0.05	1.00

   (c) 0.80
   (d) 0.85

9. (a) 3.66
   (b) 2.7644
   (c) 1.6626

10. (a) 3.05
    (b) 2.0475, 1.431

11. (a) 0.26, 0.577
    (b) $390

12. (a) 4.14
    (b) 3.70, 1.924

13. (a) 3.84
    (b) 0.61196

15. (a) 0.3456
    (b) 0.0778
    (c) 0.0102

16. (a) 0.0819
    (b) 0.0154
    (c) 0.2621

17.

x	p
0	0.3430
1	0.4410
2	0.1890
3	0.0270

19. (a) 0.9
    (b) 0.794

20. (a)

x	p
0	0.6561
1	0.2916
2	0.0486
3	0.0036
4	0.0001

    (b) 0.4
    (c) 0.6

21. (a) 0.1821
 (b) 0.0388
 (c) 0.0000
 (d) 0.0002
 (e) 0.8298
 (f) 3

22. (a) 0.0000
 (b) 0.5987
 (c) 0.0746
 (d) 0.0011
 (e) 0.9138
 (f) 0.5000
 (g) 0.6892

23. (a) 0.0352
 (b) 0.0000
 (c) 0.1671
 (d) 0.1643
 (e) 3, 2.4

25. (a) 0.1171
 (b) 0.0001
 (c) 0.0212
 (d) 0.0053

26. (a) 1.5
 (b) 0.0141
 (c) 0.0656
 (d) 0.9343

27. (a) 0.0498
 (b) 0.0000
 (c) 0.0504
 (d) 0.0038
 (e) 3
 (f) 1.73

29. (a) 10/56 = 0.1786
 (b) 15/28 = 0.5357
 (c) 1/56 = 0.01786
 (d) 10/56 = 0.1786

30. (a) 12/35=0.3428571
    (b) 4/35 = 0.1143

31. 120/6,497,400 = 0.00001847

CHAPTER SIX

CONTINUOUS PROBABILITY DISTRIBUTIONS

CHAPTER OUTLINE AND REVIEW

In this chapter, you have been introduced to probability distributions of continuous random variables. Three major continuous probability distributions, the *uniform, normal,* and *exponential* distributions were introduced. More specifically, you have learned the following concepts:

A.	**Uniform Probability Distribution:**	A probability distribution of a continuous random variable, where the probability that the random variable will take on a value in any interval of equal length, is the same for each interval.
B.	**Probability Density Function:**	The function which describes the probability distribution of a continuous random variable.
C.	**Normal Probability Distribution:**	A continuous probability distribution. The normal distribution is a symmetrical distribution with a mean, μ, and a standard deviation, σ.
D.	**Standard Normal Distribution:**	A normal distribution with a mean of zero ($\mu = 0$) and a standard deviation of one ($\sigma = 1$).

E. **Continuity Correction Factor:** When a continuous distribution (such as a normal distribution) is used to approximate a discrete probability distribution (such as a binomial), a value of 0.5 is added (or subtracted) from the value of x. This factor of 0.5 is known as the continuity correction factor.

F. **Exponential Probability Distribution:** A continuous probability distribution which is useful in describing time, or space, between occurrences of an event.

CHAPTER FORMULAS

Uniform Probability Density Function for a Random Variable x

$$f(x)=\begin{cases}\dfrac{1}{b-a} & \text{for } a\le x\le b\\[2ex] 0 & \text{elsewhere}\end{cases} \tag{6.1}$$

Mean and Variance of a Uniform Continuous Probability Distribution

$$\mu=\frac{a+b}{2}$$

$$\sigma^2=\frac{(b-a)^2}{12}$$

Normal Probability Density Function

$$f(x)=\frac{1}{\sigma\sqrt{2\pi}}e^{-(x-\mu)^2/2\sigma^2} \qquad \text{for} \quad -\infty\le x\le\infty \tag{6.2}$$

where μ = mean of the random variable x

σ^2 = variance of the random variable x

π = 3.14159...

e = 2.71828...

The transformation of any Random Variable x with Mean μ and Standard Deviation σ to the Standard Normal Distribution

$$z = \frac{(x - \mu)}{\sigma} \qquad (6.3)$$

where z = the number of standard deviations

Exponential Probability Density function

$$f(x) = \frac{1}{\mu} e^{-x/\mu} \qquad (6.4)$$

Exponential Distribution Probabilities

$$p(x \le x_o) = 1 - e^{-x_o/\mu} \qquad (6.5)$$

EXERCISES

***1.** The driving time for an individual from his home to his work is uniformly distributed between 300 to 480 seconds.

(a) Give a mathematical expression for the probability density function.

Answer: The probability density function of a uniform probability distribution is given in equation 6.1 as

$$f(x)=\begin{cases}\dfrac{1}{b-a} & \text{for } a\leq x\leq b\\[2ex] 0 & \text{elsewhere}\end{cases}$$

In our example, b = 480 and a = 300. Therefore, the probability density function is

$$f(x) = \frac{1}{480 - 300} = \frac{1}{180}$$

This figure indicates that for any value of x (from 300 to 480), the f(x) is constant and its value is equal to 1/180.

(b) Compute the probability that the driving time will be less than or equal to 435 seconds.

Answer: In this part, we are interested in determining the $p(300 \leq x \leq 435)$. This probability is the area under a uniform distribution with a height of 1/180 and a width of 135 seconds (i.e., 435 - 300 = 135). Hence, the area is (135)(1/180)= 0.75. This indicates that $p(300 \leq x \leq 435) = 0.75$. The shaded area in Figure 6.1 (on the following page) represents this probability.

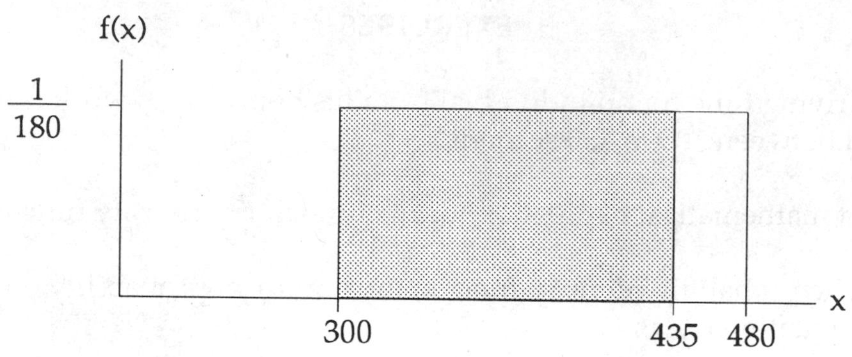

Figure 6.1

(c) Determine the expected driving time and its standard deviation.

Answer: The expected value of a uniform continuous random variable is

$$E(x) = \frac{a + b}{2}$$

Hence, in our example, the expected value of x is

$$E(x) = \frac{480 + 300}{2} = 390$$

Furthermore, the variance is given by

$$\text{Variance } (x) = \frac{(b - a)^2}{12}$$

Therefore

$$\text{Variance } (x) = \frac{(480 - 300)^2}{12} = 2700$$

Hence, the standard deviation is

$$\sigma = \sqrt{2700} = 51.96$$

2. The Body Paint, an automobile body paint shop, has determined that the painting time of automobiles is uniformly distributed and that the required time ranges between 45 minutes to $1\frac{1}{2}$ hours.

(a) Give a mathematical expression for the probability density function.

(b) What is the probability that the painting time will be less than or equal to one hour?

(c) What is the probability that the painting time will be more than 50 minutes?

(d) Determine the expected painting time and its standard deviation.

3. The length of time patients must wait to see a doctor in a local clinic is uniformly distributed between 15 minutes and $2\frac{1}{2}$ hours.

(a) Give a mathematical expression for the probability density function.

(b) What is the probability that a patient would have to wait between 45 minutes and 2 hours?

(c) Compute the probability that a patient would have to wait over 2 hours.

(d) Determine the expected waiting time and its standard deviation.

***4.** The *standard normal distribution* represents the probability distribution of a random variable which is normally distributed and has a mean of zero ($\mu = 0$) and a standard deviation of one ($\sigma = 1$). Let z represent the number of standard deviations from the mean.

(a) What is the probability that z will be between -1.2 and +1.2?

Answer: The table of probabilities for the standard normal distribution provides us with the probabilities associated with a specified value for z. In Figure 6.2, the shaded area represents the probability that z will fall between -1.2 and +1.2.

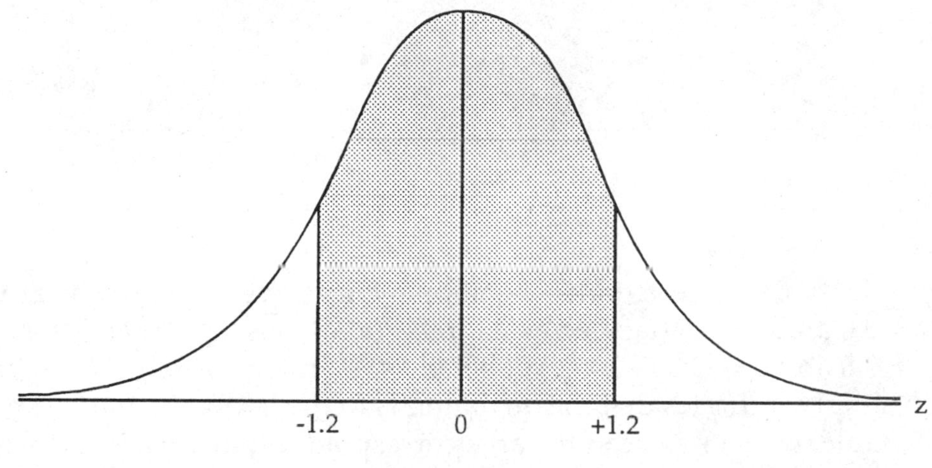

Figure 6.2

Referring to the table of areas under the normal curve (in your textbook), we note that the area between the mean (i.e., z = 0.0) to z = 1.2 is 0.3849. Since the area to the left of the mean is the mirror image of the area to the right of the mean, we conclude that the area between z = 0.0 and z = -1.2 is also 0.3849. Hence, the area between z = -1.2 to z = 1.2 is 0.3849 + 0.3849 = 0.7698, which is the probability that z will be between -1.2 and +1.2.

(b) What is the probability that a z value will be between z = -2.5 and z= 1.8?

Answer: The shaded area in Figure 6.3 shows the desired area.

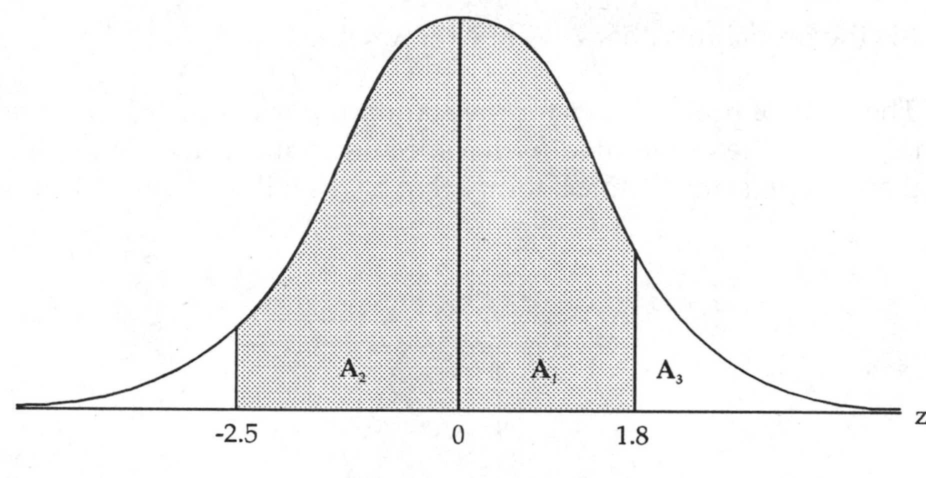

Figure 6.3

Referring to the table of areas under the normal curve, we note that the area between z = 0.0 and z = 1.8 (labeled as A_1 in Figure 6.3) is 0.4641, and the area between z = 0.0 and z = -2.5 is 0.4938 (labeled as A_2 in Figure 6.3). Note that the minus sign simply indicates that the deviation is to the left of the mean; and in reading the table, we simply read the area corresponding to z = 2.5. Hence, the area between z = -2.5 and z = 1.8 is the sum of the two areas (A_1 and A_2), and its value is 0.4641 + 0.4938 = 0.9579.

(c) What is the probability that z will have a value larger than 1.8?

Answer: Referring to Figure 6.3, you will note that the requested area is labeled as A_3. Since we know that one half of the entire area (or 0.5) falls above the mean and since in Part b we have already determined that the area between z = 0.0 and z = 1.8 is 0.4641, we can determine the required area as

$$\begin{aligned} A_3 &= 0.5 - A_1 \\ &= 0.5 - 0.4641 \\ &= 0.0359 \end{aligned}$$

(d) What is the probability that z will have a value between z = 1.4 and z = 1.9?

Answer: The shaded area in Figure 6.4 shows the required area.

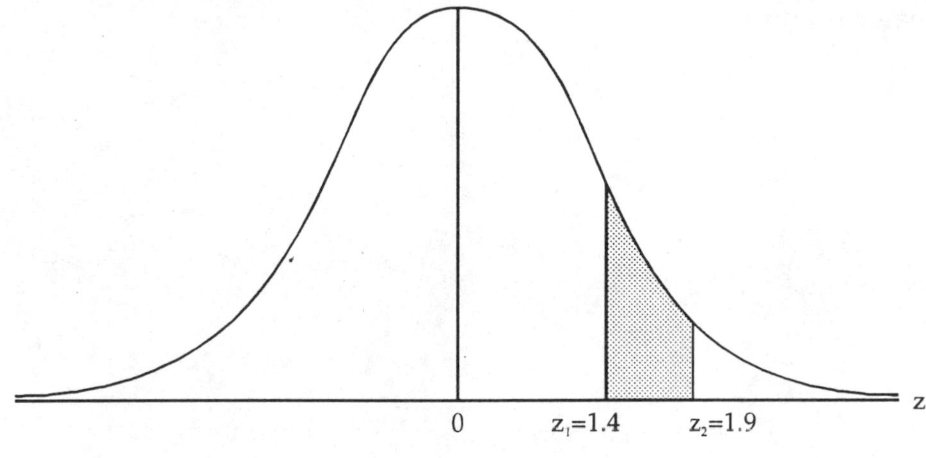

Figure 6.4

To determine this probability, we first need to determine the area from the mean to $z_2 = 1.9$, then find the area from the mean to $z_1 = 1.4$ and, finally, subtract the area corresponding to 1.4 standard deviations from the area which corresponds to 1.9 standard deviations from the mean. From the table of areas under the normal curve (in your textbook), we note that the area from the mean to $z_2 = 1.9$ is 0.4713, and the area from the mean to $z_1 = 1.4$ is 0.4192. Therefore, the required probability is 0.4713 - 0.4192 = 0.0521.

5. For a standard normal distribution, determine the probabilities of obtaining the following z values. It is helpful to draw a normal distribution for each case and show the corresponding area.

(a) Greater than zero.

(b) Between -2.4 and -2.0

(c) Less than 1.6.

(d) Between -1.9 to 1.7.

(e) Between 1.5 and 1.75.

***6.** Sun Love grapefruit growers have determined that the diameters of their grapefruits are normally distributed with a mean of 4.5 inches and a standard deviation of 0.3 inches.

(a) What is the probability that a randomly selected grapefruit will have a diameter of at least 4.14 inches?

Answer: The required area is shown by the shaded area in Figure 6.5, which is the sum of the areas labeled as A_1 and A_2.

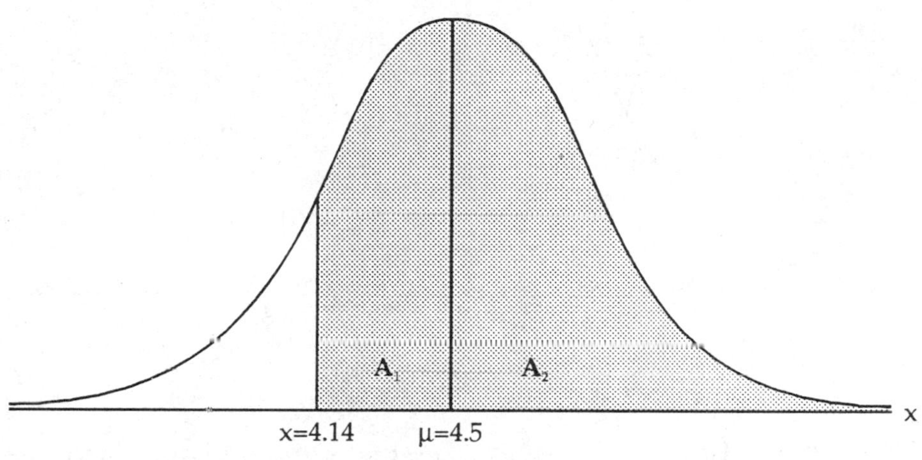

Figure 6.5

Since we have a normal distribution, the area to the right of the mean (A_2) represents 50% of the total area. Therefore, we need to determine the area labeled as A_1. To determine this area, we first need to find the number of standard deviations from the point of x = 4.14 to the mean. The number of standard deviations is computed as

$$z = \frac{(x - \mu)}{\sigma} = \frac{4.14 - 4.5}{0.3} = -1.2$$

Therefore, there are 1.2 standard deviations between the point of x = 4.14 to the mean. Looking in the table of areas under the normal curve, we note an area of 0.3849 corresponds to 1.2 standard deviations. Therefore, the total area, or the probability of a grapefruit having a diameter greater than 4.14, is $A_1 + A_2 =$ 0.3849 + 0.5 = 0.8849.

(b) What percentage of the grapefruits have a diameter between 4.8 and 5.04 inches?

Answer: You should realize that the entire area under the normal curve represents 100% of the grapefruits. The percentage of the area which falls between 4.8 and 5.04 inches is shown in Figure 6.6 by the shaded area.

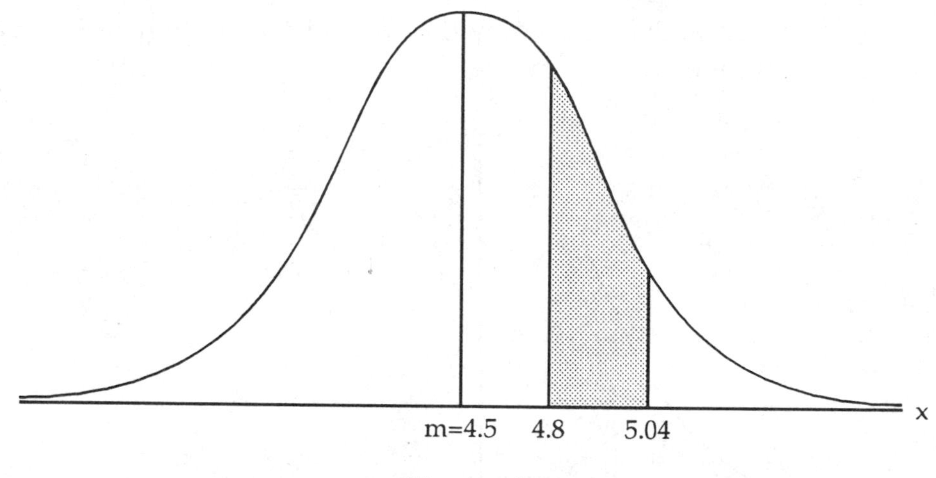

Figure 6.6

To find this area, we first need to determine the number of standard deviations from the mean to x = 5.04 as

$$z = \frac{(x - \mu)}{\sigma} = \frac{5.04 - 4.5}{0.3} = 1.8$$

Then read the area which corresponds to 1.8 standard deviations in the table of areas under the normal curve. From the table, we read this area to be 0.4641. Then we need to determine the number of standard deviations from the mean to x = 4.8 as

$$z = \frac{(x - \mu)}{\sigma} = \frac{4.8 - 4.5}{0.3} = 1.0$$

From the table of areas under the normal curve, the area which corresponds to 1.0 standard deviations is 0.3413. Finally, the area between x = 4.8 and x = 5.04 is the difference between the two areas which we just determined, or 0.4641 - 0.3413 = 0.1228. Hence, we can conclude that 12.28% of the grapefruits have diameters between 4.8 and 5.04 inches.

(c) Sun Love packs their largest grapefruits in special packages called the *super pack*. If 5% of all their grapefruits are packed as *super packs*, what is the smallest diameter of the grapefruits which are in the *super packs*?

Answer: The shaded area in Figure 6.7 represents the percentage of the area representing the *super pack* grapefruits, which is 5%. Therefore, the area from the mean to this point is 0.45 of the entire area. Now looking in the body of the table of areas under the normal curve, we see that the area of 0.45 occurs approximately 1.64 standard deviations above the mean. Hence, to find the smallest diameter of the grapefruits in the *super pack* corresponding to z = 1.64, we calculate x as follows:

$$z = \frac{(x - \mu)}{\sigma}$$

$$1.64 = \frac{x - 4.5}{0.3}$$

$$x = 4.5 + (1.64)(0.3) = 4.992$$

This figure indicates that the smallest grapefruit in the *super pack* will have a diameter of 4.992 inches.

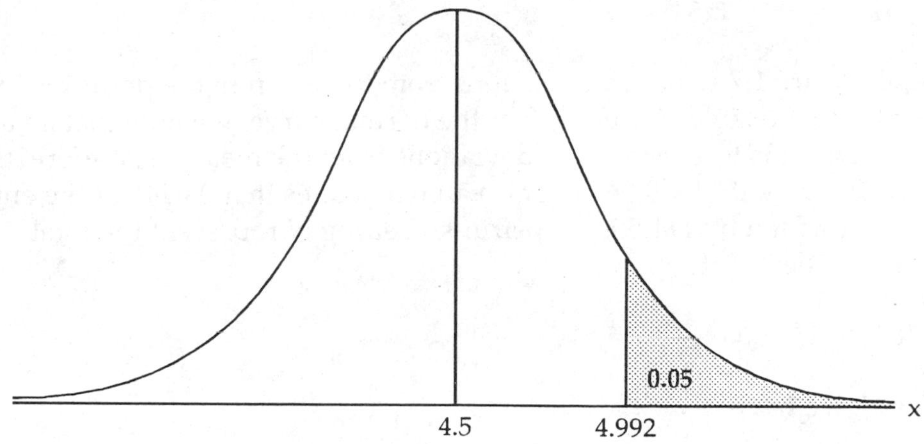

Figure 6.7

(d) In this year's harvest, there were 111,500 grapefruits which had a diameter over 5.01 inches. How many grapefruits has Sun Love harvested this year?

Answer: The shaded area in Figure 6.8 represents the portion of the entire area which satisfies the above condition.

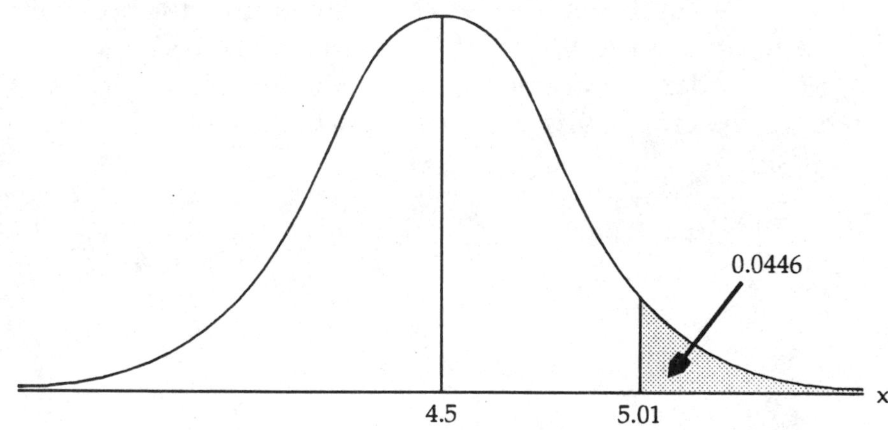

Figure 6.8

To find this area, we will first find the area from the mean to the point of 5.01 inches and then subtract the result from 0.5.

$$z = \frac{(x - \mu)}{\sigma} = \frac{5.01 - 4.5}{0.3} = 1.7$$

Therefore, there are 1.7 standard deviations from the mean to the point x = 5.01. Now referring to the table of areas under the normal curve, we note that an area of 0.4554 corresponds to 1.7 standard deviations from the mean. Therefore, the shaded area is 0.5 - 0.4554 = 0.0446. This value indicates that 0.0446 of the entire harvest is represented by 111,500 grapefruits. Letting N represent the total harvest, we can then write

$$0.0446\,N = 111{,}500$$

Solving for N, we have

$$N = \frac{111{,}500}{0.0446} = 2{,}500{,}000$$

Therefore, Sun Love has harvested 2,500,000 grapefruits this year.

7. The life expectancy of a particular brand of hair dryer is normally distributed with a mean of four years and a standard deviation of eight months.

(a) What is the probability that a hair dryer will be in working condition more than five years?

(b) The company has a three year warranty period on their hair dryers. What percentage of their hair dryers will be in operating condition after the warranty period?

(c) What is the minimum and the maximum life expectancy of the middle 90% of the hair dryers?

(d) Ninety-five percent of the hair dryers will have a life expectancy of at least how many months?

8. Duckworth Drug Company is a large manufacturer of various kinds of liquid vitamins. The quality control department has noted that the bottles of vitamins marked 6 ounces vary in content with a standard deviation of 0.3 ounces.

(a) What percentage of all bottles produced contain more than 6.51 ounces of vitamins?

(b) What percentage of all bottles produced contain less than 5.415 ounces?

(c) What percentage of bottles produced contain between 5.46 and 6.495 ounces?

(d) Ninety-five percent of the bottles will contain at least how many ounces?

(e) What percentage of the bottles contain between 6.3 and 6.6 ounces?

9. A professor at a local community college noted that the grades of his students were normally distributed with a mean of 74 and a standard deviation of 10. The professor has informed us that 6.3 percent of his students received A's while only 2.5 percent of his students failed the course and received F's.

(a) What is the minimum score needed to make an A?

(b) What is the maximum score among those who received an F?

(c) If there were 5 students who did not pass the course, how many students took the course?

10. In grading shrimp into small, medium, and large, the Globe Fishery packs the shrimp that weigh more than 2 ounces in packages marked ***large*** and the shrimp that weigh less than 0.75 ounces into packages marked ***small***; the remainder are packed in ***medium*** size packages. If a day's catch showed that 16.6 percent of the shrimp were ***large*** and 6.68 percent were ***small***, determine the mean and the standard deviation for the shrimp weights. Assume that the shrimps' weights are normally distributed.

11. The weekly earnings of computer systems analysts are normally distributed with a mean of $395. If only 1.1 percent of the systems analysts have a weekly income of more than $429.35, what is the value of the standard deviation of the weekly earnings of the computer systems analysts?

12. A major credit card company has determined that their customers charge an average of \$280 per month on their account with a standard deviation of \$20.

(a) What percentage of their customers charge more than \$275 per month?

(b) What percentage of their customers charge less than \$243 per month?

(c) What percentage of their customers charge between \$241 and \$301.60 per month?

***13.** The First National Mortgage Company has noted that 6% of their customers pay their mortgage payments past the due date.

(a) What is the probability that in a random sample of 150 mortgages, 7 will be late on their payments?

Answer: In this situation, we are interested in determining the binomial probability of exactly 7 successes in 150 trials, where the probability of each success is 0.06. That is, we want to find

$$P(x = 7,\ n = 150,\ p = 0.06)$$

Since n is greater than 20, we are unable to use the binomial tables (note that the binomial tables in your text give probabilities for sample sizes up to 20). However, we can approximate this binomial probability by the use of a normal distribution with a mean of $\mu = np$ and a standard deviation of

$$\sigma = \sqrt{np(1 - p)} = \sqrt{(150)(.06)(.94)} = 2.91$$

$$\mu = np = (150)(.06) = 9$$

Then to approximate the binomial probability of exactly 7 successes, we need to find the area under the normal curve between 6.5 and 7.5, which is shown by the shaded area in Figure 6.9. Note that a continuity correction factor of 0.5 is added to and subtracted from 7 in order to approximate a discrete distribution using a continuous probability distribution.

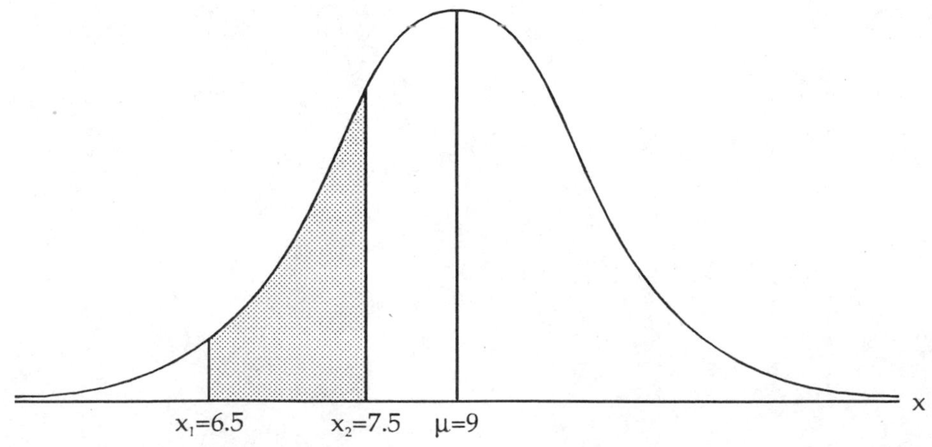

Figure 6.9

In order to find this area, we first determine the number of standard deviations from each point to the mean as

$$z_1 = \frac{x_1 - \mu}{\sigma} = \frac{6.5 - 9}{2.91} = -0.86$$

$$z_2 = \frac{x_2 - \mu}{\sigma} = \frac{7.5 - 9}{2.91} = -0.52$$

Now referring to the table of areas under the normal curve, we see that an area of 0.3051 corresponds to 0.86 standard deviations and that an area of 0.1985 corresponds to 0.52 standard deviations. Therefore, the required area is

0.3051 - 0.1985 = 0.1066

Thus, the normal approximation to the 7 successes in 150 trials is 0.1066.

(b) What is the probability that in a random sample of 150 mortgages at least 10 will be late on their payments?

Answer: In this case, we need to determine the shaded area in Figure 6.10.

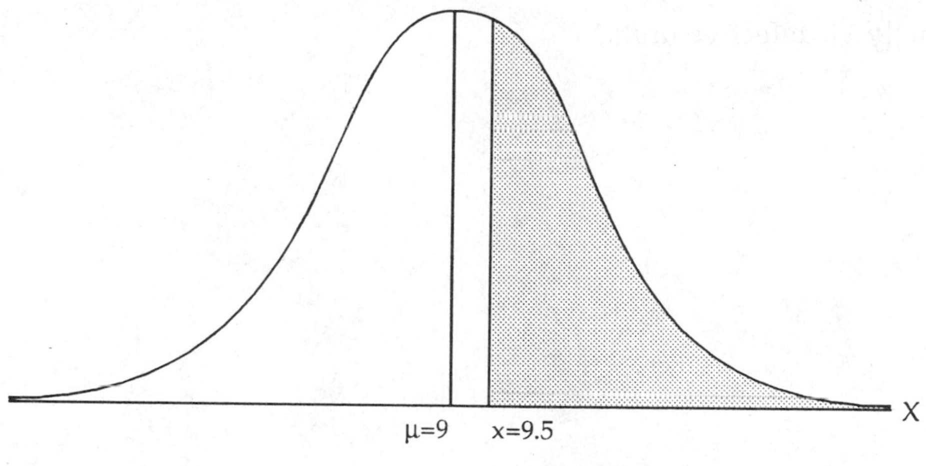

Figure 6.10

Note that the continuity correction factor has been subtracted from 10. Hence, the desired probability is represented by the area to the right of 9.5. To determine this probability, we first determine the area between 9 and 9.5 and subtract the result from 0.5. The z value corresponding to x = 9.5 is

$$z = \frac{(x - \mu)}{\sigma} = \frac{9.5 - 9}{2.91} = 0.17$$

From the table of areas under the normal curve, we determine that the area corresponding to 0.17 standard deviations is 0.0675. Therefore, the area to the right of 9.5 is

$$0.5 - 0.0675 = 0.4325$$

14. The records show that 8% of the items produced by a machine do not meet the specifications. What is the probability that a sample of 100 units contains

(a) Exactly 6 defective units?

(b) Exactly 11 defective units?

(c) Six or fewer defective units?

(d) Five or more defective units?

(e) Ten or fewer defective units?

(f) Eleven or less defective units?

15. Approximate the following binomial probabilities by the use of normal approximation.

(a) $P(x = 18,\ n = 50,\ p = 0.3)$

(b) $P(x \geq 15,\ n = 50,\ p = 0.3)$

(c) $P(x \leq 12,\ n = 50,\ p = 0.3)$

(d) $P(12 \leq x \leq 18,\ n = 50,\ p = 0.3)$

16. An airline has determined that 20% of its international flights are not on time. What is the probability that of the next 80 international flights

(a) Exactly 16 will not be on time?

(b) Fourteen or more will not be on time?

(c) Fifteen or less will not be on time?

(d) Eighteen or more will not be on time?

(e) Exactly 17 will not be on time?

17. The ticket sales for events held at the new civic center are believed to be normally distributed, with a mean of 12,000 and a standard deviation of 1,000.

(a) What is the probability of a selling more than 10,000 tickets?

(b) What is the probability of selling between 9,500 and 11,000 tickets?

(c) What is the probability of selling more than 13,500 tickets?

***18.** The time required to assemble a part of a machine follows an exponential probability distribution. The average time of assembling the part is 10 minutes.

(a) Give the appropriate probability density function.

Answer: The probability density function for an exponential distribution is given by

$$f(x) = \frac{1}{\mu} e^{-x/\mu}$$

where μ is the mean and e = 2.71828. Thus, the probability density functions for our example is

$$f(x) = \frac{1}{10} e^{-x/10}$$

(b) What is the probability that the part can be assembled in 7 minutes or less?

Answer: The probability that x is less than or equal to a specific value x_o is given by

$$p(x \leq x_o) = 1 - e^{-x_o/\mu}$$

Thus, the probability of x being less than or equal to 7 minutes is computed as

$$p(x \leq 7) = 1 - e^{-7/10} = 1 - 0.4966 = 0.5034$$

(c) Find the probability of completing the assembly in 3 to 7 minutes.

Answer: The probability of this interval (3 to 7) can be computed by $p(x \leq 7) - p(x \leq 3)$. In Part b, we computed $p(x \leq 7) = 0.5034$.

Now let us compute $p(x \leq 3)$:

$$p(x \leq 3) = 1 - e^{-3/10} = 1 - 0.7408 = 0.2592$$

Therefore, the probability of completing the assembly in 3 to 7 minutes is

$$p(x \leq 7) - p(x \leq 3) = 0.5034 - 0.2592 = 0.2442$$

19. The time between arrivals of customers at the drive-up window of a bank follows an exponential probability distribution with a mean of 14 minutes.

(a) Give the appropriate probability density function.

(b) What is the probability that the arrival time between customers is 7 minutes or less?

(c) What is the probability that the arrival time between customers is 3.5 to 7 minutes?

20. The time it takes to complete an examination follows an exponential distribution with a mean of 40 minutes.

(a) What is the probability of completing the examination in 30 minutes or less?

(b) What is the probability of completing the examination in 30 to 35 minutes?

SELF-TESTING QUESTIONS

In the following multiple choice questions, circle the correct answer. An answer key is provided following the questions.

1. For the standard normal probability distribution, the area to the left of the mean is

a) greater than 0.5
b) -0.5
c) one
d) 0.5
e) none of the above

2. In a standard normal distribution, the range of z values is

a) 0 to 1
b) -1 to 1
c) $-\infty$ to ∞
d) -3.09 to 3.09
e) none of the above

3. If a z value is to the left of the mean, then its value is

a) negative
b) positive
c) any value between $-\infty$ to ∞
d) zero
e) none of the above

In questions 4 through 8, assume z is a standard normal random variable.

4. The $p(-1.50 \leq z \leq 1.90)$ equals

a) 0.0381
b) 0.9045
c) -0.0381
d) 0.4
e) none of the above

5. The $p(-2.0 \leq z \leq -1.0)$ equals

a) 0.8185
b) 0.1469
c) 1.0000
d) 0.1359
e) none of the above

6. The $p(2.0 \leq z \leq 2.5)$ equals

a) 0.9710
b) 0.0166
c) 0.5000
d) 4.5000
e) none of the above

7. The $p(-2.54 \leq z \leq 2.54)$ equals

a) 0.4945
b) 0.0000
c) 0.5400
d) 0.9890
e) none of the above

8. The $p(2.32 \leq z \leq 3.05)$ equals

a) 0.4989
b) 0.9887
c) 0.0091
d) 0.7300
e) none of the above

In questions 9 through 11, assume z is the standard normal random variable.

9. If the area between zero and z is 0.4115, then the z value is

a) 2.70
b) 1.35
c) 1.00
d) 0.2077
e) none of the above

10. If the area to the right of z is 0.8413, then z is

a) -1.0
b) 1.0
c) 2.0
d) -2.0
e) none of the above

11. If the area to the right of z is 0.0668, then z is

a) 0.17
b) 2.00
c) 1.50
d) 1.00
e) none of the above

Answer questions 12 through 14 based on the above information.

The travel time for a businesswoman traveling between Dallas and Fort Worth is uniformly distributed between 40 and 90 minutes.

12. The probability that she will finish her trip in 80 minutes or less is

a) 0.02
b) 0.2
c) 0.8
d) 1.0
e) none of the above

13. The probability that her trip will take longer than 60 minutes is

a) 0.4
b) 0.6
c) 0.02
d) 1.00
e) none of the above

14. The probability that her trip will take exactly 50 minutes is

a) 1.0
b) 0.02
c) 0.06
d) 0.2
e) almost zero

15. A standard normal distribution is a normal distribution

a) with a mean of 1 and a standard deviation of 0
b) with any mean and any standard deviation
c) with a mean of 0 and any standard deviation
d) with a mean of 0 and a standard deviation of 1
e) none of the above

16. A normal probability distribution

a) is a discrete probability distribution
b) is a continuous probability distribution
c) can be either continuous or discrete
d) must always have a mean of zero
e) none of the above

Answer questions 17 through 19 based on the following information.

The life expectancy of a particular brand of tire is normally distributed with a mean of 40,000 miles and a standard deviation of 5,000 miles.

17. What is the probability that a randomly selected tire will have a life of at least 30,000 miles?

a) 0.4772
b) 0.9772
c) 0.0228
d) none of the above

18. What is the probability that a randomly selected tire will have a life of at least 47,500 miles?

a) 0.4332
b) 0.9332
c) 0.0668
d) none of the above

19. What percentage of tires will have a life of 34,000 to 46,000 miles?

a) 38.49%
b) 76.98%
c) 50%
d) none of the above

20. An exponential probability distribution

a) is a continuous distribution
b) is a discrete distribution
c) can be either continuous or discrete
d) none of the above

ANSWERS TO THE SELF-TESTING QUESTIONS

1. d
2. c
3. a
4. b
5. d
6. b
7. d
8. c
9. b
10. a
11. c
12. c
13. b
14. e
15. d
16. b
17. b
18. c
19. b
20. a

ANSWERS TO CHAPTER SIX EXERCISES

2. (a)

$$f(x) = \begin{cases} \frac{1}{45} & \text{for } 45 \le x \le 90 \\ 0 & \text{elsewhere} \end{cases}$$

   (b) 0.333
   (c) 0.889
   (d) 67.5, 12.99

3. (a)

$$f(x) = \begin{cases} \frac{1}{135} & \text{for } 15 \le x \le 150 \\ 0 & \text{elsewhere} \end{cases}$$

   (b) 0.556
   (c) 0.222
   (d) 82.5, 38.97

5. (a) 0.5
   (b) 0.0146
   (c) 0.9452
   (d) 0.9267
   (e) 0.0267

7. (a) 0.0668
   (b) 93.32%
   (c) 34.88 to 61.12 months
   (d) 34.88

8. (a) 4.46%
   (b) 2.56%
   (c) 91.46%
   (d) 5.508 ounces
   (e) 13.59%

9. (a) 89.3
   (b) 54.4
   (c) 200

10. Mean = 1.5 Standard deviation = 0.5

11. 15

12. (a) 59.87%
    (b) 3.22%
    (c) 0.8343

14 (a) 0.1124
   (b) 0.0803
   (c) 0.2912
   (d) 0.9015
   (e) 0.8212
   (f) 0.9015

15. (a) 0.0805
    (b) 0.5596
    (c) 0.2206
    (d) 0.7198

16. (a) 0.1114
    (b) 0.7580
    (c) 0.4443
    (d) 0.3372
    (e) 0.1071

17. (a) 0.9772
    (b) 0.1525
    (c) 0.0668

19. (a) $f(x) = \frac{1}{14} e^{-x/14}$
    (b) 0.3935
    (c) 0.1723

20. (a) 0.5276
    (b) 0.0555

CHAPTER SEVEN

SAMPLING AND SAMPLING DISTRIBUTIONS

CHAPTER OUTLINE AND REVIEW

In Chapter 1, you were informed that one can make inferences about the characteristics of a population based on the sample information. In this chapter, you have studied the concept of sampling, how samples can be taken, and the characteristics of various sampling distributions. The objective of this chapter has been to prepare you for future chapters where you will learn how one can use these sampling distributions in order to make inferences about a population's characteristics. The following is an outline of the main points which you should have learned.

A. Parameter: A descriptive measure of a population, such as a population mean (μ), a population standard deviation (σ), and a population proportion (p).

B. Statistic: A descriptive measure of a sample, such as a sample mean ($\overline{X}$), a sample standard deviation (S), or a sample proportion ($\overline{p}$). Each descriptive measure from the sample is used to estimate the value of the population parameter.

C. Simple Random Sampling: The process of selecting a sample from a population where each individual element of the population has an equal chance of being selected.

D. Sampling Distribution: A probability distribution showing all possible values that a sample statistic (such as a sample mean, a sample standard deviation, or a sample proportion) can assume.

E. Sampling Without Replacement: The process of selecting items for a sample from a population and not returning them to the population.

F. Sampling With Replacement: The process of selecting items for a sample from a population and returning them to the population so that the same item may be selected again.

G. Point Estimate: A single numerical value used for estimating a population parameter.

H. Point Estimator: A sample statistic, such as a sample mean ($\overline{X}$), a sample standard deviation (S), or a sample proportion ($\overline{p}$), which is used to estimate a population parameter.

I. Standard Error: The standard deviation of a point estimator, such as the standard error of the mean ($\sigma_{\overline{x}}$) and the standard error of the proportion ($\sigma_{\overline{p}}$).

J. Finite Population Correction Factor: When a population is finite (as a rule of thumb, consider a population finite if the sample size is more than or equal to 5% of the population, i.e., $n/N \geq 0.05$), the multiplier $\sqrt{(N - n)/(N - 1)}$, known as the finite correction factor, is used to estimate $\sigma_{\overline{x}}$ and $\sigma_{\overline{p}}$.

K. Central Limit Theorem: A theorem which states that when samples of size n are selected from a population with mean μ and standard deviation σ, the distribution of sample means ($\overline{X}$) will approach a normal distribution with mean μ and standard deviation $\sigma/\sqrt{n}$ as the sample size increases.

L. Unbiased: If the expected value of a poin
the population parameter whi
point estimator is known as an
and the point estimate as an u

M. Consistency: A property of a point estimato
sizes yield point estimates closer to the population parameter.

N. Relative Efficiency: The ratio of the variances of two point estimators of a given parameter. The smaller the variance, the more efficient the estimator.

O. Sufficiency: A sufficient point estimator is one that uses all of the information available in the sample to develop the point estimate of the population parameter.

P. Probability Sample: Stratified random sampling, cluster sampling, systematic sampling, and simple random sampling are examples of methods leading to probability samples. In each of these sampling methods, each element in the population has a known probability of being included in the sample.

Q. Nonprobability Sample: When a sample is selected in a manner that the probability of each element being included in the sample is unknown (such as convenience and judgment samples), it is called a nonprobability sample.

R. Convenience Sampling: A nonprobabilistic method of sampling whereby elements are selected on the basis of convenience.

S. Judgment Sampling: A nonprobabilistic method of sampling whereby the element selected is based on the judgment of the person doing the study.

T. Stratified Simple Random Sampling: A method of selecting a sample in which the population is first divided into strata and a simple random sample is then taken from each stratum.

Cluster Sampling: A probabilistic method of sampling in which the population is first divided into clusters and then one or more clusters is selected for sampling. In single-stage cluster sampling, every element in each selected cluster is sampled; in two-stage cluster sampling, a sample of the elements in each selected cluster is collected.

V. **Systematic Sampling:** A method of choosing a sample by randomly selecting the first element and then selecting every k^{th} element thereafter.

CHAPTER FORMULAS

The number of different simple random samples of size n that can be selected from a finite population of size N

$$\frac{N!}{n!(N-n)!}$$

FINITE POPULATION | *INFINITE POPULATION*

Expected Value of $\overline{X}$

$E(\overline{X}) = \mu$ $\qquad$ $E(\overline{X}) = \mu$ $\qquad$ (7.1)

where: $E(\overline{X})$ = the expected value of the random variable $\overline{X}$
μ = the population mean

Standard Deviation of the Distribution of $\overline{X}$ Values (The Standard Error of the Mean)

$\sigma_{\bar{x}} = \sqrt{\frac{N-n}{N-1}} \cdot \frac{\sigma}{\sqrt{n}}$ $\qquad$ $\sigma_{\bar{x}} = \frac{\sigma}{\sqrt{n}}$ $\qquad$ (7.2)

Expected Value of $\overline{p}$

$E(\overline{p}) = p$ $\qquad$ $E(\overline{p}) = p$ $\qquad$ (7.4)

where: $E(\overline{p})$ = the expected value of the random variable $\overline{p}$
p = the population proportion

CHAPTER FORMULAS
(continued)

Standard Deviation of the Distribution of $\bar{p}$ Values
(The Standard Error of the Proportion)

$$\sigma_{\bar{p}} = \sqrt{\frac{N - n}{N - 1}} \cdot \sqrt{\frac{p(1 - p)}{n}} \qquad \sigma_{\bar{p}} = \sqrt{\frac{p(1 - p)}{n}} \qquad (7.5)$$

Unbiasedness

The sample statistic $\hat{\theta}$ is an unbiased estimator of the population parameter θ if

$$E(\hat{\theta}) = \theta \qquad (7.6)$$

where: $E(\hat{\theta})$ = expected value of the sample statistic $\hat{\theta}$

EXERCISES

***1.** A small firm has 97 employees, and we want to select a random sample of 6 employees to represent the firm. Explain how we can select this random sample using the table of random numbers.

Answer: Table 8 of Appendix B in your text provides us with random numbers, where digits 1 through 9 have an equal probability of appearing in any position. Before using the table, we need to assign numbers 1 through 97 to the employees and then draw two-digit random numbers from the table, each number representing the employee to be selected. Let us use the fifth row of the random numbers of Table 8 (we could have selected any row). The numbers in the fifth row are

55363   07449   34835   . . .

A simple way of selecting two-digit random numbers is to make a two-digit grouping of the random numbers given above as

55   36   30   74   49   34

Hence, employees with numbers 55, 36, 30, 74, 49, and 34 are selected.

2. Use the first row of the table of random numbers to select a random sample of 10 employees.

3. Assume that in a small town there are 9,832 eligible people for jury duty. Using row 11 of Table 8 in Appendix B (of your textbook), identify the 12 random jurors to be selected.

***4.** A simple random sample of 6 computer programmers in Houston, Texas revealed the sex of the programmers and the following information about their weekly incomes.

Programmer	Weekly Income	Sex
A	$250	M
B	270	M
C	285	F
D	240	M
E	255	M
F	290	F

(a) What is the point estimate for the average weekly income of all the computer programmers in Houston?

Answer: Recall that a point estimate is a single numerical value used for estimating the population parameter. In this case, we are interested in the average. Hence, we need to find the average income of the individuals in the sample:

$$\overline{X} = \frac{\Sigma X_i}{n} = \frac{250 + 270 + 285 + 240 + 255 + 290}{6} = 265$$

Therefore, the point estimate for the average weekly income of all the computer programmers in Houston is $265.

(b) What is the point estimate for the standard deviation for the population?

Answer: The point estimate for the standard deviation for the population is determined by finding the standard deviation of the sample which is

$$S = \sqrt{\frac{\Sigma(X_i - \overline{X})^2}{n-1}}$$

The $\Sigma(X_i - \overline{X})^2$ is calculated as

X_i	$(X_i - \overline{X})^2$
250	225
270	25
285	400
240	625
255	100
290	625

$$\Sigma(X_i - \overline{X})^2 = 2000$$

Therefore, the standard deviation of the sample is

$$S = \sqrt{\frac{\Sigma(X_i - \overline{X})^2}{n-1}} = \sqrt{\frac{2000}{6 - 1}} = 20$$

Hence, the point estimate for the standard deviation of the population is $20.

(c) Determine a point estimate for the proportion of all programmers in Houston who are female.

Answer: In the sample, there are 2 female programmers. Therefore, the proportion of female programmers in the sample is

$$\overline{p} = \frac{2}{6} = 0.33$$

The above figure (0.33) is used as a point estimate for the population proportion.

5. In exercise 1 of Chapter 3, you were given the gasoline prices from a sample of 9 gasoline stations in Chattanooga, Tennessee. The following is the price per gallon:

Gas Station #	Price Per Gallon (x)
1	$1.14
2	1.19
3	1.25
4	1.21
5	1.17
6	1.19
7	1.22
8	1.24
9	1.19

(a) What is the point estimate for the prices of all gasoline stations in Chattanooga?

(b) What is the point estimate for the standard deviation of the population?

6. A random sample of 15 telephone calls in an office showed the duration of each call and whether it was a local or a long distance call.

Call Number	Duration (In Minutes)	Type of Call
1	2	local
2	12	long distance
3	10	local
4	3	local
5	5	long distance
6	6	local
7	3	local
8	5	local
9	8	local
10	4	local
11	5	local
12	4	local
13	5	local
14	4	local
15	9	long distance

(a) What is the point estimate for the average duration of all calls?

(b) What is the point estimate for the standard deviation of the population?

(c) What is the point estimate for the proportion of all calls which were long distance?

7. A random sample of 10 examination papers in a course, which was given on a pass or fail basis, showed the following scores:

Paper Number	Grade	Status
1	65	Pass
2	87	Pass
3	92	Pass
4	35	Fail
5	79	Pass
6	100	Pass
7	48	Fail
8	74	Pass
9	79	Pass
10	91	Pass

(a) What is the point estimate for the mean of the population?

(b) What is the point estimate for the standard deviation of the population?

(c) What is the point estimate for the proportion of all students who passed the course?

***8.** Consider a population of four weights identical in appearance but weighing 2, 4, 6, and 8 grams. If we select samples of size 2 **with replacement**, there will be a total of 16 different samples.

(a) List all the possible samples of size 2 and determine the mean weight of each sample.

Answer: The 16 possible samples of size 2, and the average weight of each sample is shown below:

Possible Sample	Sample Mean ($\overline{X}$)
2 and 2	2
2 and 4	3
2 and 6	4
2 and 8	5
4 and 2	3
4 and 4	4
4 and 6	5
4 and 8	6
6 and 2	4
6 and 4	5
6 and 6	6
6 and 8	7
8 and 2	5
8 and 4	6
8 and 6	7
8 and 8	8

(b) List all possible sample means, determine the frequency of the appearance of each mean, and draw a histogram of the sampling distribution of $\overline{X}$.

Answer: From Part a of this exercise, we can determine the frequency of each mean as follows:

Possible Sample Mean	Frequency
2	1
3	2
4	3
5	4
6	3
7	2
8	1

Therefore, the histogram of the sampling distribution can be shown as follows.

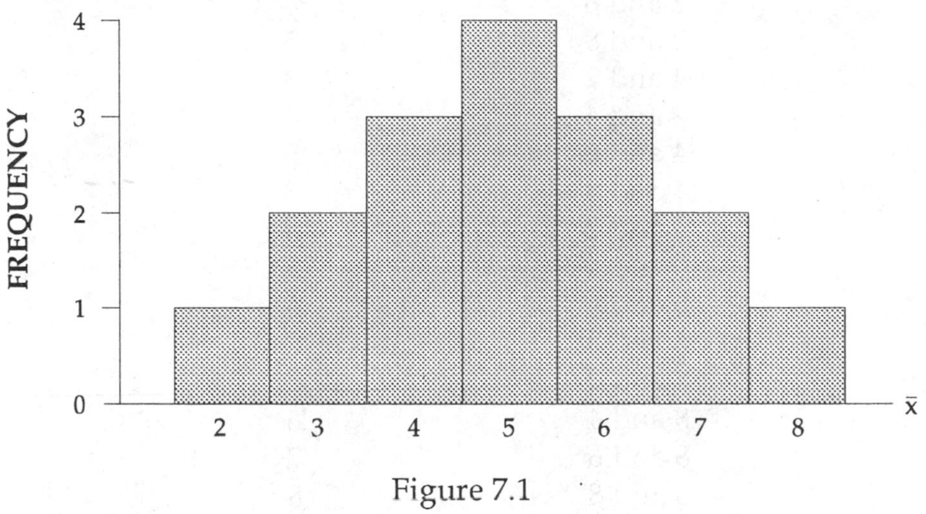

Figure 7.1

As you note, the sampling distribution of the sample means ($\overline{X}$), as shown in Figure 7.1, is approximately normally distributed.

***9.** Now let us consider a situation where sampling is done **without replacement**. Assume a population consists of 5 weights identical in appearance but weighing 1, 3, 5, 7, and 9 ounces. If we select samples of size 2 **without replacement**, there will be a total of 10 different samples.

(a) List all the possible samples of size 2 and determine the mean weight of each sample.

Answer: The 10 possible samples of size 2, and the average weight of each sample is shown below.

Possible Sample	Sample Mean ($\overline{X}$)
1 and 3	2
1 and 5	3
1 and 7	4
1 and 9	5
3 and 5	4
3 and 7	5
3 and 9	6
5 and 7	6
5 and 9	7
7 and 9	8

(b) List all possible sample means; determine the frequency of the appearance of each mean, and draw a histogram of the sampling distribution of $\overline{X}$.

Answer: From Part a of this exercise, we can determine the frequency of each mean:

Possible Sample Mean	Frequency
2	1
3	1
4	2
5	2
6	2
7	1
8	1

Therefore, the histogram of the sampling distribution can be shown as presented in Figure 7.2..

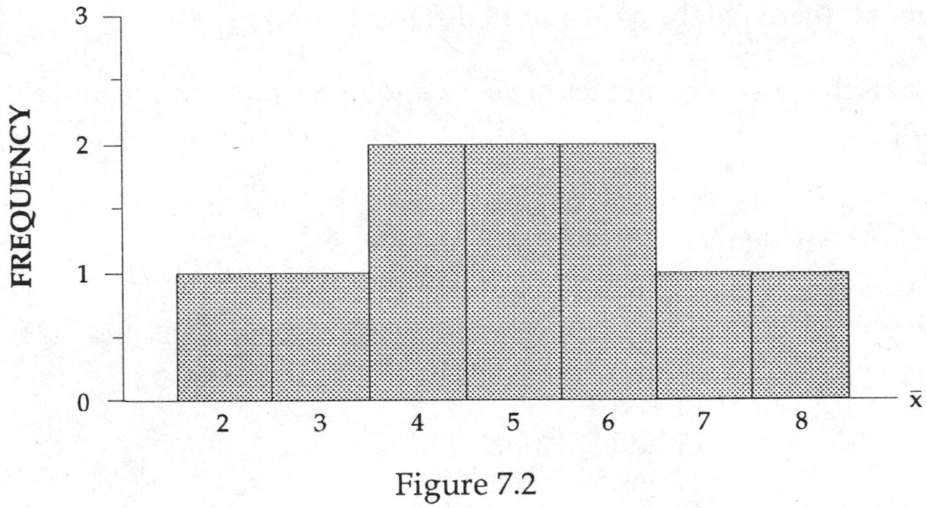

Figure 7.2

As you note, the sampling distribution of the sample means ($\overline{X}$), as shown in Figure 7.2, is approximately normally distributed.

10. Consider a population of five families with the following data representing the number of children in each family.

Family	Number of Children
A	2
B	6
C	4
D	3
E	1

(a) There are ten possible samples of size 2 (**sampling without replacement**). List the 10 possible samples of size 2, and determine the mean of each sample.

2, 6 4
2, 4 3
2, 3 2.5
2, 1 1.5
6, 4 5
6, 3 4.5
6, 1 3.5
4, 3 3.5
4, 1 2.5
3, 1 2

(b) List all the possible sample means and determine the frequency of the appearance of each mean.

4 – 1
3 – 1
2.5 – 2
1.5 – 1
5 – 1
4.5 – 1
3.5 – 2
2 – 1

11. Refer to exercise 10. Sampling **without replacement** with a sample size of 3 results in a total of 10 possible samples.

(a) List the 10 possible samples of size 3.

ABC
ABD
ABE
ACD
ACE
ADE

(b) Determine the mean of each sample.

***12.** Refer to exercise 8 in which a population consisted of 4 weights of 2, 4, 6, and 8 grams.

(a) Determine the mean and the variance of the population.

Answer: The mean of the population is

$$\mu = \frac{\Sigma X_i}{N} = \frac{2+4+6+8}{4} = 5$$

and the variance of the population can be determined as

$$\sigma^2 = \frac{\Sigma(X_i - \mu)^2}{N}$$

Then, calculating the value of the numerator, we have

X_i	$(X_i - \mu)^2$
2	9
4	1
6	1
8	9

$$\Sigma(X_i - \mu)^2 = 20$$

Therefore, the variance of the population is

$$\sigma^2 = \frac{\Sigma(X_i - \mu)^2}{N} = \frac{20}{4} = 5$$

(b) Sampling **with replacement** from the above population with a sample size of 2 produces sixteen possible samples, which have been shown in Part a of exercise 8. Using the sixteen $\overline{X}$ values, determine the mean and the variance of $\overline{X}$.

Answer: The sixteen sample means as previously shown are

Possible Means ($\overline{X}$)	$(\overline{X}-\mu)^2$
2	9
3	4
4	1
5	0
3	4
4	1
5	0
6	1
4	1
5	0
6	1
7	4
5	0
6	1
7	4
8	9
$\Sigma\overline{X} = 80$	$\Sigma(\overline{X}-\mu)^2 = 40$

Then, the mean of the above sixteen sample means is

$$E(\overline{X}) = \frac{80}{16} = 5$$

As you will note, the mean of the sample means is 5, which is equal to the mean of the population (as shown in Part a). In other words, we have just seen that $E(\overline{X}) = \mu$.

To determine the variance of $\overline{X}$, we need to determine $\Sigma(\overline{X}-\mu)^2$; the calculation of which is shown in the second column above. Therefore, the variance of the 16 sample means is

$$\sigma_{\overline{x}}^2 = \frac{\Sigma(\overline{X}-\mu)^2}{N} = \frac{40}{16} = 2.5$$

(c) Use equation 7.2 to determine the variance of $\overline{X}$.

Answer: Since sampling is done **with replacement** the following form of equation 7.2 is used

$$\sigma_{\bar{x}} = \frac{\sigma}{\sqrt{n}}$$

Therefore,

$$\sigma^2_{\bar{x}} = \frac{\sigma^2}{n}$$

In Part a of this exercise, we calculated the variance of the population. The value of which was equal to 5. Now using equation 7.2, we can determine the variance of $\overline{X}$ as

$$\sigma^2_{\bar{x}} = \frac{\sigma^2}{n} = \frac{5}{2} = 2.5$$

As you can see, we can determine the variance of $\overline{X}$ by direct computation as shown in Part b or simply use equation 7.2 to arrive at the same value.

***13.** Now let us apply the above procedure to sampling **without replacement.** Refer to exercise 9 in which a population consisted of 5 weights of 1, 3, 5, 7, and 9 ounces.

(a) Determine the mean and the variance of the population.

Answer: The mean of the population is

$$\mu = \frac{\Sigma X}{N} = \frac{1+3+5+7+9}{5} = 5$$

and the variance of the population can be determined as

$$\sigma^2 = \frac{\Sigma(X-\mu)^2}{N}$$

Then, calculating the value of the numerator, we have

X	$(X-\mu)^2$
1	16
3	4
5	0
7	4
9	16

$$\Sigma(X-\mu)^2 = 40$$

Therefore, the variance of the population is

$$\sigma^2 = \frac{\Sigma(X-\mu)^2}{N} = \frac{40}{5} = 8$$

(b) Sampling **without replacement** from the above population with a sample size of 2 produces ten possible samples, which have been shown in Part a of exercise 9. Using the ten $\overline{X}$ values, determine the mean and the variance of the $\overline{X}$.

Answer: The ten sample means, as previously shown, are

Possible Means ($\overline{X}$)	$(\overline{X}-\mu)^2$
2	9
3	4
4	1
5	0
4	1
5	0
6	1
6	1
7	4
8	9

$$\Sigma\overline{X} = 50 \qquad \Sigma(\overline{X}-\mu)^2 = 30$$

Then, the mean of the above ten sample means is

$$E(\overline{X}) = \frac{50}{10} = 5$$

As you will note, the mean of the sample means is 5, which is equal to the mean of the population (as shown in Part a). In other words, we have just seen that $E(\overline{X}) = \mu$.

To determine the variance of $\overline{X}$, we need to compute $\Sigma(\overline{X} - \mu)^2$; the calculation of which is shown in the second column on the previous page. Therefore, the variance of the 10 sample means is

$$\sigma^2_{\overline{x}} = \frac{\Sigma(\overline{X} - \mu)^2}{n} = \frac{30}{10} = 3$$

(c) Use equation 7.2 to determine the variance of the $\overline{X}$.

Answer: Equation 7.2 states

$$\sigma_{\overline{x}} = \sqrt{\frac{N-n}{N-1}}\left(\frac{\sigma}{\sqrt{n}}\right)$$

Therefore,

$$\sigma^2_{\overline{x}} = \frac{N-n}{N-1} \cdot \frac{\sigma^2}{n}$$

In Part a of this exercise, we calculated the variance of the population, the value of which was equal to 8. Now using equation 7.2, we can determine the variance of $\overline{X}$ as

$$\sigma^2_{\overline{x}} = \frac{N-n}{N-1} \cdot \frac{\sigma^2}{n} = \frac{5-2}{5-1} \cdot \frac{8}{2} = 3$$

As you can see, we can determine the variance of the $\overline{X}$ by direct computation as shown in Part b or simply use equation 7.2 to arrive at the same value.

14. In exercise 10, you were given the following information regarding a population of five families and the number of children in each family.

Family	Number of Children
A	2
B	6
C	4
D	3
E	1

(a) Determine the mean and the variance of the population.

(b) Using the ten $\overline{X}$ values (computed in exercise 10), compute the mean and the variance of $\overline{X}$.

(c) Using equations 7.1 and 7.2, determine the mean and the variance of the sample means ($\overline{X}$). Compare your values to those which you determined in Part b.

***15.** The average weekly earnings of the plumbers in a city is \$750 (that is μ) with a standard deviation of \$40 (that is σ). Assume that we select a random sample of 64 plumbers.

(a) Show the sampling distribution of the sample means ($\overline{X}$).

Answer: From the central limit theorem, we know that the distribution of the $\overline{X}$ is normal with a mean equal to μ and a standard deviation equal to $\sigma_{\overline{x}}$. Therefore, the mean of the distribution is

$$E(\overline{X}) = \mu = 750$$

and the standard deviation of the sampling distribution (the standard error of the mean) is calculated as

$$\sigma_{\overline{X}} = \frac{\sigma}{\sqrt{n}} = \frac{40}{\sqrt{64}} = 5$$

(b) What is the probability that the sample mean will be greater than \$740?

Answer: This shaded area of the sampling distribution in Figure 7.3 shows this desired probability.

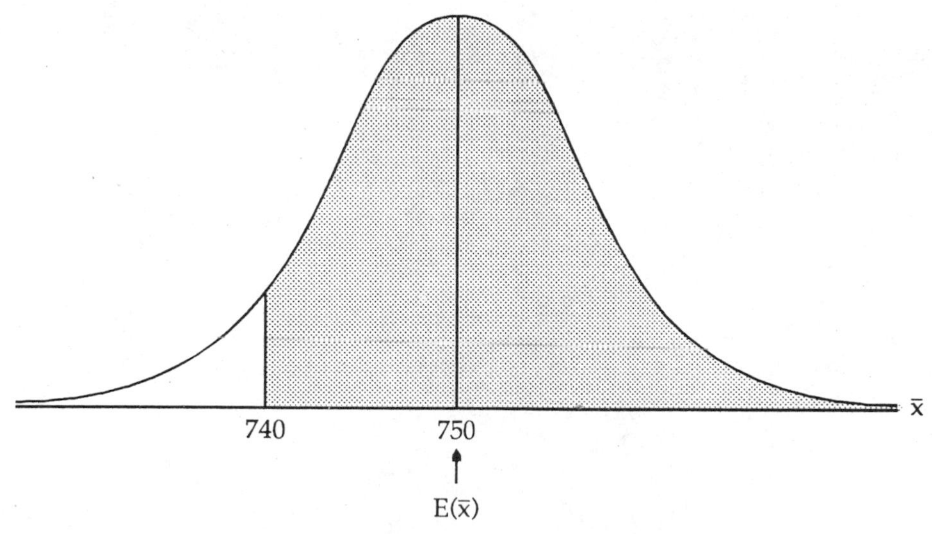

Figure 7.3

Since the sampling distribution is normal with a mean of \$750 and a standard deviation of \$5, we can use the normal distribution to calculate this probability. We calculate the number of standard errors of the mean between 740 to 750 as

$$Z = \frac{740 - 750}{5} = -2.0$$

Then from the table of areas under the normal curve, we read the area corresponding to 2 standard deviations to be 0.4772. Hence, the area to the right of 740 is 0.4772 + .5 = 0.9772. Therefore, there is a 0.9772 probability that the sample mean will be greater than \$740.

(c) If the population of plumbers consisted of 320 plumbers, what would be the standard error of the mean?

Answer: With such a population, we note that $n/N = 64/320 = 0.2$. (This means that the sample represents 20% of the population.) Since n/N is greater than 0.05, we consider the population to be a finite population. Hence, we use equation 7.2 to determine the standard error of the mean:

$$\sigma_{\bar{x}} = \sqrt{\frac{N-n}{N-1}} \cdot \frac{\sigma}{\sqrt{n}}$$

$$\sigma_{\bar{x}} = \sqrt{\frac{320 - 64}{320 - 1}} \cdot \frac{40}{\sqrt{64}}$$

$$= (0.896)(5)$$

$$= 4.48$$

16. An automotive repair shop has determined that the average service time on an automobile is 130 minutes with a standard deviation of 26 minutes. A random sample of 40 automotive services is selected.

(a) Show the sampling distribution of the $\overline{X}$.

$$\sigma_{\bar{x}} = \frac{\sigma}{\sqrt{n}} = \frac{26}{\sqrt{40}} = \frac{26}{6.325} = 4.111$$

(b) What is the probability that the sample of 40 automotive services will have a mean service time greater than 136 minutes?

130 136

$$\sigma_{\bar{x}} = \sqrt{\frac{N-n}{N-1}} \cdot \frac{\sigma}{\sqrt{n}}$$

$$\sigma_{\bar{x}} = \sqrt{\frac{130-136}{40-1}} \cdot \frac{26}{\sqrt{40}} = \sqrt{\frac{-6}{39}} = -.154 \cdot \frac{26}{6.325} = .633$$

$$.0721 \quad \sqrt{.154} = .392 \cdot \frac{5}{6.325}$$

(c) Assume the population consists of 400 automotive services. Determine the standard error of the mean.

$$\sigma_{\bar{x}} = \frac{\sigma}{\sqrt{n}} = \frac{26}{\sqrt{400}} = \frac{26}{20} = 1.30$$

3.9

17. There are 8,000 students at the University of Tennessee at Chattanooga. The average age of all the students is 24 years with a standard deviation of 9 years. A random sample of 36 students is selected.

(a) Determine the standard error of the mean.

(b) What is the probability that the sample mean will be larger than 25.5?

(c) What is the probability that the sample mean will be between 21.6 and 27 years?

***18.** In a local university, 40% of the students live in the dormitories. A random sample of 80 students is selected for a particular study.

(a) What is the sampling distribution of the $\bar{p}$?

Answer: The distribution of the sample proportion is normal with an expected value of $\bar{p}$ as

$$E(\bar{p}) = p$$

where p is the population proportion. In this case, since 40% of the students live in the dormitories, p = 0.4. Therefore, $E(\bar{p}) = 0.4$.

The standard deviation of the p, known as the standard error of the proportion, is determined as

$$\sigma_{\bar{p}} = \sqrt{\frac{p\,(1 - p)}{n}}$$

$$= \sqrt{\frac{.4\ (1 - .4)}{80}} = 0.05477$$

(b) What is the probability that the sample proportion (the proportion living in the dormitories) is between 0.30 and 0.50?

Answer: The shaded area shown in Figure 7.4 represents the desired probability.

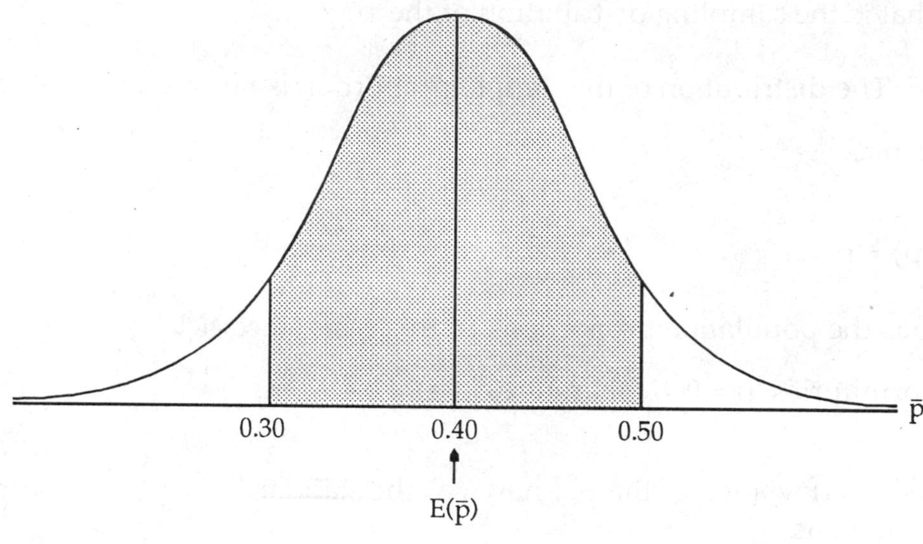

Figure 7.4

The area between 0.4 and 0.50 can be determined by calculating the number of standard errors of proportions between the two points:

$$Z = \frac{0.5 - 0.4}{0.0547} = 1.83$$

Then from the table of areas under the normal curve, we read the area corresponding to 1.83 as 0.4664, and the area between 0.3 and 0.4 as 0.4664 also. Therefore, the probability will be 0.4664 + 0.4664 = 0.9328.

19. A department store has determined that 25% of all their sales are credit sales. A random sample of 60 sales is selected.

(a) What is the sampling distribution of the $\bar{p}$?

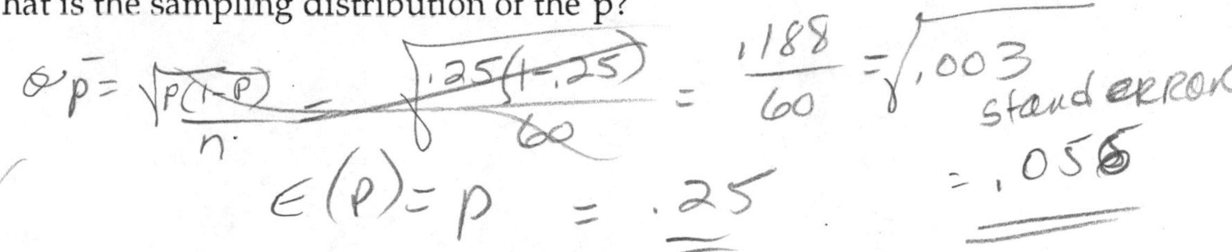

(b) What is the probability that the sample proportion will be greater than 0.30?

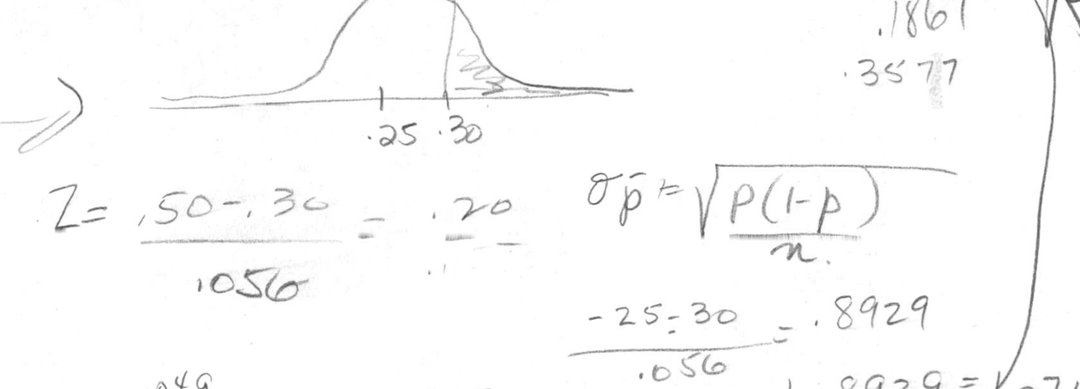

(c) What is the probability that the sample proportion will be between 0.20 to 0.30?

20. Only 4% of the items produced by a machine are defective. A random sample of 200 items is selected and checked for defects.

(a) Determine the standard error of the proportion.

(b) What is the probability that the sample contains more than 7% defective units?

21. There are 500 employees in a firm; 45% of whom are union members. A sample of 60 employees is selected randomly.

(a) Determine the standard error of the proportion. (Hint: First determine n/N.)

$\sigma_{\bar{p}} = \sqrt{\frac{p(1-p)}{n}} = \sqrt{\frac{.4(1-.4)}{60}} = \sqrt{\frac{.24}{60}} = \sqrt{.004} = .063$

(b) What is the probability that the sample proportion (proportion of union members) is between 0.40 and 0.55?

.7482

.40 .45 .55

.2852
.4441

$\frac{.40-.45}{.0603} = \frac{-.05}{.0603} = .0103$

.2939
.4515
.7454

$\frac{.45-.55}{.0603} = \frac{-.10}{.0603} = 1.66$

SELF-TESTING QUESTIONS

In the following multiple choice questions, circle the correct answer. An answer key is provided following the questions.

1. A probability distribution for all possible values of a sample statistic is known as

a) a sample statistic
b) a parameter
c) simple random sampling
d) a sampling distribution
e) none of the above

2. A population characteristic, such as a population mean is called

a) a statistic
b) a parameter
c) a sample
d) none of the above

3. A property of a point estimator that occurs whenever larger sample sizes tend to provide point estimates closer to the population parameter is known as

a) efficiency
b) unbiased sampling
c) consistency
d) none of the above

4. The ratio of the variances of two point estimators of a given parameter is called

a) the coefficient of variation
b) the average variance
c) relative efficiency
d) none of the above

5. A measure from a sample, such as a sample mean, is known as

a) a statistic
b) a parameter
c) mean deviation
d) none of the above

6. The standard deviation of a point estimator is called the

a) standard deviation
b) standard error
c) point estimator
d) none of the above

7. A single numerical value used as an estimate of a population parameter is known as

a) a parameter
b) a population parameter
c) either a or b
d) a point estimate
e) none of the above

8. The sample statistic, such as $\overline{X}$, S, and $\overline{p}$, that provides the point estimate of the population parameter is known as the

a) point estimator
b) parameter
c) population parameter
d) nonc of the above

9. A theorem that allows us to use the normal probability distribution to approximate the sampling distribution of $\overline{X}$ and $\overline{p}$ whenever the sample size is large is known as the

a) approximation theorem
b) normal probability theorem
c) central limit theorem
d) none of the above

10. A property of a point estimator that occurs whenever the expected value of the point estimator is equal to the population parameter it estimates is known as

a) consistency
b) the expected value
c) the estimator
d) unbiased
e) none of the above

11. Given two unbiased point estimators of the same population parameter, the point estimator with the smaller variance is said to have

a) smaller relative efficiency
b) greater relative efficiency
c) smaller consistency
d) none of the above

12. Whenever the estimation process summarizes **all** of the information a sample has about a population parameter, the point estimator has the property of

a) relative consistency
b) full consistency
c) sufficiency
d) insufficiency
e) none of the above

13. The number of different simple random samples of size 4 that can be selected from a population of size 6 is

a) 24
b) 30
c) 15
d) 4
e) none of the above

14. Since the sample size is always smaller than the size of the population, then the sample mean

a) must always be smaller than the population mean
b) must be larger than the population mean
c) must be equal to the population mean
d) could be larger, smaller, or equal to the mean of the population
e) none of the above

15. As the sample size increases

a) the standard deviation of the population decreases
b) the population mean increases
c) the standard error of the mean decreases
d) the standard error of the mean increases
e) none of the above

16. The point estimation of μ is

a) S
b) $\overline{X}$
c) $\overline{p}$
d) σ
e) none of the above

17. The point estimator of σ is

a) S
b) $\overline{X}$
c) X
d) μ
e) none of the above

18. Which of the following is a point estimator?

a) σ
b) $\overline{X}$
c) μ
d) none of the above

19. In computing the standard error of the mean, the finite population correction factor is used when

a) $n/N > 30$
b) $N/n \leq 0.05$
c) $n/N \leq 0.05$
d) $n/N \geq 0.05$
e) none of the above

20. As the sample size becomes larger, the sampling distribution of the sample mean approaches a

a) binomial distribution
b) normal distribution
c) hypergeometric distribution
d) chi-square distribution
e) none of the above

ANSWERS TO THE SELF-TESTING QUESTIONS

1. d
2. b
3. c
4. c
5. a
6. b
7. d
8. a
9. c
10. d
11. b
12. c
13. c
14. d
15. c
16. b
17. a
18. b
19. d
20. b

ANSWERS TO CHAPTER SEVEN EXERCISES

5. (a) $1.20
   (b) 0.03428

6. (a) 5.67
   (b) 2.85
   (c) 0.20

7. (a) 75
   (b) 20.48
   (c) 0.8

10. (a) AB, AC, AD, AE
    BC, BD, BE
    CD, CE
    DE

    (b)

Possible Sample Mean	Frequency
1.5	1
2.0	1
2.5	2
3.0	1
3.5	2
4.0	1
4.5	1
5.0	1

11. (a) ABC, ABD, ABE, ACD, ACE, ADE
BCD, BCE, BDE
CDE

(b)

Sample	Sample Mean
ABC	4
ABD	3.667
ABE	3
ACD	3
ACE	2.33
ADE	2
BCD	4.33
BCE	3.67
BDE	3.33
CDE	2.67

14. (a) 3.2, 2.96
(b) 3.2, 1.11
(c) 3.2, 1.11

16. (a) $E(\overline{X}) = 130$ Standard Error = 4.11
(b) 0.0721
(c) 3.9

17. (a) 1.5
(b) 0.1587
(c) 0.9224

19. (a) $E(\overline{p}) = 0.25$ Standard Error = .056 (rounded)
(b) 0.1867
(c) 0.6266

20. (a) 0.0139 (rounded)
(b) 0.0154

21. (a) 0.0603
(b) 0.7482

CHAPTER EIGHT

INTERVAL ESTIMATION

CHAPTER OUTLINE AND REVIEW

You were introduced to the concept of a point estimate in Chapter 7. As you learned in that chapter, a point estimate is a single numerical value used for estimating a population parameter. Therefore, one cannot expect a point estimate to be exactly equal to the population parameter. Furthermore, a point estimate does not give us information about how close the point estimate is to the true value of the population parameter.

In Chapter 8, you have been introduced to the concept of interval estimation, which is the determination of an interval for the population parameter based on sample information. More specifically, you have been introduced to the following concepts:

A. Interval Estimate: An estimate of a population parameter that provides an interval of values believed to contain the value of the parameter.

B. Sampling Error: The difference between the sample statistic (point estimate) and the population parameter that the statistic is estimating. In regards to the mean, the sampling error is $|\overline{X} - \mu|$; and in regards to the proportion, the sampling error is $|\overline{p} - p|$.

C. Confidence Level: As was mentioned earlier, the interval estimate is believed to contain the value of the parameter. The confidence that is placed on the ability of an interval estimate to contain the value of the parameter is called the confidence level.

D. t Distribution: In order to develop an interval estimate for the mean of the population, where the standard deviation of the population is not known, a family of probability distributions, known as the "t distribution," is used. The number of degrees of freedom for the computation of an interval estimate for the mean is n - 1, where n is the sample size.

CHAPTER FORMULAS

Sampling Error (Mean) $= | \overline{X} - \mu |$ (8.1)

Interval Estimation of a Population Mean

I. Large samples ($n \geq 30$)

A. When the standard deviation of the population σ is known,

$$\mu = \bar{x} \pm Z_{\alpha/2} \cdot \sigma_{\bar{x}} \quad (8.2)$$

where $\sigma_{\bar{x}} = \dfrac{\sigma}{\sqrt{n}}$

B. When the standard deviation σ is unknown, the sample standard deviation S is used to approximate σ, therefore the equation becomes

$$\mu = \bar{x} \pm Z_{\alpha/2} \cdot S_{\bar{x}} \quad (8.3)$$

where $S_{\bar{x}} = \dfrac{S}{\sqrt{n}}$

II. Small samples ($n < 30$)

A. When the population has a normal probability distribution and the standard deviation of the population σ is known, use equation 8.2.

B. When the population has a normal probability distribution and the standard deviation of the population σ is unknown,

$$\mu = \bar{x} \pm t_{\alpha/2} \cdot S_{\bar{x}} \quad (8.4)$$

where $S_{\bar{x}} = \dfrac{S}{\sqrt{n}}$ (Where S is the standard deviation of the sample)

CHAPTER FORMULAS
(Continued)

Sample Size for An Interval Estimate of a Population Mean

$$n = \frac{(Z_{\alpha/2})^2 \cdot \sigma^2}{E^2} \qquad (8.5)$$

where E = the size of the sampling error

Interval Estimation of a Population Proportion

$$p = \bar{p} \pm Z_{\alpha/2} \cdot S_{\bar{p}} \qquad (8.8)$$

where $S_{\bar{p}} = \sqrt{\frac{\bar{p}(1-\bar{p})}{n}}$

Sample Size for An Interval Estimate of a Population Proportion

$$n = \frac{(Z_{\alpha/2})^2 \cdot p(1-p)}{E^2} \qquad (8.9)$$

If the value of p in equation 8.9 is not known and a good estimate of p is not available, use p=0.50.

EXERCISES

***1.** In order to estimate the average electric usage per month, a sample of 81 houses was selected, and the electric usage was determined.

(a) Assume a population standard deviation of 450 kilowatt hours. Determine the standard error of the mean.

Answer: When the standard deviation of the population is known, the standard error of the mean is

$$\sigma_{\bar{x}} = \frac{\sigma}{\sqrt{n}} = \frac{450}{\sqrt{81}} = 50$$

(b) With a 0.95 probability, what can be said about the size of the sampling error?

Answer: The probability of 0.95 indicates that $\alpha = 0.05$, which means that the area in the upper tail of the normal distribution is $\alpha/2 = 0.025$. Thus, the area from the mean to the upper tail is 0.475. Now looking in the body of Table 1 of Appendix B in your text (Standard Normal Distribution Table), we note that the Z value corresponding to an area of 0.475 is 1.96. Therefore, we can state that 95% of the sample means will lie within plus or minus 1.96 standard deviations of the mean. Since $1.96\sigma_{\bar{x}} = (1.96)(50) = 98$, we can state that there is a 0.95 probability that the value of the sample mean will result in a sampling error of 98 KWH or less.

(c) If the sample mean is 1858 KWH, what is the 95% confidence interval estimate of the population mean?

Answer: We have already determined that $\sigma_{\bar{x}} = 50$ and $Z_{.025} = 1.96$. When the standard deviation of the population is known, the interval estimate of the population mean is

$$\mu = \bar{x} \pm Z_{\alpha/2} \cdot \sigma_{\bar{x}} = 1858 \pm (1.96)(50)$$

$$= 1{,}858 \pm 98$$

Therefore, the confidence interval of the mean is from 1,760 to 1,956 KWH.

2. A random sample of 81 credit sales in a department store showed an average sale of $68.00. From past data, it is known that the standard deviation of the population is $27.00.

(a) Determine the standard error of the mean.

(b) With a 0.95 probability, what can be said about the size of the sampling error?

(c) What is the 95% confidence interval of the population mean?

3. In order to determine the average weight of carry-on luggage by passengers in airplanes, a sample of 25 pieces of carry-on luggage was collected and weighed. The average weight was 18 pounds. Assume that we know the standard deviation of the population to be 7.5 pounds.

(a) Determine a 98% confidence interval estimate for the mean weight of the carry-on luggage.

(b) Determine an 80% confidence interval estimate for the mean weight of the carry-on luggage.

4. A small stock brokerage firm wants to determine the average daily sales (in dollars) of stocks to their clients. A sample of the sales for 36 days revealed average sales of $139,000. Assume that the standard deviation of the population is known to be $12,000. Provide a 95% and a 99% confidence interval estimate for the average daily sales.

***5.** A random sample of 64 children with working mothers showed that they were absent from school an average of 5.3 days per term with a standard deviation of 1.8 days. Provide a 96% confidence interval for the average number of days absent per term for all the children.

Answer: In this situation, we want to determine a confidence interval for the mean where the standard deviation of the population is not known. Therefore, we determine the standard error of the mean by the use of the standard deviation of the sample as

$$S_{\bar{x}} = \frac{S}{\sqrt{n}} = \frac{1.80}{\sqrt{64}} = 0.225$$

Then the interval estimate for the mean for a large sample ($n \geq 30$) is

$$\mu = \bar{x} \pm Z_{\alpha/2} . S_{\bar{x}}$$

A 96% confidence indicates that $\alpha = 0.4$, which means $\alpha/2 = 0.02$. Thus, the area from the mean to the upper tail is 0.48. Now looking in the standard normal distribution table, we note that a Z value corresponding to an area of 0.48 (or the closest value to it, which is actually 0.4798) is $Z_{.02} = 2.05$. Therefore, the interval estimate will be

$$\mu = \bar{x} \pm Z_{\alpha/2} . S_{\bar{x}} = 5.3 \pm (2.05)(0.225)$$

$$= 5.3 \pm 0.461$$

Hence, the 96% confidence interval estimate for the population mean is 4.838 days to 5.761 days.

6. The Highway Safety Department wants to study the driving habits of individuals. A sample of 196 cars traveling on the highway revealed an average speed of 67 miles per hour with a standard deviation of 7 miles per hour. Determine a 99% confidence interval estimate for the speed of all cars.

7. To determine how many hours per week freshmen college students watch television, a random sample of 225 students was selected. It was determined that the students in the sample spent an average of 35 hours with a standard deviation of 12 hours watching TV per week. Provide a 95% confidence interval estimate for the average number of hours that all college freshmen spend watching TV per week.

8. Computer Services, Inc. wants to determine a confidence interval for the average CPU time of their teleprocessing transactions. A sample of 144 transactions yielded a mean of 0.20 seconds with a standard deviation of 0.05. Determine a 92% confidence interval for the average CPU time.

***9.** The proprietor of a boutique in New York wanted to determine the average age of his customers. A random sample of 25 customers revealed an average age of 32 years with a standard deviation of 8 years. Determine a 95% confidence interval estimate for the average age of all his customers.

Answer: Since we do not know the standard deviation of the population, we can determine the standard error of the mean using the standard deviation of the sample as

$$S_{\bar{x}} = \frac{S}{\sqrt{n}} = \frac{8}{\sqrt{25}} = 1.6$$

Then we note that the sample is small ($n < 30$); hence, we determine an interval for the mean of the population as

$$\mu = \bar{x} \pm t_{\alpha/2} . S_{\bar{x}}$$

To read the appropriate t value, we note that the sample size is 25. Therefore, there are 24 degrees of freedom ($n - 1$); and at 95% confidence, $\alpha = .05$ and $\alpha/2 = .025$. Thus, we can read t at 24 degrees of freedom from Table 2 of Appendix B of your text (t Distribution Table) as $t_{.025} = 2.064$. Therefore, the interval estimate will be

$$\mu = \bar{x} \pm t_{\alpha/2} . S_{\bar{x}} = 32 \pm (2.064)(1.6)$$

$$= 32 \pm 3.3024$$

Hence, at 95% confidence, the interval estimate for the average age of all his customers is from 28.6976 to 35.3024 years.

In exercises 10 through 17, assume the populations are normally distributed.

10. A sample of 16 patients in a doctor's office showed that they had to wait an average of 43 minutes, with a standard deviation of 8 minutes, before they could see the doctor. Provide a 95% and a 99% confidence interval estimate for the average waiting time of all the patients who visit this doctor.

11. Refer to exercise 6. Assume that a sample of 20 cars was taken. Determine a 98% confidence interval estimate for the speed of all cars.

12. Refer to exercise 7. Assume that a sample of 28 students was selected. Provide a 90% confidence interval estimate for the average number of hours that all college freshmen spend watching TV per week.

13. The owner of a restaurant wants to determine the average number of customers who eat lunch at his restaurant each day. A sample of 16 days showed an average of 130 lunches were served daily. The standard deviation of the sample was 12. Provide a 90% confidence interval estimate for the average number of lunches served per day.

***14.** In order to determine the life expectancy of the picture tubes of a particular brand of portable televisions, a sample of 6 tubes was selected randomly. The sample revealed the following life expectancies:

Picture Tube	Life Expectancy (X_i) (In Thousands of Hours)	$(X_i - \bar{X})^2$
1	8.2	0.04
2	7.5	0.25
3	9.5	2.25
4	6.5	2.25
5	8.5	0.25
6	7.8	0.04
	$\Sigma X_i = 48.0$	$\Sigma(X_i - \bar{X})^2 = 5.08$

Provide a 90% confidence interval estimate for the life expectancy of all picture tubes for this brand of portable televisions.

Answer: From the above data, we first must determine the mean of the sample as

$$\bar{X} = \frac{\Sigma X_i}{n} = \frac{48}{6} = 8$$

Since we do not know the standard deviation of the population, we calculate the standard deviation of the sample (S) and use it as a point estimator for the standard deviation of the population (σ).

$$S = \sqrt{\frac{\Sigma(X_i - \bar{X})^2}{n-1}} = \sqrt{\frac{5.08}{6 - 1}} = 1.008$$

Then, we can estimate the standard error of the mean as

$$S_{\bar{x}} = \frac{S}{\sqrt{n}} = \frac{1.008}{\sqrt{6}} = 0.4115$$

Since the standard deviation of the population is not known and since the sample size is small, we determine an interval estimate by

$$\mu = \bar{x} \pm t_{\alpha/2} . S_{\bar{x}}$$

From Table 2 of Appendix B of your textbook (t Distribution Table), at 90% confidence and 5 degrees of freedom (n - 1), we read

$$t_{\alpha/2} = t_{.05} = 2.015$$

Hence, the interval estimate becomes

$$\mu = \bar{x} \pm t_{\alpha/2} . S_{\bar{x}} = 8 \pm (2.015)(0.4115)$$

$$= 8 \pm 0.829$$

Thus, the 90% confidence interval estimate of the population mean is from 7,171 hours to 8,829. Note that the units for life expectancy are in thousands.

15. A fluorescent light bulb manufacturer selected a sample of 10 fluorescent bulbs and gathered the following data:

Light Bulbs	Life Expectancy (In Hundreds of Hours)
1	39
2	42
3	41
4	38
5	40
6	43
7	36
8	37
9	43
10	41

Determine a 98% confidence interval estimate for the life expectancy of all the bulbs produced by this manufacturer.

16. The following data represents the hourly wages in a sample of 12 carpenters in the city of Seattle, Washington.

Carpenter	Hourly Wages (In Dollars)
1	8.17
2	9.23
3	10.50
4	7.55
5	9.45
6	8.65
7	8.20
8	10.35
9	12.80
10	9.60
11	6.10
12	7.40

Provide a 90% confidence interval estimate for all the carpenters in the city of Seattle, Washington.

17. A local university administers a comprehensive examination to the candidates for B.S. degrees in Business Administration. Five examinations are selected at random and scored. The scores are shown below.

Grade
94
72
93
54
77

Develop a 98% confidence interval estimate for the mean of the population.

***18.** The monthly starting salaries of students who receive business degrees have a standard deviation of \$600. What size sample should be selected so that there is 0.95 probability of estimating the mean monthly income within \$150 or less?

Answer: From the above, it is indicated that the sampling error E = \$150. Furthermore, $\alpha = 0.05$ and $\alpha/2 = 0.025$ (note that $1 - \alpha$ is 0.95). Thus, we read the $Z_{\alpha/2} = Z_{.025} = 1.96$. Then we determine the sample size as

$$n = \frac{(Z_{\alpha/2})^2 \cdot \sigma^2}{E^2} = \frac{(1.96)^2 \ (600)^2}{(150)^2} = 61.4656$$

Since the computed n is not an integer, we **round up** to the next integer value and conclude that the desired sample size is 62 students.

19. A coal company wants to determine a 95% confidence interval estimate for the average daily tonnage of coal which they mine. Assuming that the company reports that the standard deviation of daily output is 80 tons, how many days should they sample so that the sampling error is 20 tons or less?

E = \$20

Z = 1.96

$$n = \frac{(1.96)^2 (80)^2}{(20)^2} = \frac{3.8416 \cdot 6400}{400} = \frac{245862.40}{400} = 624.$$

20. If the standard deviation of vacuum cleaners lifetime is estimated to be 400 hours, how large a sample must be taken in order to be 93% confident that the sampling error will not exceed 50 hours?

E = 50 .93 = .07/2 .035

***21.** In the last presidential election, a sample of 120 registered voters in Washington, D. C. showed that 30 of them voted for the incumbent president. Develop a 98% confidence interval estimate for the proportion of all Washington registered voters who voted for the incumbent president.

Answer: In the sample of 120 registered voters, there were 30 who voted for the incumbent president. Thus, the point estimate of the proportion of voters who voted for the incumbent president is $\bar{p} = 30/120 = 0.25$. Now we can estimate the standard error of the proportion using this sample proportion as

$$S_{\bar{p}} = \sqrt{\frac{\bar{p}(1-\bar{p})}{n}} = \sqrt{\frac{0.25\ (1\ -\ 0.25)}{120}} = 0.039$$

At 98% confidence, $Z_{\alpha/2} = Z_{.01} = 2.33$. Therefore, the interval estimate can be determined as

$$p = \bar{p} \pm Z_{\alpha/2} \cdot S_{\bar{p}} = 0.25 \pm (2.33)\ (0.039)$$

$$= 0.25 \pm 0.09$$

Hence, we note that at a 98% confidence level the interval estimate for the proportion of all registered voters who voted for the incumbent president is from 0.16 to 0.34.

22. A health club is considering the addition of a running track to its current facilities. The owner of the club selects a random sample of 200 of members and determines that 160 have indicated that they would use the track. Determine a 96% confidence interval for the proportion of all the club members who would use the new track.

$\bar{p} = 160/200 = .80$ $\quad .96/2 = .4800 = 2.06$

$S_{\bar{p}} = \sqrt{\frac{\bar{p}(1-\bar{p})}{n}} = \sqrt{\frac{.80(1-.80)}{200}} = \sqrt{\frac{.16}{200}} = .028$

$.80 \pm (2.06)(.028) =$

$.80 \pm .0577 =$ ~~.85~~

$.74 - .85$

23. A local health care facility noted that in a sample of 200 patients, 180 were referred to them by the local hospital. Provide a 99% confidence interval for all the patients who are referred to this facility by the hospital.

24. In a random sample of 150 residents of New Hampshire, 90 residents indicated that they voted for the Republican candidate in the last presidential election. Develop a 95% confidence interval estimate for the proportion of all New Hampshire residents who voted for the Republican candidate.

***25.** Refer to exercise 23. What size sample would be required to estimate the proportion of hospital referrals with a sampling error of 0.08 or less at 95% confidence?

Answer: To determine this sample size, we use

$$n = \frac{(Z_{\alpha/2})^2 \cdot p(1-p)}{E^2}$$

But since p is not known, we use the sample proportion:

$$\bar{p} = \frac{180}{200} = 0.9$$

Therefore,

$$n = \frac{(1.96)^2\,(0.9)(0.1)}{(0.08)^2} = 54.02$$

Rounding up, we realize that the required sample size is 55.

26. The manager of a grocery store wants to determine what proportion of people who enter his store are his regular customers. What size sample should he take so that at 95% confidence the sampling error will not be more than 0.1? Hint: Since p is not known, you need to determine the sample size by using $p = 0.5$.

27. A local weight loss clinic is interested in the effectiveness of its weight reduction program. In a random sample of 50 participants, 35 lost weight. How large a sample size is necessary for the sampling error to be no more than 10% at 95% confidence?

SELF-TESTING QUESTIONS

In the following multiple choice questions, circle the correct answer. An answer key is provided following the questions.

1. An estimate of a population parameter that provides an interval of values believed to contain the value of the parameter is known as the

a) confidence level
b) interval estimate
c) parameter value
d) population estimate

2. The difference between the point estimate, such as the sample mean $\overline{X}$, and the value of the population parameter it estimates, such as the population mean μ, is known as the

a) confidence level
b) sampling error
c) parameter estimate
d) interval estimate

3. If an interval estimate is said to be constructed at the 90% confidence level, the confidence coefficient would be

a) 0.1
b) 0.95
c) 0.9
d) none of the above

4. Whenever the population standard deviation is unknown and the population has a normal or near-normal distribution, which distribution is used in developing an interval estimation?

a) standard distribution
b) chi-square distribution
c) beta distribution
d) t distribution

5. As the number of degrees of freedom in a t distribution increases, the difference between the t distribution and the standard normal distribution

a) becomes larger
b) becomes smaller
c) varies according to the confidence coefficient
d) none of the above

6. The t distribution is applicable whenever

a) the sample is considered small ($n < 30$)
b) the population is normal and the sample standard deviation is used to estimate the population standard deviation
c) both a and b
d) none of the above

7. In developing an interval estimate, if the population standard deviation is unknown

a) it is impossible to develop an interval estimate
b) the standard deviation is arrived at using historical data
c) the sample standard deviation can be used
d) none of the above

Answer questions 8 through 10 based on the following information.

In order to estimate the average time spent on the computer terminals per student at a local university, data were collected from a sample of 81 business students over a one week period. Assume the population standard deviation is 1.2 hours.

8. The standard error of the mean is

a) 7.5
b) 0.014
c) 0.160
d) 0.133
e) none of the above

9. With a 0.95 probability, the sampling error is approximately

a) 0.26
b) 1.96
c) 0.21
d) 1.64
e) none of the above

10. If the sample mean is 9 hours, then the 95% confidence interval is

a) 7.04 to 110.96 hours
b) 7.36 to 10.64 hours
c) 7.80 to 10.20 hours
d) 8.74 to 9.26 hours
e) none of the above

11. The absolute value of the difference between the point estimate and the population parameter it estimates is

a) the standard error
b) the sampling error
c) precision
d) the error of confidence
e) none of the above

12. If we want to provide a 95% confidence interval for the proportion of a population, the Z value is

a) 0.45
b) 0.90
c) 1.96
d) 1.645
e) none of the above

13. The Z value for a 97% confidence interval estimation is

a) 1.88
b) 1.96.
c) 2.00
d) 2.17
e) none of the above

14. The t value for a 95% confidence interval estimation with 28 degrees of freedom is

a) 2.467
b) 2.052
c) 2.048
d) 2.473
e) none of the above.

15. A 90% confidence interval for a population mean is determined to be 800 to 900. If the confidence coefficient is increased to 0.95, the interval for μ

a) becomes narrower
b) becomes 0.05
c) does not change
d) becomes wider
e) none of the above

ANSWERS TO THE SELF-TESTING QUESTIONS

1. b
2. b
3. c
4. d
5. b
6. c
7. c
8. d
9. a
10. d
11. b
12. b
13. d
14. c
15. d

ANSWERS TO CHAPTER EIGHT EXERCISES

2. (a) 3.0
   (b) 5.88
   (c) $62.12 to $73.88

3. (a) 14.505 to 21.495
   (b) 16.08 to 19.92

4. $135,080 to $142,920
   $133,840 to $144,160

6. 65.71 to 68.29

7. 33.432 to 36.568

8. 0.1927 to 0.2073

10. 38.738 to 47.262
    37.106 to 48.894

11. 63.03 to 70.97

12. 31.14 to 38.86

13. 125 to 135 (rounded)

15. 37.813 to 42.186

16. $8.09 to $9.91

17. 50.29 to 105.71

19. 62 (rounded)

20. 210 (rounded)

22. 0.7426 to 0.8574

23. 0.8453 to 0.9546

24. 0.5213 to 0.6787

26. 97 (rounded)

27. 81 (rounded)

CHAPTER NINE

HYPOTHESIS TESTING

CHAPTER OUTLINE AND REVIEW

This chapter has introduced you to another form of statistical inference, namely, testing of hypothesis. You have learned that one can state a hypothesis about the characteristic of a population and then, based on the sample information, decide whether or not the hypothesis is rejected. The main topics covered in this chapter are the following.

A.	**Hypothesis:**	An assumption made about the value of a population parameter.
B.	**Null Hypothesis:**	The hypothesis which is tentatively assumed to be true in hypothesis testing. Based on sample information, we can decide whether to reject or not reject the null hypothesis.
C.	**Alternative Hypothesis:**	The hypothesis which is assumed to be true when the null hypothesis is rejected.
D.	**Type I Error:**	When the null hypothesis is in fact true, but the sample information has resulted in the rejection of the true null hypothesis, a Type I error has been made.
E.	**Type II Error:**	When the null hypothesis is false, but the sample information has resulted in the acceptance of the false null hypothesis, a Type II error has been made.

F. **Critical Value:** A value which is compared with the test statistic in order to determine whether or not the null hypothesis is to be rejected.

G. **Level of Significance:** The maximum probability of a Type I error that the decision maker will tolerate in the test of hypothesis procedure.

H. **One-Tailed Test:** A testing situation in which the null hypothesis is rejected for the values of the point estimator only on one side (tail) of the sampling distribution.

I. **Two-Tailed Test:** A testing situation in which the null hypothesis is rejected for values of the point estimator in either tail of the sampling distribution.

J. **Power:** The probability of rejecting the null hypothesis.

K. **Power Curve:** A graph of the probability of the rejection of the null hypothesis for all possible values of the population parameter.

L. **P-Value:** The smallest value of α for which H_0 would be rejected given sample results. The null hypothesis should be rejected only if the p-value is less than the level of significance for the test. That is,
Reject H_0 if p-value $< \alpha$.

CHAPTER FORMULAS

Standardized Testing Procedure

I. Hypothesis Tests about a Population Mean: Large Sample (n ≥30)

Test Statistic: $Z = \dfrac{\overline{X} - \mu_o}{\sigma_{\overline{X}}}$ (9.6)

where $\sigma_{\overline{X}} = \dfrac{\sigma}{\sqrt{n}}$

(If σ is not known, substitute S for σ in computing Z statistic.)

A. Lower one-tailed test of the form

H_0: $\mu \geq \mu_0$

H_a: $\mu < \mu_0$

Reject H_0 if $Z < -Z_\alpha$

B. Upper one-tailed test of the form

H_0: $\mu \leq \mu_0$

H_a: $\mu > \mu_0$

Reject H_0 if $Z > Z_\alpha$

C. Two-tailed test of the form

H_0: $\mu = \mu_0$

H_a: $\mu \neq \mu_0$

Reject H_0 if $Z < -Z_{\alpha/2}$ or $Z > Z_{\alpha/2}$

CHAPTER FORMULAS
(Continued)

II. Hypothesis Tests about a Population Mean: Small Samples (n < 30) and Unknown σ

Standardized Test Statistic: $$t = \frac{\overline{X} - \mu_o}{S_{\overline{X}}} \quad (9.7)$$

where $S_{\bar{x}} = \frac{S}{\sqrt{n}}$

The decision rules are the same as those shown in Part I with the t statistic substituted for the Z statistic.

III. Hypothesis Tests about a Population Proportion

Standardized Test Statistic: $$Z = \frac{\overline{p} - p_o}{\sigma_{\overline{p}}} \quad (9.8)$$

where $$\sigma_{\overline{p}} = \sqrt{\frac{p_o\,(1 - p_o)}{n}} \quad (9.9)$$

The decision rules are the same as those shown in Part I with p substituted for μ.

P-Value Criterion for Hypothesis Testing

Do not reject H_0 if p-value $\geq \alpha$

Reject H_0 if p-value $< \alpha$

CHAPTER FORMULAS
(Continued)

Determining the Sample Size for a Hypothesis Test about a Population Mean

Recommended Sample Size for a One-Tailed Hypothesis Test

$$n = \frac{(Z_\alpha + Z_\beta)^2 \sigma^2}{(\mu_o - \mu_a)^2} \tag{9.12}$$

where Z_α = Z value associated with an area of α in one tail of the distribution

Z_β = Z value associated with an area of β in one tail of the distribution

σ = the standard deviation of the population

μ_o = the value of the population mean in the null hypothesis

μ_α = the actual value of the population mean in the statement about a Type II error

In a two-tailed test, replace Z_α with $Z_{\alpha/2}$.

EXERCISES

***1.** The average gasoline price of one of the major oil companies has been $1.25 per gallon. Recently, the company has undertaken several efficiency measures in order to reduce prices. Management is interested in determining whether their efficiency measures have actually **reduced** prices. That is, they are interested in determining whether or not the current average price is significantly less than $1.25. A random sample of 49 of their gas stations is selected and the average price is determined to be $1.20. Furthermore, assume that the standard deviation of the population (σ) is $0.14.

(a) At 95% confidence, test to determine whether the measures were effective in reducing the average price.

Answer: The first step is to formulate the null and the alternative hypotheses.

H_0: $\mu \geq 1.25$

H_a: $\mu < 1.25$ (There was a reduction in the average price.)

In this case, the standard deviation of the population is known; and we have a lower one-tailed hypothesis testing situation.

The null hypothesis is rejected if the test statistic $Z < -Z_\alpha$, where the test statistic Z is

$$Z = \frac{\overline{X} - \mu_o}{\sigma_{\overline{X}}} = \frac{\overline{X} - \mu_o}{\sigma/\sqrt{n}} = \frac{1.20 - 1.25}{.02} = -2.5$$

At a 0.05 level of significance, we read the value of Z_α from the table of areas under the normal curve as $Z_{.05} = 1.64$. As you can see in Figure 9.1 for any value of $Z < -1.64$, the null hypothesis is rejected. In this case, since $-2.5 < -1.64$, the null hypothesis is rejected, and it can be concluded that the efficiency measures were effective in reducing the average price.

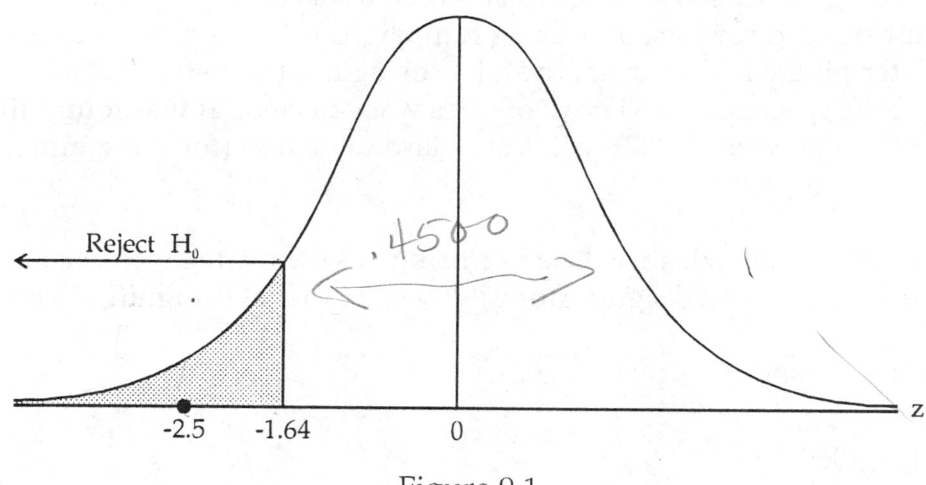

Figure 9.1

(b) What is the p-value associated with the above sample results, and what conclusions can be drawn based on the p-value?

Answer: The p-value is the smallest value of α for which we cannot reject the null hypothesis, based on the sample results.

In this problem the p-value is the area in the lower tail of the sampling distribution. Thus, we can compute the p-value as

$$Z = \frac{\overline{X} - \mu_0}{\sigma_{\overline{X}}} = \frac{1.20 - 1.25}{0.02} = -2.5$$

The area between the mean and Z = -2.5 is 0.4938. (See the table for the standard normal distribution.) Thus the area in the lower tail of the sampling distribution is 0.5 - 0.4938 = 0.0062. Therefore, the p-value is 0.0062.

The null hypothesis is rejected if the p-value < α. In this problem, the p-value is 0.0062 which is smaller than α of 0.05, thus the null hypothesis is rejected.

***2.** The average daily sales of a grocery store has been $8,000 per day. The store has introduced several advertising campaigns in order to increase sales. In order to determine whether or not the advertising has been effective in increasing sales, a sample of 64 days of sales was selected. It was found that the average daily sales were $8,250 with a standard deviation (for the sample) of $1,200.

(a) Test to determine whether the advertising has been effective. That is, has the average sales increased significantly? Use a 0.01 level of significance.

Answer: The H_o and H_a are stated as

H_o: $\mu \leq 8000$

H_a: $\mu > 8000$ (Manager's belief is true)

In this case, we have an upper one-tailed hypothesis testing situation. We can read the Z_α from the table of areas under the normal curve as $Z_{.01} = 2.33$. Since the standard deviation of the population is not known, but the sample size is large, we can determine $S_{\bar{x}}$ as

$$S_{\bar{x}} = \frac{S}{\sqrt{n}} = \frac{1200}{\sqrt{64}} = 150$$

In this case, the null hypothesis will be rejected if $Z > Z_\alpha$.

We have the value of $Z_\alpha = Z_{.01} = 2.33$. Then, we calculate

$$Z = \frac{\overline{X} - \mu_o}{S_{\overline{X}}} = \frac{8250 - 8000}{150} = 1.67$$

Since 1.67 < 2.33, the null hypothesis is not rejected. Therefore, there is not sufficient evidence to conclude that sales have increased. The region where the null hypothesis is rejected is shown in Figure 9.2.

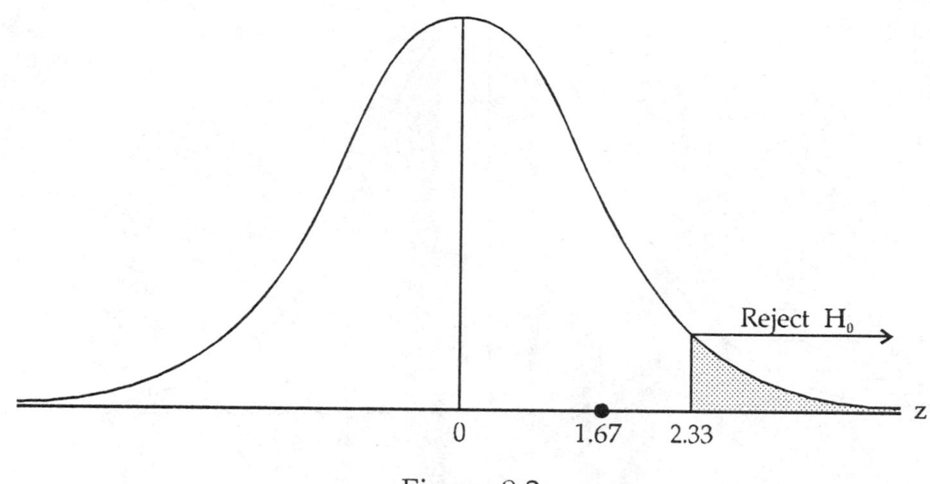

Figure 9.2

***3.** A sample of 64 account balances of a credit company showed an average balance of $1,040 with a standard deviation of $200. Test to determine if the mean of all account balances (i.e., population mean) is significantly different from $1,000. Use a 0.05 level of significance.

Answer: Since in this situation we are interested in determining whether the average balance is significantly different (larger or smaller) from $1,000, a two-tailed test is needed. The null and the alternative hypotheses are formulated as

H_o: $\mu = 1000$

H_a: $\mu \neq 1000$

The level of significance, which is the probability of a Type I error (i.e., the probability of rejecting the null hypothesis when, in fact, it is true), is given as 0.05. In the situation of a two-tailed test, half of the level of significance or, in this case, half of 0.05 or 0.025 will be in each tail of the sampling distribution as shown in Figure 9.3. From Table 1 of Appendix B of your textbook, $Z_{\alpha/2}$ corresponding to an area of 0.025 in the upper tail is read as $Z_{.025} = 1.96$. Since we are dealing with a two-tailed test, there will be a corresponding $Z_{.025} = -1.96$ in the lower tail. (See Figure 9.3)

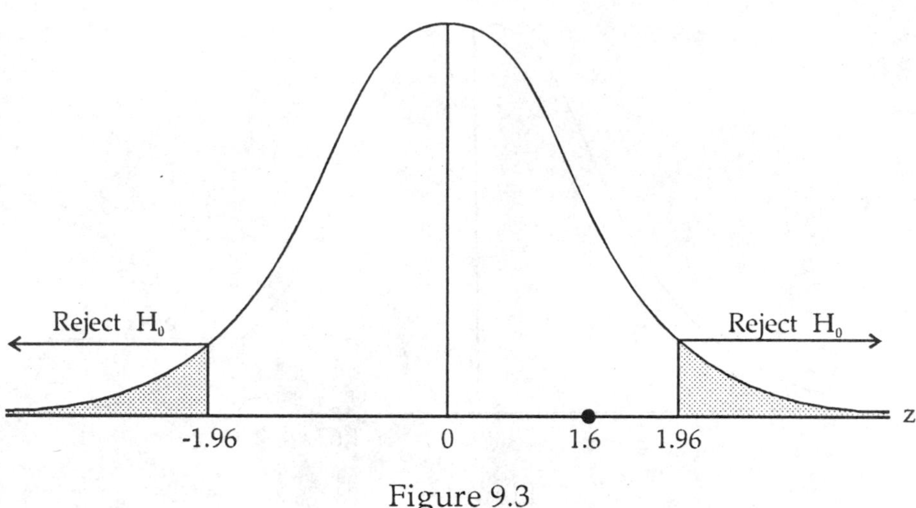

Figure 9.3

Now the decision rule can be stated:

Reject H_0 if $Z < -1.96$ or $Z > 1.96$,

where the Z statistic is computed as

$$Z = \frac{\overline{X} - \mu_0}{S / \sqrt{n}} = \frac{1040 - 1000}{200 / \sqrt{64}} = 1.6$$

Since 1.6 is between -1.96 and 1.96, the null hypothesis cannot be rejected; and we conclude that there is no evidence that the mean is significantly different from $1,000.

***4.** From a population of cereal boxes marked "12 ounces," a sample of 16 boxes is selected and the contents of each box is weighed. The sample revealed a mean of 11.7 ounces with a standard deviation of 0.8. Test to see if the mean of the population is at least 12 ounces. Use a 0.05 level of significance.

Answer: The null and the alternative hypotheses are formulated as

H_0: $\mu \geq 12$

H_a: $\mu < 12$

Since the standard deviation of the population is not known and the sample size is less than 30, the t distribution is used for the hypothesis test. The rejection region for this test is located in the lower tail of the sampling distribution. The value of t with n - 1 = 16 - 1 = 15 degrees of freedom can be read from Table 2 of Appendix B as $t_{.05} = -1.753$. (Note that the actual value which is read from the table is 1.753; but since we are dealing with the lower tail of the distribution, t is actually -1.753.)

Now the decision rule is

Reject H_0 if $t < -1.753$

where the t statistic is computed as

$$t = \frac{\overline{X} - \mu_o}{S / \sqrt{n}} = \frac{11.7 - 12}{0.8 / \sqrt{16}} = -1.5$$

Since -1.5 is greater than -1.753, the null hypothesis is not rejected. (See Figure 9.4.) Therefore, at a 0.05 level of significance, there is not sufficient evidence to conclude that the mean content is not at least 12 ounces.

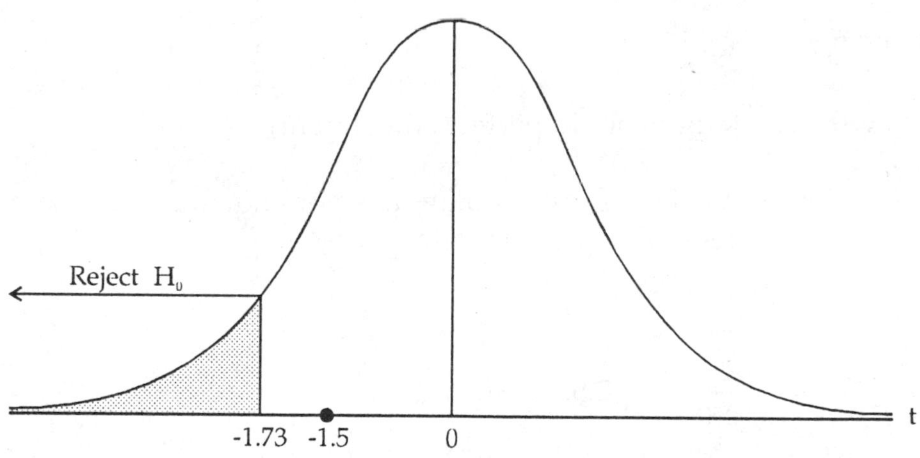

Figure 9.4

***5.** A lathe is set to cut steel bars into lengths of 6 centimeters. The lathe is considered to be in perfect adjustment if the average length of the bars it cuts is 6 centimeters. A sample of 25 bars is selected randomly, and the lengths are measured. It is determined that the average length of the bars in the sample is 6.1 centimeters with a standard deviation of 0.2 centimeters.

(a) Compute the standard error of the mean.

Answer: Since the standard deviation of the population is not known, the standard deviation of the sample is substituted in it's place and the standard error is computed:

$$S_{\bar{x}} = \frac{S}{\sqrt{n}} = \frac{0.2}{\sqrt{25}} = 0.04$$

(b) Test to determine whether or not the lathe is in perfect adjustment. Use a 0.05 level of significance.

Answer: In this situation, if the bars are cut too large or too small, the machine is considered not to be in perfect adjustment. Therefore, we have a two-tailed hypothesis testing situation in which the hypotheses are

H_o: $\mu = 6$

H_a: $\mu \neq 6$ (The lathe is not in perfect adjustment)

Since the standard deviation of the population is not known and the sample size is small ($n < 30$), we determine the t statistic:

$$t = \frac{\bar{X} - \mu_o}{S_{\bar{X}}} = \frac{6.1 - 6}{0.04} = 2.5$$

The null hypothesis will be rejected if the test statistic $t > t_{\alpha/2}$ or if $t < -t_{\alpha/2}$. The value of $t_{\alpha/2}$ with 24 degrees of freedom ($n - 1$) can be read from Table 2 of Appendix B in your text as $t_{.025} = 2.064$. Since 2.5 is larger than 2.064, the null hypothesis is rejected; and we conclude that the machine is not perfectly adjusted. Figure 9.5 shows the value of $t = 2.5$, where the null hypothesis is rejected.

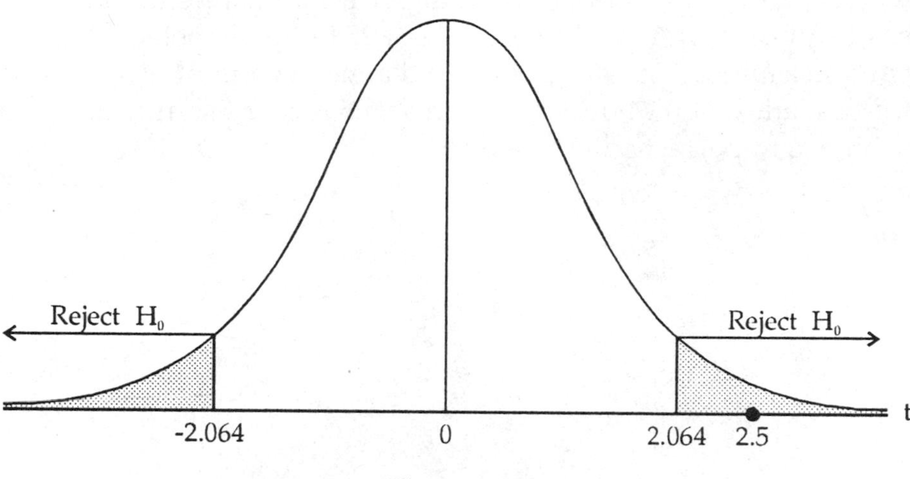

Figure 9.5

6. Automobiles manufactured by the Efficiency Company have been averaging 42 miles per gallon of gasoline in highway driving. It is believed that their new automobiles average **more** than 42 miles per gallon. An independent testing service road-tested 25 of their automobiles. The sample showed an average of 43.2 miles per gallon with a standard deviation of 1.25 miles per gallon.

(a) With a 0.05 level of significance, test to determine whether or not the new automobiles actually do average more than 42 miles per gallon.

(b) What is the p-value associated with the sample results? What is your conclusion based on the p-value?

7. A soft drink filling machine, when in perfect adjustment, fills bottles with 12 ounces of soft drink. A random sample of 25 bottles is selected, and the contents are measured. The sample yielded a mean content of 11.88 ounces with a standard deviation of 0.24 ounces. With a 0.05 level of significance, test to see if the machine is in perfect adjustment.

8. The McCollough Corporation, a producer of various kinds of batteries, has been producing "D" size batteries with a life expectancy of 87 hours. Due to an improved production process, the management believes that there has been an increase in the life expectancy of their "D" size batteries. A sample of 36 batteries showed an average life of 88.5 hours. Assume from past information that it is known that the standard deviation of the population is 9 hours.

(a) Use a 0.01 level of significance and test the management's belief.

(b) What is the p-value associated with the sample results? What is your conclusion based on the p-value?

9. In the past, the average age of evening students at a local college has been 21. A sample of 49 evening students was selected in order to determine whether the average age of the evening students has increased. The average age of the students in the sample was 23 with a standard deviation of 3.5. Determine whether or not there has been an increase in the average age of the evening students. Use a 0.1 level of significance.

10. A tire manufacturer has been producing tires with an average life expectancy of 26,000 miles. Now the company is advertising that their **new** tires' life expectancy has increased. In order to test the legitimacy of their advertising campaign, an independent testing agency tested a sample of 6 of their tires and has provided the following data:

Life Expectancy (In Thousands of Miles)
28
27
25
28
29
25

Use a 0.01 level of significance and test to determine whether or not the tire company is using legitimate advertising.

11. In order to determine the average price of hotel rooms in Small Town, U.S.A., a sample of 64 hotels was selected. It was determined that the average price of the rooms in the sample was $43 with a standard deviation of $10. Use a 0.05 level of significance and determine whether or not the average room price is significantly different from $40.

12. Refer to exercise 11. Assume that the average price of the hotel rooms in the sample was $38 with a standard deviation of $10. Use a 0.05 level of significance to determine whether or not the average price is significantly less than $40.

***13.** A recent survey of 80 accounting firms in Atlanta showed that only 12 of them were using any form of advertising. There are indications that nationally 16% of all accounting firms are advertising. Use a 0.05 level of significance in order to determine whether or not the percentage of the firms in Atlanta who are advertising is significantly less than the national percentage.

Answer: The null and the alternative hypotheses are

H_o: $p \geq 0.16$

H_a: $p < 0.16$ (The proportion is significantly less than 16%)

The standard error of the proportion can be determined as

$$\sigma_{\bar{p}} = \sqrt{\frac{p\,(1 - p)}{n}} = \sqrt{\frac{0.16\,(1 - 0.16)}{80}} = 0.041$$

The null hypothesis will be rejected if $Z < -Z_\alpha$, where

$$Z = \frac{\bar{p} - p}{\sigma_{\bar{p}}}$$

In the sample of 80 firms, only 12 were advertising. Therefore, the sample proportion is

$$\bar{p} = \frac{12}{80} = 0.15$$

Now we can compute the test statistic Z as

$$Z = \frac{\bar{p} - p}{\sigma_{\bar{p}}} = \frac{0.15 - 0.16}{0.041} = -0.24$$

and compare its value with $-Z_\alpha$. From Table 1 of Appendix B of your textbook, we read $Z_\alpha = Z_{.05} = -1.64$. Since $-0.24 > -1.64$, the null hypothesis is not rejected. Therefore, we conclude that there is not sufficient evidence to verify that the percentage of the firms in Atlanta who are advertising is significantly less than the national percentage.

Figure 9.6 shows the region where the null hypothesis would be rejected.

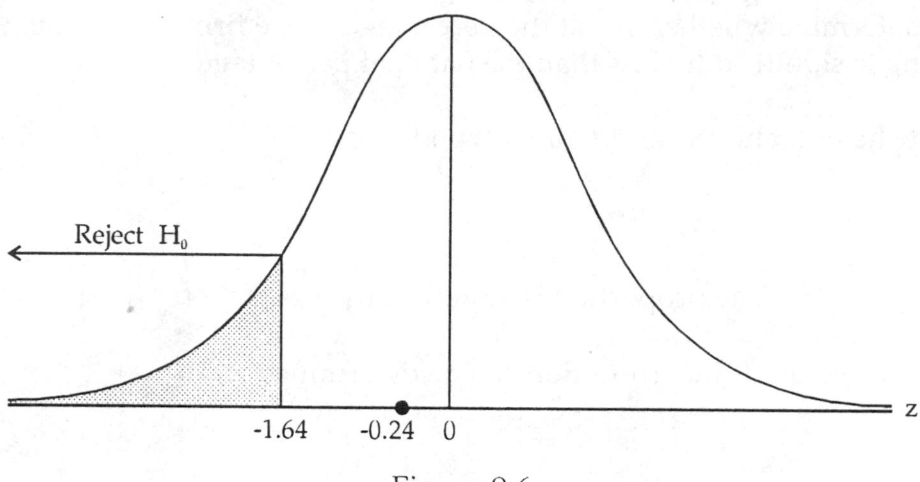

Figure 9.6

14. An insurance company which currently only carries automobile insurance is planning to introduce homeowners insurance to their customers. The management has indicated that they will introduce the homeowners insurance if more than 85% of their current customers indicate that they will purchase the new insurance. A random sample of 400 customers was selected and 348 customers indicated that they will purchase the homeowners insurance. Using a 0.02 level of significance, do you recommend that the company introduce the homeowners insurance?

15. One thousand numbers are selected randomly; 440 were odd numbers. At 0.05 level of significance, determine whether the proportion of odd numbers is significantly different from 50%.

16. In the last presidential election, a national survey company claimed that more than 56% of all registered voters voted for the Republican candidate. In a random sample of 200 registered voters, 116 voted for the Republican candidate. Use a 0.05 level of significance and test the survey company's claim.

17. An automobile manufacturer stated that he would be willing to mass produce electrically powered cars if more than 30% of potential buyers indicate they would purchase the newly designed electric cars. In a sample of 500 potential buyers, 160 indicated that they would buy such a product. Should the manufacturer produce the new electrically powered cars? Use a 0.05 level of significance.

18. On graduation day in a large university among a random sample of 60 graduates, 12 indicated that they had changed their major at least one time during their course of study. The records office believes that less than 25% of all their graduates change their major. Do the sample results support the records office's belief? Use a 0.01 level of significance.

19. It is said that more than 50% of registered voters actually vote in a national election. A research organization selected a random sample of 300 registered voters and reported that 165 of those actually voted. Based on the sample results, can you conclude that more than 50% of the registered voters actually vote? Use a 0.05 level of significance.

***20.** Refer to exercise 1 of this chapter.

(a) What is the probability of committing a Type II error if the actual price per gallon is $1.19?

Answer: A Type II error refers to the error of not rejecting H_0 when it is actually false. The probability of making a Type II error β can be computed by the following step-by-step procedure.

STEP 1. Formulate the null and the alternative hypotheses. In exercise 1, the hypotheses were formulated as

$$H_0: \mu \geq 1.25$$

$$H_a: \mu < 1.25$$

STEP 2. Read the Z value (from the table of areas under the normal curve) associated with the specified level of significance α. Since our test was conducted at a 0.05 level of significance, we read the Z_α to be 1.64. This Z_α establishes the rejection rule. That is, the null hypothesis will be rejected if the *test statistic* Z is less than -1.64.

Note: If we wanted to conduct a two tailed test, we would read $Z_{\alpha/2}$.

STEP 3. Using the Z value from Step 2, solve for the sample mean which identifies the rejection region as

$$Z = \frac{\overline{X} - 1.25}{0.14 / \sqrt{49}} < -1.64$$

Solving for $\overline{x}$ in the above expression yields

$$\overline{X} < 1.25 - (1.64)\left(\frac{0.14}{\sqrt{49}}\right) = 1.2172$$

which indicates the null hypothesis will be rejected if $\overline{X} < 1.2172$.

STEP 4. The results of Step 3 indicate that the null hypothesis will not be rejected for sample mean values of $\overline{X} \geq 1.2172$.

STEP 5. Now we can determine the probability of a Type II error (β) by finding the shaded area of Figure 9.7. This shaded area shows the probability of committing a Type II error when the mean is actually \$1.19.

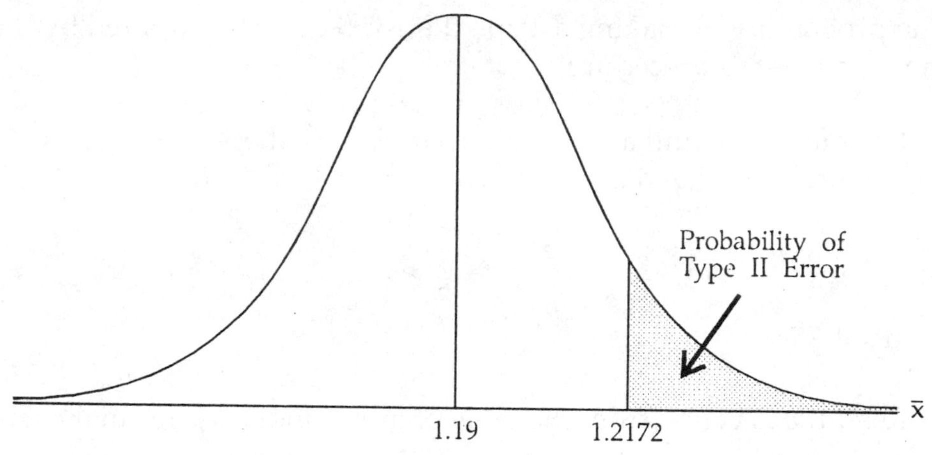

Figure 9.7

To find this area, we first find the area between 1.19 and 1.2172 and then subtract the result from 0.5. Therefore,

$$Z = \frac{1.2172 - 1.19}{.02} = 1.36$$

Thus, the area between 1.19 and 1.2172 (from Table 1 of Appendix B) is 0.4131. Subtracting this value from 0.5 yields 0.5 - 0.4131 = 0.0869, which is β, or the probability of committing a Type II error.

(b) Compute the *power* of the test and explain its meaning.

Answer: The power of the test refers to the probability of correctly rejecting the null hypothesis. Therefore, the power of the test is $1 - \beta = 1 - 0.0869 = 0.9131$.

21. Refer to exercise 2 of this chapter.

(a) What is the probability of committing a Type II error if the actual daily sales are $8,400?

(b) Compute the power of the test and explain its meaning.

***22.** Refer to exercise 1 of this chapter. What size sample should be taken in order to limit the probability of a Type I error (α) to 0.05 when the null hypothesis is true, $\beta = 0.3$, and the mean is actually 1.19?

Answer: The required sample size can be determined:

$$n = \frac{(Z_\alpha + Z_\beta)^2 \sigma^2}{(\mu_o - \mu_a)^2}$$

where: Z_α = Z value associated with an area of α in one tail of the distribution. In our problem, $Z_\alpha = Z_{.05} = 1.64$.

Z_β = Z value associated with an area of β in one tail of the distribution. In our problem, $Z_\beta = Z_{0.3} = 0.52$.

σ = the standard deviation of the population which is given to be 0.14.

μ_o = the value of the population mean in the null hypothesis. In exercise 1, the mean was hypothesized to be \$1.25.

μ_a = the actual value of the population mean in the statement about a Type II error. This value in this problem is stated to be \$1.19.

With the above information, the sample size can be computed:

$$n = \frac{(Z_\alpha + Z_\beta)^2 \sigma^2}{(\mu_o - \mu_a)^2} = \frac{(1.64 + 0.52)^2 (0.14)^2}{(1.25 - 1.19)^2} = 25.4$$

Rounding up, the required sample size is 26.

SELF-TESTING QUESTIONS

In the following multiple choice questions, circle the correct answer. An answer key is provided following the questions.

1. In hypothesis testing, the hypothesis which is tentatively assumed to be true is called the

a) correct hypothesis
b) null hypothesis
c) alternative hypothesis
d) level of significance

2. When the null hypothesis has been true, but the sample information has resulted in the rejection of the null, a ____________________ has been made.

a) level of significance
b) Type II error
c) critical value
d) Type I error

3. The maximum probability of a Type I error that the decision maker will tolerate is called the

a) level of significance
b) critical value
c) decision value
d) probability value

4. Which notation is used to represent the null hypothesis?

a) H_1
b) H_a
c) H_o
d) none of the above

5. A Type II error is the error of

a) accepting H_o when it is false
b) accepting H_o when it is true
c) rejecting H_o when it is false
d) rejecting H_o when it is true

6. For setting the decision rule in a small-sample case, if it is reasonable to assume that the population is normal, we use

a) the Z distribution
b) the t distribution with n - 1 degrees of freedom
c) the t distribution with n + 1 degrees of freedom
d) none of the above

7. A hypothesis test in which rejection of the null hypothesis occurs for values of the point estimator in either tail of the sampling distribution is called

a) the null hypothesis
b) the alternative hypothesis
c) a one-tailed test
d) a two-tailed test

Use the following information for questions 8 through 10.

The ABC Company claims that the batteries it produces have useful lives of more than 100 hours, with a known standard deviation (of the population) of 20 hours. A test is undertaken to test the validity of this claim.

8. The correct set of hypotheses for this test is

a) H_o: $\mu = 100$
H_a: $\mu \neq 100$

b) H_o: $\mu \geq 100$
H_a: $\mu < 100$

c) H_o: $\mu \leq 100$
H_a: $\mu > 100$

d) none of the above

9. A sample of 64 batteries had an average useful life of 110 hours. The test statistic has a value of

a) -4
b) 32
c) 4
d) 10
e) none of the above

10. With a 0.05 level of significance, the proper decision is

a) do not reject H_o and conclude claim is correct
b) do not reject H_o and conclude claim is false
c) reject H_o and conclude claim is correct
d) reject H_o and conclude claim is false

11. The average monthly income of recent business major graduates was reported to be $2,800. It is hypothesized that a recession has **reduced** the average income. The correct set of hypotheses is

a) H_o: $\mu < 2800$
H_a: $\mu \geq 2800$

b) H_o: $\mu \geq 2800$
H_a: $\mu < 2800$

c) H_o: $\mu > 2800$
H_a: $\mu \leq 2800$

d) H_o: $\mu \leq 2800$
H_a: $\mu \geq 2800$

e) none of the above

12. The average life expectancy of Strong tires has been 30,000 miles. Because of improved processing, it is believed that the average life has **increased**. The correct set of hypotheses for testing this belief is

a) H_0: $\mu \leq 30{,}000$
H_a: $\mu > 30{,}000$

b) H_0: $\mu \geq 30{,}000$
H_a: $\mu < 30{,}000$

c) H_0: $\mu < 30{,}000$
H_a: $\mu \geq 30{,}000$

d) H_0: $\mu > 30{,}000$
H_a: $\mu \leq 30{,}000$

e) none of the above

13. The level of significance is the

a) maximum allowable probability of Type II error
b) maximum allowable probability of Type I error
c) same as the confidence coefficient
d) same as the P-value
e) none of the above

14. Type II error is committed when

a) a true alternative hypothesis is mistakenly rejected
b) a true null hypothesis is mistakenly rejected
c) the sample size has been too small
d) not enough information has been available
e) none of the above

15. The error of rejecting a true null hypothesis is

a) a Type I error
b) a Type II error
c) can be either a or b, depending on the situation
d) committed when not enough information is available
e) none of the above

16. In hypothesis testing, α is

a) the probability of committing a Type II error
b) the probability of committing a Type I error
c) the probability of either a Type I or Type II, depending on the hypothesis to be tested
d) none of the above

17. In hypothesis testing, β is

a) the probability of committing a Type II error
b) the probability of committing a Type I error
c) the probability of either a Type I or Type II, depending on the hypothesis to be tested
d) none of the above

18. When testing the following hypotheses at α level of significance

H_0: $p \leq 0.7$
H_a: $p > 0.7$

The null hypothesis will be rejected if the test statistic Z is

a) $Z > Z_\alpha$
b) $Z < Z_\alpha$
c) $Z < -Z_\alpha$
d) none of the above

19. Which of the following does **not** need to be known in order to compute the P-value?

a) knowledge of whether the test is one-tailed or two-tailed
b) the value of the test statistic
c) the level of significance
d) All of the above are needed.
e) none of the above

20. If the level of significance of a hypothesis test is raised from .01 to .1, the probability of a Type II error

a) will increase from .01 to .05
b) will not change
c) will decrease
d) Not enough information is given to answer this question.
e) none of the above

ANSWERS TO THE SELF-TESTING QUESTIONS

1. b
2. d
3. a
4. c
5. a
6. b
7. d
8. c
9. c
10. c
11. b
12. a
13. b
14. a
15. a
16. b
17. a
18. a
19. c
20. c

ANSWERS TO CHAPTER NINE EXERCISES

6. (a) Ho: $\mu \leq 42$

   H_a : $\mu > 42$

   Since t = 4.8 > 1.711, reject H_o and conclude that the new cars average more than 42 miles per gallon.

   (b) p-value = 0 < 0.05, therefore reject H_o (area to the right of Z = 4.8 is almost zero)

7. H_o: $\mu = 12$

   H_a: $\mu \neq 12$

   Since t = -2.5 < -2.064, reject H_o and conclude that the machine is not perfectly adjusted.

8. (a) H_o: $\mu \leq 87$

   H_a: $\mu > 87$

   Since Z = 1 < 2.33, do not reject H_o and conclude that there is insufficient evidence to support the corporation's claim.

   (b) p-value = 0.1587 > .01, therefore do not reject H_o

9. H_o: $\mu \leq 21$

   H_a: $\mu > 21$

   Since Z = 4 > 1.28, reject H_o and conclude that there has been an increase in the evening students' average ages.

10. H_o: $\mu \leq 26000$

    H_a: $\mu > 26000$

    Since t = 1.47 < 3.365, do not reject H_o and conclude that there is insufficient evidence to support the manufacturer's claim.

11. H_o: $\mu = 40$

    H_a: $\mu \neq 40$

    Since Z = 2.4 > 1.96, reject H_o and conclude that the average room price is significantly different from $40.

12. H_o: $\mu \geq 40$
H_a: $\mu < 40$
Since $Z = -1.6 > -1.64$, do not reject H_0 and conclude that there is insufficient evidence to show that the average room price is significantly less than \$40.

14. H_o: $p \leq 0.85$
H_a: $p > 0.85$
Since $Z = 1.12 < 2.05$, do not reject H_0 and, therefore, you should not recommend introducing homeowners insurance.

15. H_o: $p = 0.5$
H_a: $p \neq 0.5$
Since $Z = -3.79 < -1.96$, reject H_0 and conclude that the proportion of odd numbers is significantly different from 50%.

16. H_o: $p \leq 0.56$
H_a: $p > 0.56$
Since $Z = 0.57 < 1.64$, do not reject H_0 and conclude that there is insufficient evidence to support the survey company's claim.

17. H_o: $p \leq 0.3$
H_a: $p > 0.3$
Since $Z = 0.98 < 1.64$, do not reject H_0 and conclude that there is insufficient evidence to support the production of the electric cars.

18. H_o: $p \geq 0.25$
H_a: $p < 0.25$
Since $Z = -0.89 > -2.33$, do not reject H_0 and, therefore, there is insufficient evidence to support the records office's belief.

19. H_o: $p \leq 0.5$
H_a: $p > 0.5$
Since $Z = 1.73 > 1.64$, reject H_0 and conclude that more than 50% of the registered voters actually vote.

21. (a) 0.3669
(b) 0.6331 The probability of correctly rejecting the null hypothesis.

CHAPTER TEN

STATISTICAL INFERENCE ABOUT MEANS AND PROPORTIONS WITH TWO POPULATIONS

CHAPTER OUTLINE AND REVIEW

In this chapter, you have learned how to determine an interval estimate and also how to test a hypothesis where two populations are involved. The main topics of study have been inferences about the differences between means and also the differences between proportions of two populations. Some of the new terminology which you have learned includes

A.	**Pooled Variance Estimate:**	When we are faced with two populations with assumed equal variances, we estimate the variance based on the combination of the two sample variances, known as the pooled variance estimate.
B.	**Independent Samples:**	When the elements of a sample from one population are selected independently of the elements of a second sample from a second population, the samples are known as independent samples.
C.	**Matched Samples:**	When each data value in one sample is matched with a corresponding data value in the second sample, the samples are known as matched samples.

CHAPTER FORMULAS

MEANS

Point Estimate for the Difference Between the Means of Two Populations:

$$\bar{x}_1 - \bar{x}_2 \tag{10.1}$$

Sampling Distribution of $(\bar{x}_1 - \bar{x}_2)$

Expected Value: $E(\bar{x}_1 - \bar{x}_2) = \mu_1 - \mu_2$ (10.2)

Standard Deviation: $\sigma_{\bar{x}_1 - \bar{x}_2} = \sqrt{\dfrac{\sigma_1^2}{n_1} + \dfrac{\sigma_2^2}{n_2}}$ (10.3)

Interval Estimate of the Difference Between the Means of Two Populations

I. Large-Sample Case ($n_1 \geq 30$ and $n_2 \geq 30$) with σ_1 and σ_2 Known

$$(\bar{x}_1 - \bar{x}_2) \pm Z_{\alpha/2} \cdot \sigma_{\bar{x}_1 - \bar{x}_2} \tag{10.4}$$

where $\sigma_{\bar{x}_1 - \bar{x}_2} = \sqrt{\dfrac{\sigma_1^2}{n_1} + \dfrac{\sigma_2^2}{n_2}}$ (10.3)

II. Large-Sample Case ($n_1 \geq 30$ and $n_2 \geq 30$) with σ_1 and σ_2 Unknown

$$(\bar{x}_1 - \bar{x}_2) \pm Z_{\alpha/2} \cdot S_{\bar{x}_1 - \bar{x}_2} \tag{10.6}$$

where $S_{\bar{x}_1 - \bar{x}_2} = \sqrt{\dfrac{S_1^2}{n_1} + \dfrac{S_2^2}{n_2}}$ (10.5)

CHAPTER FORMULAS
(Continued)

III. Small sample case ($n_1 < 30$ or $n_2 < 30$). Assuming both populations are normally distributed and have equal variances.

$$(\bar{x}_1 - \bar{x}_2) \pm t_{\alpha/2} \cdot S_{\bar{x}_1 - \bar{x}_2} \qquad (10.10)$$

where the t value is based on a t distribution with $n_1 + n_2 - 2$ degrees of freedom

$$\text{where } S_{\bar{x}_1 - \bar{x}_2} = \sqrt{S^2\left(\frac{1}{n_1} + \frac{1}{n_2}\right)} \qquad (10.9)$$

$$\text{and the pooled estimate of variance } S^2 = \frac{(n_1 - 1)S_1^2 + (n_2 - 1)S_2^2}{n_1 + n_2 - 2} \qquad (10.8)$$

Hypothesis Testing (Means), Independent Samples

Test Statistic for a Hypothesis Test Regarding the Difference Between the Means of Two Populations:

I. Large-sample case with $n_1 \geq 30$ and $n_2 \geq 30$:

$$Z = \frac{(\bar{x}_1 - \bar{x}_2) - (\mu_1 - \mu_2)}{\sigma_{\bar{x}_1 - \bar{x}_2}} = \frac{(\bar{x}_1 - \bar{x}_2) - (\mu_1 - \mu_2)}{\sqrt{\frac{\sigma_1^2}{n_1} + \frac{\sigma_2^2}{n_2}}} \qquad (10.11)$$

When the population variances (σ_1 and σ_2) are not known, substitute the sample variances (S_1 and S_2) in the above equation.

CHAPTER FORMULAS
(Continued)

II. Small-sample case with $n_1 < 30$ or $n_2 < 30$ and assuming both populations are normally distributed and have equal variances:

$$t = \frac{(\bar{x}_1 - \bar{x}_2) - (\mu_1 - \mu_2)}{S_{\bar{x}_1 - \bar{x}_2}} = \frac{(\bar{x}_1 - \bar{x}_2) - (\mu_1 - \mu_2)}{\sqrt{S^2\left(\frac{1}{n_1} + \frac{1}{n_2}\right)}} \qquad (10.12)$$

where Equation 10.8 is used to compute pooled estimate of S^2.

Matched Samples

Test statistic:

$$t = \frac{\bar{d} - \mu_d}{s_d/\sqrt{n}} \qquad (10.13)$$

where $S_d = \sqrt{\frac{\Sigma(d_i - \bar{d})^2}{n - 1}}$

Proportions

Point Estimate for the Difference Between the Proportions of Two Populations:

$$\bar{p}_1 - \bar{p}_2 \qquad (10.14)$$

Sampling distribution of $(\bar{p}_1 - \bar{p}_2)$

Expected Value: $E(\bar{p}_1 - \bar{p}_2) = p_1 - p_2$ (10.15)

Standard deviation: $\sigma_{\bar{p}_1 - \bar{p}_2} = \sqrt{\frac{p_1(1-p_1)}{n_1} + \frac{p_2(1-p_2)}{n_2}}$ (10.16)

CHAPTER FORMULAS
(Continued)

Interval Estimate for the Difference Between the Proportions of Two Populations

$$(\bar{p}_1 - \bar{p}_2) \pm Z_{\alpha/2} \cdot \sigma_{\bar{p}_1 - \bar{p}_2} \qquad (10.17)$$

where the estimate of $\sigma_{\bar{p}_1 - \bar{p}_2}$ is:

$$S_{\bar{p}_1 - \bar{p}_2} = \sqrt{\frac{\bar{p}_1(1-\bar{p}_1)}{n_1} + \frac{\bar{p}_2(1-\bar{p}_2)}{n_2}} \qquad (10.18)$$

Hypothesis Testing (Proportions)

Test Statistic for a Hypothesis Test Regarding the Difference Between Proportions of Two Populations:

$$Z = \frac{(\bar{p}_1 - \bar{p}_2) - (p_1 - p_2)}{S_{\bar{p}_1 - \bar{p}_2}} \qquad (10.19)$$

where $S_{\bar{p}_1 - \bar{p}_2} = \sqrt{\bar{p}(1-\bar{p})\left(\frac{1}{n_1} + \frac{1}{n_2}\right)}$ (10.21)

and assuming $p_1 = p_2$, the pooled proportion $\bar{p}$ is computed as

where $\bar{p} = \frac{n_1 \bar{p}_1 + n_2 \bar{p}_2}{n_1 + n_2}$ (10.20)

EXERCISES

***1.** The management of a department store is interested in estimating the difference between the mean credit purchases made by their customers using the store's credit card versus those customers using a national major credit card. Independent samples of credit sales are shown below.

Store's Card	Major Credit Card
$n_1 = 64$	$n_2 = 49$
$\bar{x}_1 = \$140$	$\bar{x}_2 = \$125$
$s_1 = \$10$	$s_2 = \$8$

(a) Develop a point estimate for the difference between the mean purchases of the users of the two credit cards.

Answer: The point estimate for the difference between the means of the two populations is $\bar{x}_1 - \bar{x}_2 = 140 - 125 = \$15$.

(b) Develop an interval estimate for the difference between the average purchases of the customers using the two different credit cards. Use a confidence coefficient of 0.95.

Answer: The interval estimate for the difference between the means of the two populations is (large-sample case with $n_1 \geq 30$ and $n_2 \geq 30$)

$$(\bar{x}_1 - \bar{x}_2) \pm Z_{\alpha/2} \cdot \sigma_{\bar{x}_1 - \bar{x}_2}$$

Since the standard deviations of the populations are not known, we estimate

$\sigma_{\bar{x}_1 - \bar{x}_2}$ as

$$s_{\bar{x}_1 - \bar{x}_2} = \sqrt{\frac{s_1^2}{n_1} + \frac{s_2^2}{n_2}} = \sqrt{\frac{(10)^2}{64} + \frac{(8)^2}{49}} = 1.69$$

Furthermore, $z_{\alpha/2} = z_{.025} = 1.96$. Thus, the interval estimate is

$$(140 - 125) \pm (1.96)(1.69) = 15 \pm 3.31$$

Therefore, at 95% confidence, the interval estimate for the difference in the mean purchases is from \$11.69 to \$18.31.

2. A potential investor conducted a 45 day survey in each theater in order to determine the difference between the average daily attendance at the North Mall and South Mall theaters. The North Mall Theater averaged 630 patrons per day with a standard deviation of 60; while the South Mall Theater averaged 598 patrons per day with a standard deviation of 71. Develop an interval estimate for the difference between the average daily attendance at the two theaters. Use a confidence coefficient of 0.95.

3. The business manager of a local health clinic is interested in estimating the difference between the fees for extended office visits in their center and the fees of a newly opened group practice. She gathered the following information regarding the two offices.

Health Clinic	Group Practice
n_1 = 50 visits	n_2 = 45 visits
$\bar{x}_1$ = \$21	$\bar{x}_2$ = \$19
s_1 = \$2.75	s_2 = \$3.00

Develop an interval estimate for the difference between the average fees of the two offices. Use a confidence coefficient of 0.95.

***4.** In order to estimate the difference between the average daily sales of 2 branches of a department store, the following data has been gathered.

Downtown Store	North Mall Store
$n_1 = 12$ days	$n_2 = 14$ days
$\bar{x}_1 = \$36,000$	$\bar{x}_2 = \$32,000$
$s_1 = \$1,200$	$s_2 = \$1,000$

Develop an interval estimate for the difference between the two population means with a confidence coefficient of 0.95. *Assume the two populations are normally distributed and have equal variances.*

Answer: The interval estimate for the difference between the means of the two populations is as follows. Note this is a small-sample case with $n_1 < 30$ and $n_2 < 30$.

$$(\bar{x}_1 - \bar{x}_2) \pm t_{\alpha/2} \cdot S_{\bar{x}_1 - \bar{x}_2}$$

Since the variances of the populations are not known, the variances of the two samples are combined to compute an estimate of σ^2. We have assumed the two populations are normally distributed and have equal variances.

$$s^2 = \frac{(n_1 - 1)\, s_1^2 + (n_2 - 1)\, s_2^2}{n_1 + n_2 - 2}$$

$$= \frac{(12 - 1)\,(1200)^2 + (14 - 1)\,(1000)^2}{12 + 14 - 2}$$

$$= 1,201,666.7$$

Then we determine an estimate of $\sigma_{\bar{x}_1 - \bar{x}_2}$ as

$$s_{\bar{x}_1 - \bar{x}_2} = \sqrt{s^2 \left(\frac{1}{n_1} + \frac{1}{n_2} \right)}$$

$$= \sqrt{(1,201,666.7) \left(\frac{1}{12} + \frac{1}{14} \right)}$$

$$= 431.25$$

Now, we can look up the t value, at $n_1 + n_2 - 2 = 12 + 14 - 2 = 24$ degrees of freedom, which is $t_{\alpha/2} = t_{.025} = 2.064$.

Thus, the interval estimate becomes

$$(36{,}000 - 32{,}000) \pm (2.064)(431.25) = 4000 \pm 890.1$$

which means that at 95% confidence, the interval estimate for the difference in the average sales of the two stores is from \$3109.90 to \$4890.10.

5. In order to estimate the difference between the age of computer consulting firms in the East and the West of the United States, the following information was gathered.

East	West
$n_1 = 20$	$n_2 = 22$
$\bar{x}_1 = 72$ months	$\bar{x}_2 = 78$ months
$s_1 = 6$ months	$s_2 = 8$ months

Develop an interval estimate for the difference between the average age of the firms in the East and the West. Use a confidence coefficient of 0.98.

***6.** In order to determine whether or not there is a significant difference between the hourly wages of two companies, the following data have been accumulated.

Company A	Company B
$n_1 = 80$	$n_2 = 60$
$\bar{x}_1 = \$6.75$	$\bar{x}_2 = \$6.25$
$s_1 = \$1.00$	$s_2 = \$0.95$

Is there any significant difference between the hourly wages of the two companies? Let $\alpha = 0.05$.

Answer: The hypotheses of interest are

H_0: $\mu_1 - \mu_2 = 0$

H_a: $\mu_1 - \mu_2 \neq 0$

Since the sampling distribution is approximately normal, the following test statistic is used.

$$Z = \frac{(\bar{x}_1 - \bar{x}_2) - (\mu_1 - \mu_2)}{s_{\bar{x}_1 - \bar{x}_2}} = \frac{(\bar{x}_1 - \bar{x}_2) - (\mu_1 - \mu_2)}{\sqrt{\dfrac{s_1^2}{n_1} + \dfrac{s_2^2}{n_2}}} = \frac{(6.75 - 6.25) - 0}{\sqrt{\dfrac{(1)^2}{80} + \dfrac{(.95)^2}{60}}} = 3.01$$

Using the Z table, we find that with a level of significance of 0.05 the critical values of Z are ± 1.96. Thus the decision rule is

Do not reject H_0 if $-1.96 \leq Z \leq 1.96$
Reject H_0 otherwise

Since $3.01 > 1.96$, the null hypothesis is rejected; and, therefore, we conclude there is a significant difference between the average hourly wages of the two companies.

7. The Commonwealth of the Bahamas is a collection of many islands lying in the Atlantic Ocean. Tourism has become the most important industry of the islands. Two of the most visited islands are New Province and Grand Bahamas. A random sample of 36 tourists in Grand Bahamas showed that they spent an average of $1,860 (in a week) with a standard deviation of $126, and a sample of 64 tourists in New Province showed that they spent an average of $1,935 (in a week) with a standard deviation of $138. Is there any significant difference between the average expenditures of those who visited the two islands? Use a confidence coefficient of 0.95.

8. Two universities in your state decided to administer a comprehensive examination to the recipients of M.B.A. degrees. A random sample of M.B.A. recipients was selected from each institution and were given the test. The sample sizes, the average test scores, and the standard deviations of the scores for each institution are shown below.

Central University	Northern University
$n_1 = 32$	$n_2 = 32$
$\bar{x}_1 = 83$	$\bar{x}_2 = 77$
$s_1 = 13$	$s_2 = 14$

Test at a 0.05 level of significance to determine if there is a significant difference in the average test scores of the students from the two universities.

9. A random sample of 90 days in Tennessee and Georgia revealed the following information about the rainfall.

Tennessee	Georgia
$\bar{x}_1 = 5.6$ inches	$\bar{x}_2 = 4.9$ inches
$s_1 = 0.5$ inches	$s_2 = 0.4$ inches

Test at a 0.1 level of significance to determine if there is a significant difference in the average rainfall of the two states.

***10.** Two of the major automobile manufacturers have produced compact cars with the same size engines. We are interested in determining whether or not there is a significant difference in the MPG (miles per gallon) of the two models of automobiles. A random sample of 8 cars from each manufacturer is selected, and 8 drivers are selected to drive each automobile for a specified distance. The following data show the results of the test.

Driver	MPG Model A	MPG Model B	Difference in MPG d_i	$(d_i - \bar{d})^2$
1	29	27	2	2.25
2	24	23	1	0.25
3	26	28	-2	6.25
4	24	23	1	0.25
5	25	24	1	0.25
6	27	26	1	0.25
7	30	28	2	2.25
8	25	27	-2	6.25
			$\Sigma d_i = 4$	$\Sigma(d_i - \bar{d})^2 = 18.00$

Test at a 0.05 level of significance to determine if there is a difference in the average MPG of the two models.

Answer: In this situation, we have a matched sample design, where each driver provides a pair of data values, one value for the MPG of model A and another value for the MPG of model B. Thus, we have 8 data values, namely, the differences between the MPG's of the two models, to analyze. (These differences are shown above in the fourth column marked d_i.) Now, we set our hypotheses:

H_0: $\mu_d = 0$ There is no difference between the MPG's of the two models.

H_a: $\mu_d \neq 0$ There is a difference between the MPG's of the two models.

Then, we calculate the mean for the differences as

$$\bar{d} = \frac{\Sigma d_i}{n} = \frac{4}{8} = 0.5$$

and the standard deviation for the differences is determined as

$$s_d = \sqrt{\frac{\Sigma(d_i - \bar{d})^2}{n - 1}}$$

The value of $\Sigma(d_i - \bar{d})^2 = 18$ is shown above on the bottom of the last column. Thus, the value of the standard deviation is

$$s_d = \sqrt{\frac{18}{8 - 1}} = 1.6 \text{ (rounded)}$$

The test statistic is computed as

$$t = \frac{\bar{d} - \mu_d}{s_d/\sqrt{n}} = \frac{0.5 - 0}{1.6 / \sqrt{8}} = 0.884$$

With $\alpha = .05$ and with $n - 1 = 8 - 1 = 7$ degrees of freedom, we read $t_{.025} = 2.365$.

The decision rule for the two tailed test becomes

Do not reject H_0 if $-2.365 \leq t \leq 2.365$
Reject H_0 otherwise.

Since the value of test statistic $t = 0.884$, the null hypothesis is not rejected; and, therefore, we conclude there is not sufficient evidence to indicate that there is any difference between the MPG's of the two models.

11. A large corporation wants to determine whether or not the "typing efficiency" course given at a local college can increase the typing speed of their secretaries. A sample of 6 secretaries is selected, and they are sent to take the course. The typing speeds of the secretaries in words per minute (WPM) are shown below.

Secretary	WPM Before the Course	WPM After the Course
1	65	68
2	60	62
3	61	66
4	63	66
5	64	67
6	65	67

Use $\alpha = 0.05$ and test to see if it can be concluded that taking the course will actually increase the average typing speeds of the secretaries. (Hint: This is a one-tailed test.)

***12.** Of the 40 Dallas accounting firms surveyed, 6 indicated they were advertising; while of the 50 firms surveyed in New Orleans, only 8 were advertising. Determine a 90% confidence interval estimate for the difference between the **proportions** of the firms which are advertising in the two cities.

Answer: The interval estimate for the difference between the proportion of the two populations is

$$(\bar{p}_1 - \bar{p}_2) \pm Z_{\alpha/2} \cdot s_{\bar{p}_1 - \bar{p}_2}$$

where $s_{\bar{p}_1 - \bar{p}_2}$ is

$$s_{\bar{p}_1 - \bar{p}_2} = \sqrt{\frac{\bar{p}_1(1-\bar{p}_1)}{n_1} + \frac{\bar{p}_2(1-\bar{p}_2)}{n_2}}$$

But first we need to calculate $\bar{p}_1 = \frac{6}{40} = 0.15$ and $\bar{p}_2 = \frac{8}{50} = 0.16$. Then we can calculate

$$s_{\bar{p}_1 - \bar{p}_2} = \sqrt{\frac{0.15\ (0.85)}{40} + \frac{0.16\ (0.84)}{50}} = 0.077$$

Therefore, we determine the confidence interval as

$$(0.15 - 0.16) \pm (1.64)\ (0.077) = -0.01 \pm 0.126$$

Thus, the 90% confidence interval estimate for the difference in the proportion of the firms who are advertising in the two cities is -0.136 to 0.116.

13. From production line A, a sample of 200 items is selected at random; and it is determined that 16 items are defective. While in a sample of 300 items from production process B (which produces identical items to line A), there are 21 defective items. Determine a 95% confidence interval estimate for the difference between the proportions of defectives in the two lines.

14. Among a sample of 50 M.D.s (medical doctors) in Montana, there were 10 who indicated they make house calls. While among a sample of 100 M.D.s in Idaho, 18 said they make house calls. Determine a 95% interval estimate for the difference between the proportions of doctors who make house calls in the two states.

***15.** Refer to exercise 12. Use $\alpha = 0.1$ and test the hypothesis which claims that the proportion of the firms who advertise is the same for the two cities.

Answer: The hypotheses to be tested are

H_0: $p_1 - p_2 = 0$

H_a: $p_1 - p_2 \neq 0$

The pooled proportion is computed as

$$\bar{p} = \frac{n_1 \bar{p}_1 + n_2 \bar{p}_2}{n_1 + n_2} = \frac{(40)(0.15) + (50)(0.16)}{40 + 50} = 0.156$$

Then,

$$S_{\bar{p}_1 - \bar{p}_2} = \sqrt{\bar{p}(1-\bar{p})\left(\frac{1}{n_1} + \frac{1}{n_2}\right)}$$

$$= \sqrt{(0.156)(0.844)\left(\frac{1}{40} + \frac{1}{50}\right)} = 0.077$$

With the sampling distribution approximately normal, the test statistic is computed as

$$Z = \frac{(\bar{p}_1 - \bar{p}_2) - (p_1 - p_2)}{S_{\bar{p}_1 - \bar{p}_2}} = \frac{(0.15 - 0.16) - 0}{0.077} = -0.130$$

The decision rule, with a level of significance of 0.10 is stated as

Do not reject H_0 if $-1.64 \leq z \leq 1.64$
Reject H_0 otherwise.

Since the test statistic $Z = -0.130$, the null hypothesis is not rejected; and we conclude that there is not sufficient evidence to indicate that there is a significant difference between the proportions of the firms that advertise in the two cities.

16. Refer to exercise 13. Use $\alpha = .05$ and test the hypothesis that the proportions of defectives in both lines are the same.

17. Refer to exercise 14. Use $\alpha = 0.05$ and test the hypothesis that the proportions of house calls for the two states are the same.

SELF-TESTING QUESTIONS

In the following multiple choice questions, circle the correct answer. An answer key is provided following the questions.

1. An estimate of the variance of a population based on the combination of two sample results is known as the

a) pooled standard deviation
b) matched variance
c) pooled variance estimate
d) none of the above

2. The pooled variance is appropriate whenever the two populations

a) are normally distributed
b) have equal variances
c) meet both requirements of a and b
d) none of the above

3. To construct an interval estimate of the difference between the means of two populations which are normally distributed and have equal variances, we must use a t distribution with (Let n_1 be the size of sample 1 and n_2 the size of sample 2.)

a) n_1 degrees of freedom
b) n_2 degrees of freedom
c) $n_1 + n_2$ degrees of freedom
d) $n_1 + n_2 - 2$ degrees of freedom

4. The null hypothesis which states that there is no significant difference between the means of two populations can be stated as

a) $\mu_1 - \mu_2 = 0$
b) $\mu_1 \geq \mu_2$
c) $\mu_1 - \mu_2 \leq 0$
d) none of the above

5. When each data value in one sample is matched with a corresponding data value in the other sample, the samples are known as

a) corresponding samples
b) matched samples
c) independent samples
d) none of the above

6. Assume we are interested in determining whether the proportion of voters planning to vote for Candidate A (P_C) is significantly **less** than the proportion of voters planning to vote for Candidate B (P_B). The correct set of hypotheses for testing the above is

a) H_o: $P_C - P_B \leq 0$
H_a: $P_C - P_B > 0$

b) H_o: $P_C - P_B = 0$
H_a: $P_C - P_B > 0$

c) H_o: $P_C - P_B \neq 0$
H_a: $P_C - P_B = 0$

d) H_o: $P_C - P_B \geq 0$
H_a: $P_C - P_B < 0$

e) none of the above

7. We want to test whether the average life expectancy of nonsmokers (μ_n) is significantly **more** than that of smokers (μ_s). The correct set of hypotheses for testing the above is

a) H_o: $\mu_n - \mu_s > 0$
H_a: $\mu_n - \mu_s \leq 0$

b) H_o: $\mu_n - \mu_s \geq 0$
H_a: $\mu_n - \mu_s < 0$

c) H_o: $\mu_n - \mu_s \leq 0$
H_a: $\mu_n - \mu_s > 0$

d) H_o: $\mu_n - \mu_s < 0$
H_a: $\mu_n - \mu_s \geq 0$

e) none of the above

8. If a hypothesis is rejected at the 3% level of significance:

a) it must also be rejected at the 2% level of significance
b) it must also be rejected at the 5% level of significance
c) it will sometimes be rejected and sometimes not be rejected at the 5% level of significance
d) Not enough information is given to answer this question.
e) none of the above

9. If two independent large samples are taken from two populations, the sampling distribution of the difference between the two sample means

a) can be approximated by a Poisson distribution.
b) will have a variance of one.
c) can be approximated by a normal distribution.
d) will have a mean of one.
e) none of the above.

10. Which of the following statements is **not** a required assumption for developing an interval estimate of the difference between two sample means when the samples are small?

a) Both populations have normal distributions.

b) $\sigma_1 = \sigma_2 = 1$
c) Independent random samples are selected from the two populations.
d) The variances of the two populations are equal.
e) none of the above

ANSWERS TO THE SELF-TESTING QUESTIONS

1. c
2. c
3. d
4. a
5. b
6. d
7. c
8. b
9 c
10. b

ANSWERS TO CHAPTER TEN EXERCISES

2. 4.84 to 59.16 (rounding: 5 to 60)

3. 0.8383 to 3.1617

5. -11.33 to -0.67

7. H_o: $\mu_1 - \mu_2 = 0$
   H_a: $\mu_1 - \mu_2 \neq 0$
   Since $Z = -2.76 < -1.96$, reject H_o and conclude that there is a significant difference between the average expenditures of those who visited the two islands.

8. H_o: $\mu_1 - \mu_2 = 0$
   H_a: $\mu_1 - \mu_2 \neq 0$
   Since $Z = 1.78$, do not reject H_o and conclude that there is not sufficient evidence to indicate that there is a significant difference in the average test scores from the two universities.

9. H_o: $\mu_1 - \mu_2 = 0$
   H_a: $\mu_1 - \mu_2 \neq 0$
   Since $Z = 10.37 > 1.64$, reject H_o and conclude that there is a significant difference in the average rainfall of the two states.

11. H_o: $\mu_d \leq 0$
    H_a: $\mu_d > 0$
    Since $t = 6.7 > 2.015$, reject H_o and conclude that taking the course will increase the secretaries' average typing speeds.

13. -0.037 to 0.057

14. -0.114 to 0.154

16. H_o: $p_1 - p_2 = 0$
    H_a: $p_1 - p_2 \neq 0$
    Since $Z = 0.418$, do not reject H_o and conclude that there is not sufficient evidence to indicate that there is a significant difference in the proportions of defectives in the two lines.

17. H_0: $p_1 - p_2 = 0$
H_a: $p_1 - p_2 \neq 0$
Since $Z = 0.296$, do not reject H_0 and conclude that there is not sufficient evidence to indicate that there is a significant difference in the proportions of house calls in the two states.

CHAPTER ELEVEN

INFERENCES ABOUT POPULATION VARIANCES

CHAPTER OUTLINE AND REVIEW

In this chapter, you have learned how to make inferences about population variances. You have been introduced to two new probability distributions, namely, the "chi-square" distribution and the "F" distribution. For interval estimation and hypothesis testing concerning the variance of a single normal population, you can use the chi-square distribution. When testing a hypothesis concerning the variance of two normal populations, the F distribution is used.

CHAPTER FORMULAS

Interval Estimate for a Population Variance

$$\frac{(n-1)S^2}{\chi^2_{\alpha/2}} \le \sigma^2 \le \frac{(n-1)S^2}{\chi^2_{(1-\alpha/2)}} \qquad (11.7)$$

where χ^2 values are based on a chi-square distribution with (n - 1) degrees of freedom.

Tests of Hypothesis Regarding a Single Population Variance

A. Upper one-tailed test:

H_o: $\sigma^2 \le \sigma_o^2$

H_a: $\sigma^2 > \sigma_o^2$

Test Statistic: $\chi^2 = \dfrac{(n-1)S^2}{\sigma_o^2}$

Reject H_o if $\chi^2 > \chi^2_\alpha$

B. Lower one-tailed test:

H_o: $\sigma^2 \ge \sigma_o^2$

H_a: $\sigma^2 < \sigma_o^2$

Test Statistic $\chi^2 = \dfrac{(n-1)S^2}{\sigma_o^2}$

Reject H_o if $\chi^2 < \chi^2_\alpha$

CHAPTER FORMULAS
(Continued)

C. Two-tailed test:

H_0: $\sigma^2 = \sigma_o^2$

H_a: $\sigma^2 \neq \sigma_o^2$

Test Statistic: $\chi^2 = \dfrac{(n-1)S^2}{\sigma_o^2}$

Reject H_0 if $\chi^2 < \chi^2_{(1-\alpha/2)}$ or if $\chi^2 > \chi^2_{\alpha/2}$

Inferences About the Variances of Two Populations

A. One-tailed test:

H_0: $\sigma_1^2 \leq \sigma_2^2$

H_a: $\sigma_1^2 > \sigma_2^2$

Denote the population providing the larger sample variance as population 1.

Test Statistic: $F = \dfrac{S_1^2}{S_2^2}$

Reject H_0 if $F > F_\alpha$

where value of F_α is based on an F distribution with n_1 - 1 degrees of freedom for the numerator and n_2 - 1 degrees of freedom for the denominator.

CHAPTER FORMULAS
(Continued)

B. Two-tailed Test:

H_o: $\sigma_1^2 = \sigma_2^2$

H_a: $\sigma_1^2 \neq \sigma_2^2$

Denote the population providing the larger sample variance as population 1.

Test Statistic: $F = \dfrac{S_1^2}{S_2^2}$

Reject H_o if $F > F_{\alpha/2}$

where the value of $F_{\alpha/2}$ is based on an F distribution with n_1 - 1 degrees of freedom for the numerator and n_2 - 1 degrees of freedom for the denominator.

EXERCISES

***1.** The manager of the service department of a local car dealership has noted that the service times for a sample of 15 new automobiles had a standard deviation of 4 minutes. Provide a 95% confidence interval estimate for the standard deviation of service times for all their new automobiles.

Answer: The interval estimate for a population variance is determined by

$$\frac{(n-1)S^2}{\chi^2_{\alpha/2}} \le \sigma^2 \le \frac{(n-1)S^2}{\chi^2_{(1-\alpha/2)}}$$

With a sample of 15, there are (n - 1) = (15 - 1) = 14 degrees of freedom. Thus, we can read the chi-square values from Table 3 of Appendix B in your text as

$$\chi^2_{\alpha/2} = \chi^2_{.025} = 26.12 \quad \text{(rounded)}$$

$$\chi^2_{(1-\alpha/2)} = \chi^2_{.975} = 5.63 \quad \text{(rounded)}$$

Hence, the interval estimate is determined by using the following equation.

$$\frac{(n-1)S^2}{\chi^2_{\alpha/2}} \le \sigma^2 \le \frac{(n-1)S^2}{\chi^2_{(1-\alpha/2)}}$$

$$\frac{(15-1)(4)^2}{26.12} \le \sigma^2 \le \frac{(15-1)(4)^2}{5.63}$$

Computing the above yields the following interval for the variance of the population.

$$8.58 \le \sigma^2 \le 39.79$$

Taking the square root of the above terms, we find the following 95% confidence interval estimate for the standard deviation of the population.

$$2.93 \le \sigma < 6.31$$

2. A random sample of 25 employees of a local utility firm showed that their monthly incomes had a sample standard deviation of $112. Provide a 90% confidence interval estimate for the standard deviation of the incomes for all the firm's employees.

3. A random sample of 41 scores of students taking the ACT test showed a standard deviation of 8 points. Provide a 98% confidence interval estimate for the standard deviation of all the ACT test scores.

***4.** The producer of a certain medicine claims that its bottling equipment is very accurate and that the standard deviation of all its filled bottles is 0.1 ounce or less. A sample of 20 bottles showed a standard deviation of 0.11. Does this sample result confirm the claim of the manufacturer? Use $\alpha = 0.05$.

Answer: The hypotheses to be tested are

H_o: $\sigma^2 \leq 0.01$

H_a: $\sigma^2 > 0.01$ (Claim is not true)

For an upper one-tailed test about a population variance, the null hypothesis is rejected if

$$\chi^2 = \frac{(n-1)S^2}{\sigma_o^2} > \chi_\alpha^2$$

From Table 3 of Appendix B, we read the χ_α^2 with 19 degrees of freedom as $\chi_{.05}^2 = 30.14$. With this as the critical value for the test, then the decision rule will be

Do not reject H_o if $\chi^2 \leq 30.14$

Reject H_o if $\chi^2 > 30.14$

Thus, we calculate χ^2 based on the sample information:

$$\chi^2 = \frac{(n-1)S^2}{\sigma_o^2} = \frac{(20 - 1)\,(0.11)^2}{(0.1)^2} = 22.99$$

Since 22.99 < 30.14, the null hypothesis is not rejected; and we conclude that there is not sufficient evidence to reject the company's claim.

5. A lumber company has claimed that the standard deviation for the lengths of their six-foot boards is 0.5 inches or less. To test their claim, a random sample of 17 six-foot boards is selected; and it is determined that the standard deviation of the sample is 0.43. Do the results of the sample support the company's claim? Use $\alpha = 0.1$.

6. An egg packing company has stated that the standard deviation of the weights of their grade A large eggs is 0.07 ounces or less. The sample variance for 51 eggs was 0.0065 ounces. Can this sample result confirm the company's claim? Use $\alpha = 0.1$.

***7.** Last year, the standard deviation of the ages of the students at the University of Tennessee at Chattanooga (UTC) was 1.81 years. Recently, a sample of 10 students had a standard deviation of 2.1 years. Using $\alpha = 0.05$, test to see if there has been a significant change in the standard deviation of the ages of the students at UTC.

Answer: In this case, we are interested in testing for a change in the variance. Thus, we have a two-tailed test concerning the population variance, where the hypotheses can be stated:

H_0: $\sigma^2 = 3.28$

H_a: $\sigma^2 \neq 3.28$ (There has been a change in the variance.)

For a two-tailed test, the null hypothesis is rejected if

$$\chi^2 < \chi^2_{(1-\alpha/2)} \quad \text{or if} \quad \chi^2 > \chi^2_{\alpha/2}$$

First, we can read the chi-square values from Table 3 of Appendix B (with 9 degrees of freedom) as

$$\chi^2_{(1-\alpha/2)} = \chi^2_{.975} = 2.7 \text{ (rounded)}$$

and

$$\chi^2_{\alpha/2} = \chi^2_{.025} = 19.02 \text{ (rounded)}$$

Then, we calculate chi-square from the sample information:

$$\chi^2 = \frac{(n-1)S^2}{\sigma_o^2} = \frac{(10 - 1)(2.1)^2}{3.28} = 12.1$$

Since this value is in the range of 2.7 to 19.02, the null hypothesis is not rejected. Thus, we conclude that there is not sufficient evidence to confirm that there has been any change in the standard deviation of the ages of the students at UTC.

8. The standard deviation of the daily temperatures in Honolulu last year was $\sigma = 3.2$ degrees Fahrenheit. A random sample of 19 days resulted in a standard deviation of 4 degrees Fahrenheit. Has there been a significant change in the variance of the temperatures? Use $\alpha = 0.02$.

9. Do the following data indicate that the variance of the population from which this sample has been drawn is $\sigma^2 = 17$? Use $\alpha = 0.05$.

X
12
5
9
14
10

***10.** We are interested in determining whether or not the variances of the sales at two small grocery stores are equal. A sample of 21 days of sales at Store A and a sample of 16 days of sales at Store B indicated the following:

Store A	Store B
$n_1 = 21$	$n_2 = 16$
$S_1 = \$125$	$S_2 = \$105$

Are the variances of the populations (from which these samples came) equal? Use $\alpha = 0.05$.

Answer: The hypotheses to be tested are

$$H_o: \sigma_1^2 = \sigma_2^2$$

$$H_a: \sigma_1^2 \neq \sigma_2^2$$

The sample information results in an F value of

$$F = \frac{S_1^2}{S_2^2} = \frac{(125)^2}{(105)^2} = 1.42 \text{ (rounded)}$$

Note that the population providing the larger sample variance is denoted as population 1.

For a two-tailed test regarding the variances of two populations, the null hypothesis is rejected if $F > F_{\alpha/2}$. From Table 4 of Appendix B in your text with $\alpha = 0.05$ and $(n_1 - 1) = (21 - 1) = 20$ numerator degrees of freedom and $(n_2 - 1) = (16 - 1) = 15$ denominator degrees of freedom, we read $F_{\alpha/2} = F_{.025} = 2.76$. Since $1.42 < 2.76$, the null hypothesis is not rejected; and we conclude that there is not sufficient evidence to confirm that the variances of the two populations are different.

11. At $\alpha = 0.1$, test to see if the population variances from which the following samples were drawn are equal.

Group 1	Group 2
$n_1 = 21$	$n_2 = 19$
$S_1 = 18$	$S_2 = 16$

12. The standard deviation of the ages of a sample of 16 executives from northern states was 8.2 years; while the standard deviation of the ages of a sample of 25 executives from southern states was 12.8 years. At $\alpha = 0.1$, test to see if there is any difference in the standard deviations of the ages of all northern and southern executives.

13. Student advisors are interested in determining if the variances of the grades of day students and night students are the same. The following samples are drawn:

Day	Night
$n_1 = 25$	$n_2 = 31$
$S_1 = 9.8$	$S_2 = 14.7$

Test the equality of the variances of the populations at $\alpha = 0.05$.

SELF-TESTING QUESTIONS

In the following multiple choice questions, circle the correct answer. An answer key is provided following the questions.

1. The sampling distribution used when making inferences about a single population variance is

a) the t distribution with (n - 1) degrees of freedom
b) the chi-square distribution with (n - 1) degrees of freedom
c) the F distribution with (n - 1) degrees of freedom for the numerator and (n - 1) degrees of freedom for the denominator
d) none of the above

2. The sampling distribution of the ratio of two independent sample variances extracted from normal populations with equal variances is the

a) t distribution
b) chi-square distribution
c) F distribution
d) normal distribution
e) none of the above

3. The $\chi^2_{.90}$ with 20 degrees of freedom is

a) 28.4120
b) 27.2036
c) 11.6509
d) 12.4426
e) none of the above

4. To avoid the problem of not having access to tables of F distribution with values given for the lower tail when a two-tailed test is required, let the sample with the smaller sample variance be

a) the numerator of the test statistic
b) the denominator of the test statistic
c) It makes no difference how the ratio is set up.
d) none of the above

5. A sample of 20 cans of tomato juice showed a standard deviation of 0.4 ounces. A 95% confidence interval estimate for the variance of the population is

a) 0.2313 to 0.8533
b) 0.2224 to 0.7924
c) 0.0889 to 0.3169
d) 0.0925 to 0.3413

Use the following information to answer questions 6 through 8.

A bottler of a certain soft drink claims its equipment is accurate and that the variance of all filled bottles is less than 0.05 ounces. A sample of 26 bottles had a standard deviation of 0.2.

6. The null hypothesis to test the claim would be written

a) H_0: $\sigma^2 \geq 0.05$
b) H_0: $\sigma^2 > 0.05$
c) H_0: $\sigma^2 < 0.05$
d) H_0: $\sigma^2 \leq 0.05$
e) none of the above

7. The value of the test statistic is

a) 104
b) 20.80
c) 37.65
d) 26.00
e) none of the above

8. The critical value of χ^2 at 95% confidence is

a) 14.6114
b) 15.3791
c) 37.6525
d) 38.8852
e) none of the above

9. The $F_{.05}$ value with 20 numerator degrees of freedom and 30 denominator degrees of freedom is

a) 1.93
b) 1.94
c) 2.20
d) 2.55
e) none of the above

10. A sample of 40 items from population 1 has a sample variance of 8 while a sample of 60 items from population 2 has a sample variance of 10. If we test whether the variances of the two populations are equal, the test statistic will have a value of

a) 0.8
b) 1.56
c) 1.5
d) 1.25
e) none of the above

ANSWERS TO THE SELF-TESTING QUESTIONS

1. b
2. c
3. d
4. b
5. d
6. d
7. b
8. c
9. a
10. d

ANSWERS TO CHAPTER ELEVEN EXERCISES

2. 90.93 to 147.43 (rounded)

3. 6.34 to 10.74 (rounded)

5. H_0: $\sigma^2 \leq 0.25$

   H_a: $\sigma^2 > 0.25$

   Since $\chi^2 = 11.83 < 23.54$, do not reject H_0 and conclude that there is not sufficient evidence to reject the company's claim.

6. H_0: $\sigma^2 \leq 0.0049$

   H_a: $\sigma^2 > 0.0049$

   Since $\chi^2 = 66.33 > 63.17$, reject H_0 and conclude that the sample results do **not** support the company's claim.

8. H_0: $\sigma^2 = 10.24$

   H_a: $\sigma^2 \neq 10.24$

   $\chi^2_{.01} = 34.8$ $\chi^2_{.99} = 7.01$

   Calculated $\chi^2 = 28.125$ thus, do not reject H_0 and conclude that there is not sufficient evidence to show a change in the variance of the temperature.

9. H_0: $\sigma^2 = 17$

   H_a: $\sigma^2 \neq 17$

   $\chi^2_{.025} = 11.14$ $\chi^2_{.975} = 0.48$

   Calculated $\chi^2 = 2.70$ thus, do not reject H_0 and conclude that there is not sufficient evidence to indicate that the variance of the population is significantly different from 17.

11. H_0: $\sigma_1^2 = \sigma_2^2$

    H_a: $\sigma_1^2 \neq \sigma_2^2$

    Since F = 1.26 < 2.19, do not reject H_0, therefore,, there is not sufficient evidence to conclude that the population variances are **unequal**.

12. H_o: $\sigma_1^2 = \sigma_2^2$

H_a: $\sigma_1^2 \neq \sigma_2^2$

Since F = 2.44 > 2.29, reject H_o and conclude that there is a difference in the standard deviation of the ages of northern and southern executives.

13. H_o: $\sigma_1^2 = \sigma_2^2$

H_a: $\sigma_1^2 \neq \sigma_2^2$

Since F = 2.25 > 2.21, reject H_o and conclude that the variances of the population are **not** equal.

CHAPTER TWELVE

TESTS OF GOODNESS OF FIT AND INDEPENDENCE

CHAPTER OUTLINE AND REVIEW

In this chapter, you have been introduced to the tests for goodness of fit and independence. You have learned how to use the chi-square distribution for determining whether or not an observed frequency distribution can be considered a hypothesized probability distribution. The goodness of fit has been demonstrated for multinomial, Poisson, and normal distributions. An extension of the goodness of fit test, namely, the test for independence, has been another major topic of this chapter. Thus, the two major topics of this chapter have been

A. **Goodness of Fit Test:** A statistical testing procedure for determining whether or not to reject a hypothesized probability distribution for a population.

B. **Contingency Table:** A table which is used to summarize the observed and the expected frequencies for a test of the independence of the population characteristics.

CHAPTER FORMULAS

Goodness of Fit Test

Test Statistic: $\chi^2 = \sum_{i=1}^{k} \frac{(f_i - e_i)^2}{e_i}$ (12.1)

where f_i = the observed frequency for category i
e_i = the expected frequency for category i based on the assumption that the null hypothesis is true
k = the number of categories
degrees of freedom = k - p - 1
where p = the number of population parameters estimated from the sample data

Reject H_0 if $\chi^2 > \chi^2_\alpha$

Test of Independence

Test Statistic: $\chi^2 = \sum_i \sum_j \frac{(f_{ij} - e_{ij})^2}{e_{ij}}$ (12.3)

where f_{ij} = the observed frequencies for contingency tables in row i and column j
e_{ij} = the expected frequencies for contingency tables in row i and column j (Under the assumption of independence)
degrees of freedom = (number of rows - 1) (number of columns - 1)

Note: $e_{ij} = \frac{\text{(Row i total) (Column j total)}}{\text{Sample size}}$ (12.2)

Reject H_0 if $\chi^2 > \chi^2_\alpha$

EXERCISES

***1.** Last school year, the student body of a local college consisted of 30% freshmen, 24% sophomores, 26% juniors, and 20% seniors. A sample of 300 students taken from this year's student body showed the following number of students in each classification.

Freshmen	83
Sophomores	68
Juniors	85
Seniors	64
Total	300

Has there been any significant change in the number of students in each classification between the last school year and this school year? Use $\alpha = 0.05$.

Answer: The null and the alternative hypotheses can be stated as

H_0: P(Freshmen) = 0.3, P(Sophomores) = 0.24, P(Juniors) = 0.26, and P(Seniors) = 0.20

H_a: The population proportions are not as stated in H_0.

If the sample results lead to the rejection of H_0, we can conclude that there has been a significant change in the number of students in each classification. However, if there has not been a significant change, we would expect the following number of students to fall into each classification.

		e_i
Freshmen	(300)(0.3) =	90
Sophomores	(300)(0.24) =	72
Juniors	(300)(0.26) =	78
Seniors	(300)(0.20) =	60
		300

The null hypothesis will be rejected if the chi-square determined from the data is larger than χ^2_α. (That is, reject H_0 if $\chi^2 > \chi^2_\alpha$.) Chi-square has a value of

$$\chi^2 = \sum_{i=1}^{k} \frac{(f_i - e_i)^2}{e_i}$$

Thus, to compute the value of chi-square, we can write the observed (f_i) and the expected frequencies (e_i) and complete the calculations as follows.

f_i	e_i	$(f_i - e_i)^2$	$(f_i - e_i)^2/e_i$
83	90	49	0.5444
68	72	16	0.2222
85	78	49	0.6282
64	60	16	0.2667
			$\chi^2 = 1.6615$

Now from Table 3 of Appendix B in your text, we read the value of chi-square at $\alpha = 0.05$ with $(k - 1) = (4 - 1) = 3$ degrees of freedom as $\chi^2_{.05} = 7.815$. Since the computed chi-square of 1.6615 is less than the critical value of 7.815, the null hypothesis is not rejected. Therefore, we conclude that there is not sufficient evidence to indicate that there has been a significant change from last year in the number of students in each classification.

2. In the last presidential election, before the candidates started their major campaigns, the percentages of registered voters who favored the various candidates were as follows.

	Percentages
Republicans	34%
Democrats	43%
Independents	23%

After the major campaigns began, a random sample of 400 voters showed that 172 favored the Republican candidate; 164 were in favor of the Democratic candidate; and 64 favored the Independent candidate. Test with $\alpha = 0.01$ to see if the proportion of voters who favored the various candidates had changed.

3. Before the Christmas shopping rush began, the manager of a department store noted that one-third of the customers paid for their purchases with the store's credit card; one-third used a major credit card; and one-third paid cash (that is, $P_1 = P_2 = P_3 = 1/3$). In a sample of 150 customers shopping during the Christmas rush, 46 used the store's credit card, 43 used a major credit card, and 61 paid cash. With $\alpha = 0.05$, test to see if the customers have changed their methods of payment during the Christmas rush.

4. A major automobile manufacturer claimed that the frequencies of repair on all of their five makes of cars are the same. A sample of 200 repair service receipts showed the following frequencies on the various makes of cars.

Make of Car	Frequency
A	32
B	45
C	43
D	34
E	46

At $\alpha = 0.05$, test the manufacturer's claim.

***5.** A group of 500 individuals were asked to cast their votes regarding a particular issue in the Equal Rights Amendment. The following contingency table shows the results of the votes.

Sex	Favor	Undecided	Oppose	Total
Female	180	80	40	300
Male	150	20	30	200
TOTAL	330	100	70	500

Test at $\alpha = 0.05$ to determine if the votes cast were independent of the sex of the individuals.

Answer: The null and the alternative hypotheses are

H_o: Casting of the vote is independent of the sex of the voter.

H_a: Casting of the vote is not independent of the sex of the voter.

The next step is to determine the expected frequencies under the assumption of independence between the votes cast and the sex of the individuals. We note that 330/500 = 0.66 of all the voters voted in favor of the issue; 100/500 = 0.2 were undecided; and 70/500 = 0.14 opposed the issue. Therefore, if the independence assumption is valid, the same fractions must be applicable to both male and female voters. Since there were 300 female and 200 male voters, we expect the following frequencies to exist.

Sex	Favor	Undecided	Oppose
Female	(0.66)(300) = 198	(0.2)(300) = 60	(0.14)(300) = 42
Male	(0.66)(200) = 132	(0.2)(200) = 40	(0.14)(200) = 28

The above table shows the expected frequencies under the assumption of independence. Then, we calculate the value of chi-square as

$$\chi^2 = \sum_i \sum_j \frac{\left(f_{ij} - e_{ij}\right)^2}{e_{ij}}$$

Thus, the chi-square value will be calculated as

$$\chi^2 = \frac{(180 - 198)^2}{198} + \frac{(80 - 60)^2}{60} + \ldots + \frac{(30 - 28)^2}{28} = 20.99$$

Now, we can read the chi-square value with (number of rows - 1) x (number of columns - 1) = (2 - 1)(3 - 1) = 2 degrees of freedom from Table 3 of Appendix B in your text as $\chi^2_{.05} = 5.99$. Since 20.99 > 5.99, we reject the null hypothesis and conclude that the votes cast were dependent upon the sex of the voters.

6. Among 1,000 managers with degrees in business administration, the following data have been accumulated as to their fields of concentration.

Major	Top Management	Middle Management	TOTAL
Management	300	200	500
Marketing	200	0	200
Accounting	100	200	300
TOTAL	600	400	1000

Test at $\alpha = 0.01$ to determine if their position in management is independent of their field (major) of concentration.

7. From a poll of 800 television viewers, the following data have been accumulated as to their levels of education and their preference of television stations.

	Educational Level			
	High School	Bachelor	Graduate	TOTAL
Public Broadcasting	150	150	100	400
Commercial Stations	50	250	100	400
TOTAL	200	400	200	800

Test at $\alpha = 0.05$ to determine if the selection of a TV station is dependent upon the level of education.

***8.** The number of emergency calls per day at a hospital over a period of 120 days are shown below.

Number of Emergency Calls (x)	Observed Frequency (f)	$f \cdot x$
0	9	0
1	12	12
2	30	60
3	27	81
4	22	88
5	13	65
6	7	42
	$n = \Sigma f = 120$	$\Sigma f \cdot x = 348$

Use $\alpha = 0.05$ and test to see if the data have a Poisson probability distribution.

Answer: The null and the alternative hypotheses are

H_0: The number of emergency calls have a Poisson distribution.

H_a: The number of emergency calls do not have a Poisson distribution.

The first step is to determine the mean of the distribution:

$$\mu = \frac{\Sigma f \,.\, x}{n}$$

The calculation of $\Sigma f \cdot x$ is shown above in the last column. Thus, the mean is

$$\mu = \frac{348}{120} = 2.9$$

Now, from Table 7 of Appendix B, of your textbook, we can read the Poisson probabilities with a mean $\mu = 2.9$ as shown below.

Number of Emergency Calls (x)	Poisson Probability f (x)	Expected Number of Emergency Calls
0	0.0550	6.600
1	0.1596	19.152
2	0.2314	27.768
3	0.2237	26.844
4	0.1622	19.464
5	0.0940	11.280
6	0.0455	5.460
7 or more	0.0286	3.432

The expected frequencies can be determined by multiplying the Poisson probability values by 120. The expected frequencies are shown above in the last column. Since the expected frequency of the last category is less than 5, we combine the last two categories into a single category. Thus, the observed and the expected frequencies can be written as shown below.

Number of Emergency Calls	Observed Freq. (f_i)	Expected Freq. (e_i)	Difference ($f_i - e_i$)	$(f_i - e_i)^2/e_i$
0	9	6.600	2.400	0.873
1	12	19.152	-7.152	2.671
2	30	27.768	2.232	0.179
3	27	26.844	0.156	0.001
4	22	19.464	2.536	0.330
5	13	11.280	1.720	0.262
6 or more	7	8.892	-1.892	0.403

Then, the value of chi-square is determined by summing the values of the last column shown above.

$$\chi^2 = \sum \frac{(f_i - e_i)^2}{e_i} = 0.873 + 2.671 + \ldots + 0.403 = 4.719$$

From Table 3 of Appendix B, we can now read the value of chi-square to be 11.07. Note that there are k - p - 1 = 7 - 1 - 1 = 5 degrees of freedom, where k is the number of categories, in this case 7, and p is the number of population parameters estimated from the sample data. (In this case, the sample was used to estimate the mean, therefore p = 1.) Since the chi-square which was calculated above (4.719) is less than $\chi^2_{.05} = 11.07$, the null hypothesis is not rejected; and we conclude that the number of emergency calls have a Poisson distribution.

9. An insurance company has gathered the following information regarding the number of accidents reported per day over a period of 100 days.

Accidents Per Day	Number of Days (f_i)
0	5
1	18
2	25
3	24
4	20
5	8

Use $\alpha = 0.05$ and test to see if the above data have a Poisson distribution.

***10.** The following data show the grades of a sample of 40 students who have taken statistics.

98	64	96	69
45	94	58	59
63	49	88	83
85	87	68	77
56	63	86	89
84	73	52	63
64	80	69	68
79	73	78	79
72	82	78	88
83	76	66	76

Use $\alpha = 0.1$ and conduct a goodness of fit test to determine if the sample comes from a population which has a normal distribution.

Answer: The mean and the standard deviation for the above data can be determined as

$$\bar{x} = \frac{\Sigma x_i}{n} = \frac{2960}{40} = 74$$

$$S = \sqrt{\frac{\Sigma(x_i - \bar{x})^2}{n - 1}} = \sqrt{\frac{6409}{40 - 1}} = 12.82$$

Then, the hypotheses are stated as

H_0: The population of the examination scores is normal with a mean of 74 and a standard deviation of 12.82.

H_a: The population of the examination scores does not have a normal distribution with a mean of 74 and a standard deviation of 12.82.

Now, we divide the sample of 40 into 8 categories, and each category will contain 5 test scores, or 12.5% of the test scores. (Recall that the rule of thumb requires that at least 5 elements be included in each expected frequency category.) We can determine the boundaries of each 12.5% of the test scores. For example, the lowest 12.5% of the test scores will have an upper limit of

$$x = 74 - (1.15)(12.82) = 59.51$$

where the z value of 1.15 is read from Table 1 of Appendix B and corresponds to an area of 0.375. Working through the normal distribution in a similar manner, we can determine the following category limits.

$$\begin{aligned}
(74) - (0.67)(12.82) &= 65.41 \\
(74) - (0.31)(12.82) &= 70.02 \\
(74) - (0.00)(12.82) &= 74.00 \\
(74) + (0.31)(12.82) &= 77.97 \\
(74) + (0.67)(12.82) &= 82.59 \\
(74) + (1.15)(12.82) &= 88.74
\end{aligned}$$

Thus, we determine the observed and the expected frequencies as

Interval of Examination Scores	Observed Frequency (f_i)	Expected Frequency (e_i)	$(f_i - e_i)$	$(f_i - e_i)^2/e_i$
less than 59.26	6	5	1	0.2
59.26 to 65.41	5	5	0	0.0
65.41 to 70.02	5	5	0	0.0
70.02 to 74.00	3	5	-2	0.8
74.00 to 77.97	3	5	-2	0.8
77.97 to 82.59	6	5	-1	0.2
82.59 to 88.74	8	5	3	1.8
over 88.74	4	5	-1	0.2
				4.0

Therefore, the value of chi-square calculated from the sample as shown at the bottom of the last column is 4. Now, we can read the chi-square value with $k - p - 1 = 8 - 2 - 1 = 5$ degrees of freedom as $\chi^2_{0.10} = 9.24$. Since $4 < 9.24$, the null hypothesis is not rejected; and, therefore, there is insufficient evidence to conclude that the examination scores do **not** have a normal distribution with a mean of 74 and a standard deviation of 12.82.

11. Use $\alpha = 0.05$ to determine if the following sample comes from a normal distribution:

105	260	314	400	520
300	306	115	200	208
418	110	410	312	360
310	314	418	316	412
516	480	490	504	518
280	270	516	419	520
420	438	511	708	300
420	519	702	690	518
510	700	650	670	612
460	600	680	692	600

SELF-TESTING QUESTIONS

In the following multiple choice questions, circle the correct answer. An answer key is provided following the questions.

1. A population where each element of the population is assigned to one and only one of several classes or categories is a

a) multinomial population
b) Poisson population
c) normal population
d) none of the above

2. The sampling distribution for a goodness of fit test is

a) the Poisson distribution
b) the t distribution
c) the normal distribution
d) the chi-square distribution
e) any of the above

3. A goodness of fit test is always conducted as

a) a lower tail test
b) an upper tail test
c) either a and b
d) none of the above

4. An important application of the chi-square distribution is

a) making inferences about a single population variance
b) testing for goodness of fit
c) testing for the independence of two variables
d) all of the above

5. The number of degrees of freedom for the appropriate chi-square distribution in a test of independence is

a) n - 1
b) k - 1
c) the number of rows minus 1 times number of columns minus 1
d) a chi-square distribution is not used

6. In order not to violate the requirements necessary to use the chi-square distribution, each expected frequency in a goodness of fit test must be

a) at least 5
b) at least 10
c) no more than 5
d) the number does not matter

7. A statistical test conducted to determine whether to reject or not reject a hypothesized probability distribution for a population is known as a

a) contingency test
b) probability test
c) goodness of fit test
d) none of the above

8. The degrees of freedom for a contingency table with 21 rows and 7 columns is

a) 20
b) 27
c) 26
d) 147
e) 120

ANSWERS TO THE SELF-TESTING QUESTIONS

1. a
2. d
3. b
4. d
5. c
6. a
7. c
8. e

ANSWERS TO CHAPTER SEVENTEEN EXERCISES

2. Chi-square = 18.42 > 9.21 Reject H_0, the proportion has changed.

3. Chi-square = 3.72 < 5.99 Do not reject H_0, the method of payment has not changed.

4. Chi-square = 4.25 < 9.487 Do not reject H_0, there is no difference in the frequencies of repair.

6. Chi-square = 222.2 > 9.21 Reject H_0, the position is not independent of the major.

7. Chi-square = 75 > 5.99 Reject H_0, the selection of a TV station is not independent of the level of education.

9. Mean = 2.6
   Chi-square = 5.018 < 9.487 Do not reject H_0, the data have a Poisson distribution.

11. $\bar{x}$ = 440.42
    S = 163.21
    Hint: divide the distribution into 10 equal intervals. Since computed χ^2 = 23.2 is greater than $\chi^2_{.05}$ = 14.067, the null hypothesis is rejected, thus concluding that the distribution is not normal.

CHAPTER THIRTEEN

ANALYSIS OF VARIANCE AND EXPERIMENTAL DESIGN

CHAPTER OUTLINE AND REVIEW

In Chapter 10, you learned how to test whether or not the means of two populations are equal. In this chapter, you have been introduced to the analysis of variance (ANOVA) procedure. You have learned this procedure (ANOVA) for determining whether or not the means of more than two populations are equal. The specific concepts which you should have learned in this chapter are

A. **Analysis of Variance (ANOVA) Procedure:** A statistical approach for determining whether or not the means of several different populations are equal.

B. **Factor:** Another word for the variable of interest in an ANOVA procedure.

C. **Treatment:** Different levels of a factor.

D. **Single-Factor Experiment:** An experiment involving only one factor with k populations or treatments.

E. **Experimental Units:** The objects of interest in the experiment.

F. **Completely Randomized Design:** An experimental design where the experimental units are randomly assigned to the treatments.

G. Mean Square: The sum of squares divided by its corresponding degrees of freedom. This quantity is used in the F ratio to determine if significant differences in means exist or not.

H. ANOVA Table: A table used to summarize the analysis of variance computations and results. It contains columns showing the source of variation, the degrees of freedom, the sum of squares, the mean squares, and the F value.

I. Partitioning: The process of allocating the total sum of squares and degrees of freedom into the various components.

J. Blocking: The process of using the same, or similar, experimental units for all treatments. The purpose of blocking is to remove a source of variation from the error term and, hence, provide a sharper test of the difference in population or treatment means.

K. Randomized Block Design: An experimental design employing blocking. The experimental unit(s) within a block is (are) assigned randomly or ordered for the treatments.

L. Factorial Experiments: An experimental design that permits statistical conclusions about two or more factors. All levels of each factor are considered with all levels of the other factors in order to specify the experimental conditions for the experiment.

M. Replication: The number of times each experimental condition is observed in a factorial design. It is the sample size associated with each treatment combination.

N. Main Effect: The response produced by the different factors in factorial design.

O. Interaction: The response produced when the treatments of one factor interact with the treatments of another in influencing the response variable.

CHAPTER FORMULAS

Mean of the jth sample

$$\bar{x}_j = \frac{\sum_{i=1}^{n_j} x_{ij}}{n_j} \qquad (13.1)$$

Variance of the jth sample

$$S_j^2 = \frac{\sum_{i=1}^{n_j} \left(x_{ij} - \bar{x}_j\right)^2}{n_j - 1} \qquad (13.2)$$

Where

x_{ij} = The ith observation in the jth sample

n_j = size of the jth sample

The overall sample mean

$$\bar{\bar{x}} = \frac{\sum_{j=1}^{k} \sum_{i=1}^{n_j} x_{ij}}{n_T} \qquad (13.3)$$

where

$$n_T = n_1 + n_2 + \ldots + n_k \qquad (13.4)$$

The overall sample mean when sample sizes are equal

$$\bar{\bar{x}} = \frac{\sum_{j=1}^{k} \bar{x}_j}{k} \qquad (13.5)$$

CHAPTER FORMULAS
(Continued)

Mean Square Between Treatments

$$MSB = \frac{SSB}{k-1} \tag{13.7}$$

where

$$SSB = \sum_{j=1}^{k} n_j \left(\bar{x}_j - \bar{\bar{x}}\right)^2 \tag{13.8}$$

Mean Square Within Treatments

$$MSW = \frac{SSW}{n_T - k} \tag{13.10}$$

where

$$SSW = \sum_{j=1}^{k} (n_j - 1) s_j^2 \tag{13.11}$$

I. Analysis of Variance for Testing the Equality of the Means of K Populations (General Form)

H_0: $\mu_1 = \mu_2 = \ldots = \mu_k$

H_a: Not all the population means are equal

where

μ_j = the mean of the jth population
k = the number of populations or treatments

$$\text{Decision rule: Reject } H_0 \text{ if } F = \frac{MSB}{MSW} > F_\alpha \tag{13.12}$$

CHAPTER FORMULAS
(Continued)

Total Sum of Squares

$$SST = \sum_{j=1}^{k} \sum_{i=1}^{n_j} \left(x_{ij} - \bar{\bar{x}} \right)^2 \qquad (13.13)$$

or

$$SST = SSB + SSW \qquad (13.14)$$

MULTIPLE COMPARISON

Fisher's LSD (Least Significance Difference) Procedure

H_0: $\mu_i = \mu_j$

H_a: $\mu_i \neq \mu_j$

Test Statistic

$$t = \frac{\bar{x}_1 - \bar{x}_2}{\sqrt{MSW\left(\frac{1}{n_i} + \frac{1}{n_j} \right)}} \qquad (13.16)$$

Reject H_0 if $t < -t_{\alpha/2}$ or $t > t_{\alpha/2}$ (Degrees of freedom = n_T - k)

CHAPTER FORMULAS
(Continued)

Fisher's LSD Procedure Based Upon the Test Statistic $\bar{x}_1 - \bar{x}_2$

H_0: $\mu_i = \mu_j$

H_a: $\mu_i \neq \mu_j$

Test Statistic

$\bar{x}_1 - \bar{x}_2$

Reject H_0 if $|\bar{x}_i - \bar{x}_j| > LSD$

where

$$LSD = t_{\alpha/2}\sqrt{MSW\left(\frac{1}{n_i} + \frac{1}{n_j}\right)} \qquad (13.17)$$

Confidence Interval Estimate of the Difference Between Two Populations Means Using Fisher's LSD Procedure

$$(\bar{x}_i - \bar{x}_j) \pm LSD \qquad (13.18)$$

where

$$LSD = t_{\alpha/2}\sqrt{MSW\left(\frac{1}{n_i} + \frac{1}{n_j}\right)} \qquad (13.19)$$

Degrees of freedom = n_T - k

CHAPTER FORMULAS
(Continued)

II. Analysis of Variance for Completely Randomized Designs

H_0: $\mu_1 = \mu_2 = \ldots = \mu_k$

H_a: Not all the population means are equal

where

μ_j = the mean of the jth population
k = the number of populations or treatments

Decision rule: Reject H_0 if $F = \dfrac{MSTR}{MSE} > F_\alpha$

To determine MSTR and MSE, the following computations are needed.

The Sum of Squares Between (Due) to Treatments

$$SSTR = \sum_{j=1}^{k} n_j \left(\bar{x}_j - \bar{\bar{x}}\right)^2$$

where

n_j = the sample size for the jth treatment

$\bar{x}_j$ = the sample mean for the jth treatment

SSTR can also be computed as

$$SSTR = \sum_{j=1}^{k} \frac{T_j^2}{n_j} - \frac{T^2}{n_T} \qquad \text{(Appendix 13.1)}$$

where

T_j = the sum of all observations in treatment j
n_j = the sample size for the ith treatment

CHAPTER FORMULAS
(Continued)

Mean Square Between (Due to) Treatments

$$MSTR = \frac{SSTR}{k-1} = \frac{\sum_{j=1}^{k} n_j \left(\bar{x}_j - \bar{\bar{x}}\right)}{k-1} \qquad (13.20)$$

The Sum of Squares Within Treatments (Due to Error)

$$SSE = \sum_{j=1}^{k} \left(n_j - 1\right) s_j^2$$

Mean Square Within Treatments (Due to Error)

$$MSE = \frac{SSE}{n_T - k} = \frac{\sum_{j=1}^{k} \left(n_j - 1\right) s_j^2}{n_T - k} \qquad (13.21)$$

The Total Sum of Squares

$$SST = SSTR + SSE \qquad (13.22)$$

or

$$SST = \sum_{j=1}^{k} \sum_{i=1}^{n_j} x_{ij}^2 - \frac{T^2}{n_T} \qquad \text{(Appendix 13.1)}$$

The F Ratio

$$F = \frac{MSTR}{MSE} \qquad (13.23)$$

CHAPTER FORMULAS
(Continued)

III. Analysis of Variance for the Randomized Block Design

The Total Sum of Squares

The following notations are used for the *randomized block design*:

x_{ij} = the value of the observation under treatment i in block j
$T_{i.}$ = the total of all observations in treatment i
$T_{.j}$ = the total of all observations in block j
T = the total of all observations
k = the number of treatments
b = the number of blocks
n_T = the total sample size, $n_T = k\,b$
$\bar{x}_{.j}$ = the sample mean of the jth treatment
$\bar{x}_{i.}$ = the sample mean for the jth block
$\bar{x}$ = the overall sample mean

The Total Sum of Squares

$$SST = SSTR + SSBL + SSE \qquad (13.24)$$

where

SST = Sum of squares total
SSTR = Sum of squares due to treatments
SSBL = Sum of squares due to blocks
SSE = Sum of squares due to error

SST is computed as

$$SST = \sum_{i=1}^{b}\sum_{j=1}^{k}\left(x_{ij} - \bar{\bar{x}}\right)^2 \qquad (13.25)$$

SST can also be computed as

$$SST = \sum_{i=1}^{b}\sum_{j=1}^{k} x_{ij}^2 - \frac{T^2}{n_T} \qquad \text{(Appendix 13.2)}$$

CHAPTER FORMULAS
(Continued)

The Sum of Squares Between (Due to) Treatments

$$SSTR = b\sum_{j=1}^{k}(\bar{x}_{.j} - \bar{\bar{x}})^2 \qquad (13.26)$$

SSTR can also be computed as

$$SSTR = \frac{\sum_{j=1}^{k} T_{.j}^2}{b} - \frac{T^2}{n_T} \qquad \text{(Appendix 13.2)}$$

degrees of freedom = k - 1

The Sum of Squares Due to Blocks

$$SSBL = k\sum_{i=1}^{b}\left(\bar{x}_{i.} - \bar{\bar{x}}\right)^2 \qquad (13.27)$$

SSBL can also be computed as

$$SSBL = \frac{\sum_{i=1}^{b} T_{i.}^2}{k} - \frac{T^2}{n_T} \qquad \text{(Appendix 13.2)}$$

Degrees of freedom = b - 1

The Sum of Squares Due to Error

$$SSE = SST - SSTR - SSBL \qquad (13.28)$$

Degrees of freedom = (k - 1) (b - 1)

CHAPTER FORMULAS
(Continued)

IV. Analysis of Variance for Factorial Experiments

The following notations are used for the *factorial experiments*:

a = the number of levels of factor A
b = the number of levels of factor B
r = the number of replications
n_T = the total number of observations taken in the experiment
x_{ijk} = the observation corresponding to the kth replicate taken from treatment i of factor A and treatment j of factor B
$T_{i.}$ = the total of all observations in treatment i (factor A)
$T_{.j}$ = the total of all observations in treatment j (factor B)
T_{ij} = the total of all observations in the combination of treatment i (factor A) and treatment j (factor B)
T = the total of all observations
$\bar{x}_{i.}$ = the sample mean for the observations in treatment i (factor A)
$\bar{x}_{.j}$ = the sample mean for the observations in treatment j (factor B)
$\bar{x}_{ij}$ = the sample mean for the observations in the combination of treatment i (factor A) and treatment j (factor B)
$\bar{\bar{x}}$ = the overall sample mean

The Total Sum of Squares

$$SST = SSA + SSB + SSAB + SSE \qquad (13.29)$$

or

$$SST = \sum_{i=1}^{a}\sum_{j=1}^{b}\sum_{k=1}^{r}(x_{ijk} - \bar{\bar{x}})^2 \quad \text{degrees of freedom} = n_T - 1 \qquad (13.30)$$

or

$$SST = \sum_{i=1}^{a}\sum_{j=1}^{b}\sum_{k=1}^{r}x_{ijk}^2 - \frac{T^2}{n_T} \qquad \text{(Appendix 13.3)}$$

CHAPTER FORMULAS
(Continued)

The Sum of Squares for Factor A

$$SSA = br\sum_{i=1}^{a}(\bar{x}_{i.} - \bar{\bar{x}})^2 \qquad \text{degrees of freedom} = a - 1 \qquad (13.31)$$

or

$$SSA = \frac{\sum_{i=1}^{a} T_{i.}^2}{br} - \frac{T^2}{n_T} \qquad \text{(Appendix 13.3)}$$

The Sum of Squares for Factor B

$$SSB = ar\sum_{j=1}^{b}(\bar{x}_{.j} - \bar{\bar{x}})^2 \qquad \text{degrees of freedom} = b - 1 \qquad (13.32)$$

or

$$SSB = \frac{\sum_{j=1}^{b} T_{.j}^2}{ar} - \frac{T^2}{n_T} \qquad \text{(Appendix 13.3)}$$

The Sum of Squares for the Interaction

$$SSAB = r\sum_{i=1}^{a}\sum_{j=1}^{b}\left(\bar{x}_{ij} - \bar{x}_{i.} - \bar{x}_{.j} + \bar{\bar{x}}\right)^2 \qquad (13.33)$$

or

$$SSAB = \frac{\sum_{i=1}^{a}\sum_{j=1}^{b} T_{ij}^2}{r} - \frac{T^2}{n_T} - SSA - SSB \qquad \text{(Appendix 13.3)}$$

Degrees of freedom = (a - 1) (b - 1)

CHAPTER FORMULAS
(Continued)

The Sum of Squares Due to Error

SSE = SST - SSA - SSB - SSAB (13.34)

Degrees of freedom = a b (r - 1)

EXERCISES

***1.** M. B. Shultz, a manufacturer of foam rubber sofas, wants to determine whether or not the type of work schedule her employees have has any effect on their productivity. She has selected 12 production employees at random and has randomly assigned 4 employees to each of the 3 proposed work schedules. The proposed work schedules are shown below.

1. 4 days - 40 hours per week
2. flexible time - 40 hours per week
3. standard 5 days - 40 hours per week

The following table shows the units of production (per week) under each of the work schedules.

Work Schedule (Treatment)		
4-Day Program	Flexible Time	5-Day Program
32	33	26
30	35	34
26	30	28
28	38	32

(a) The analysis of variance procedure is based on two major assumptions. Fully explain these assumptions.

Answer: The two assumptions are stated below.

1. The variable of interest for each population has a normal probability distribution. In this example, we assume that the variable of interest, that is, units of production is normally distributed for each of the 3 work schedules.

2. The variance associated with the variable must be the same for each population. In this example, we assume that the variance of the units of production is the same for the employees in each of the 3 types of work schedules.

(b) Use the analysis of variance (Completely randomized design) procedure with $\alpha = 0.05$ to determine if there is a significant difference in the mean weekly units of production for the three types of work schedules.

Answer: The analysis of variance (ANOVA) procedure tests the following hypotheses.

H_0: $\mu_1 = \mu_2 = \ldots = \mu_k$

H_a: Not all the population means are equal

where μ_j = the mean of the jth population
k = the number of populations or treatments

The means of each of the three random samples are computed by

$$\bar{x}_j = \frac{\sum_{i=1}^{n_j} x_{ij}}{n_j}$$

which results in the following sample means.

$\bar{x}_1 = 29$

$\bar{x}_2 = 34$

$\bar{x}_3 = 30$

Then an overall sample mean, $\bar{\bar{x}}$, is computed as the estimate of μ.

$$\bar{\bar{x}} = \frac{\sum_{j=1}^{k}\sum_{i=1}^{n_j} x_{ij}}{n_T}$$

where x_{ij} = the ith observation corresponding to the jth treatment,
n_T = the total sample size for the experiment.

Thus, the overall sample mean becomes

$$\bar{\bar{x}} = \frac{32 + 30 + 26 + \ldots + 28 + 32}{12} = 31$$

Then, we need to determine two independent estimates of the variance of the population σ^2. The first estimate is based upon the differences **between** the treatment means and the overall sample mean and is termed *mean square between treatments* (MSTR). The second estimate is based upon the differences of observations **within** each treatment from the corresponding treatment mean and is termed *mean square within treatments* or *mean square due to error* (MSE). By comparing these two estimates of the population variance, we will be able to conclude whether or not the population means are equal.

The first estimate of σ^2 or MSTR can be written as

$$\text{MSTR} = \frac{\text{SSTR}}{k - 1}$$

where $\text{SSTR} = \sum_{j=1}^{k} n_j \left(\bar{x}_j - \bar{\bar{x}}\right)^2$

Thus, SSTR can be computed as

$$\text{SSTR} = 4\,[(29 - 31)^2 + (34 - 31)^2 + (30 - 31)^2] = 4\,(4 + 9 + 1) = 56$$

Now we can compute MSTR as

$$\text{MSTR} = \frac{\text{SSTR}}{k - 1} = \frac{56}{3 - 1} = 28$$

The second estimate of σ^2 (MSE) is given by

$$\text{MSE} = \frac{\text{SSE}}{n_T - k}$$

where $\text{SSE} = \sum_{j=1}^{k} \left(n_j - 1\right) s_j^2$

SSE is referred to as the sum of squares within treatments or sum of squares due to error. First we need to compute the variance of each sample as

$$S_j^2 = \frac{\sum_{i=1}^{n_j}\left(x_{ij} - \bar{x}_j\right)^2}{n_{j-1}}$$

The variance for the first sample (4 day program) is computed as

$$S_1^2 = \frac{(32-29)^2 + (30-29)^2 + (26-29)^2 + (28-29)^2}{4-1} = \frac{20}{3} = 6.67$$

Similarly, the variances of the other two samples are computed, and their values are

$$S_2^2 = \frac{34}{4-1} = 11.34$$

$$S_3^2 = \frac{40}{4-1} = 13.33$$

Now that the variances have been computed, we can compute SSE as

$$SSE = \sum_{j=1}^{k}\left(n_j - 1\right)S_j^2 = (4 - 1)(6.67) + (4 - 1)(11.34) + (4 - 1)(13.33) = 94$$

Then MSE is computed as

$$MSE = \frac{SSE}{n_T - k} = \frac{94}{12 - 3} = 10.44$$

Now we compute an F value as

$$F = \frac{MSTR}{MSE} = \frac{28}{10.44} = 2.68$$

Then the decision rule for this problem is

Do not reject H_o if $F \le F_\alpha$

Reject H_o if $F > F_\alpha$

From Table 4 of Appendix B, we read the critical value of $F_{.05}$ with $k - 1 = 3 - 1 = 2$ degrees of freedom for the numerator and $n_T - k = 12 - 3 = 9$ degrees of freedom for the denominator as $F_{.05} = 4.26$. (Recall our computed value of F = MSTR/MSE = 2.68.) Since $2.68 < 4.26$, there is not sufficient statistical evidence to reject the null hypothesis. Thus, indicating there is not sufficient evidence to conclude that there exists a statistically significant difference in the three population means at the 0.05 level.

2. A random sample of six automobile tires from each of the three major manufacturers showed the following life expectancies All of the figures are in thousands of miles.

Manufacturer A	Manufacturer B	Manufacturer C
41	47	47
38	42	41
44	44	42
42	34	49
36	42	40
39	43	45

Test at $\alpha = 0.05$ to determine if there is a significant difference in the average lives of the three brands of tires.

*3. Use the results of exercise 1 and set up a complete ANOVA table.

Answer: The general from of the ANOVA table is shown below.

Source of Variation	Sum of Squares	Degrees of Freedom	Mean Square	F
Between Treatments	SSTR	K - 1	MSTR	$\frac{MSTR}{MSE}$
Error (Within Treatments)	SSE	n_t - K	MSE	
Total	SST	n_t - 1		

Referring to exercise 1 and filling in the required information, the ANOVA table will be as follows.

Source of Variation	Sum of Squares	Degrees of Freedom	Mean Square	F
Between Treatments	56	2	28	2.68
Error (Within Treatments)	94	9	10.44	
Total	150	11		

4. The heating bills for a selected sample of houses using various forms of heating are given below. (Values are in dollars.)

Natural Gas	Central Electric	Heat Pump
84	95	85
64	60	93
93	89	90
88	96	92
71	90	80

At $\alpha = 0.05$, test to see if there is a significant difference among the average heating bills of the homes.

5. Use the results of exercise 4 and fill in the blanks in the following ANOVA table. Use the space below to show your work.

Source of Variation	Sum of Squares	Degrees of Freedom	Mean Square	F
Between Treatments	____?	____?	____?	____?
Error (Within Treatments)	____?	____?	____?	
Total	____?	____?		

*6. Three universities in your state have decided to administer the same comprehensive examination to the recipients of MBA degrees from the three institutions. From each institution, a random sample of MBA recipients has been selected and given the test. The following table shows the scores of the students from each university.

Northern University	Central University	Southern University
56	62	94
85	97	72
65	91	93
86	82	78
93		54
		77
$T_1 = 385$	$T_2 = 332$	$T_3 = 468$

At $\alpha = 0.01$, test to see if there is any significant difference in the average scores of the students from the three universities. Solve this problem following the step-by-step procedure introduced to you in *Appendix 13.1* of your text.(Note that the sample sizes are not equal.)

Answer: The hypotheses to be tested are

H_0: $\mu_1 = \mu_2 = \ldots = \mu_k$

H_a: Not all means are equal

Now let us solve this problem following the step-by-step procedure introduced to you in *Appendix 13.1* of your text.

STEP 1: Compute the sum of squares about the mean (SST) by

$$SST = \sum_{j=1}^{k}\sum_{i=1}^{n_j} x_{ij}^2 - \frac{T^2}{n_T}$$

where x_{ij} = the value of the ith observation under treatment j,
T = the sum of all observations,
n_T = the total sample size for the experiment.

In this example then,

$$\sum_{j=1}^{k}\sum_{i=1}^{n_j} x_{ij}^2 = [(56)^2 + (85)^2 + \ldots + (93)^2] + [(62)^2 + (97)^2 + \ldots + (82)^2] + [(94)^2 + (72)^2 + \ldots + (77)^2] = 96{,}487$$

and $T = [56 + \ldots] + [62 + \ldots] + [94 + \ldots] = 1185$

and $n_T = 15$

Thus, the value of $\dfrac{T^2}{n_T} = \dfrac{(1185)^2}{15} = 93{,}615$

Now we can compute SST as

$$SST = \sum_{j=1}^{k}\sum_{i=1}^{n_j} x_{ij}^2 - \frac{T^2}{n_T} = 96{,}487 - 93{,}615 = 2{,}872$$

STEP 2: Compute the sum of squares due to treatments (SSTR) by

$$SSTR = \sum_{j=1}^{k} \frac{T_j^2}{n_j} - \frac{T^2}{n_T}$$

where T_j = the sum of all observations in treatment j,
n_j = the sample size for treatment j.

First let us compute

$$\sum_{j=1}^{k} \frac{T_j^2}{n_j} = \frac{(385)^2}{5} + \frac{(332)^2}{4} + \frac{(468)^2}{6} = 93{,}705$$

Thus, SSTR is computed as

$$SSTR = \sum_{j=1}^{k} \frac{T_j^2}{n_j} - \frac{T^2}{n_T} = 93{,}705 - 93{,}615 = 90$$

STEP 3: Compute the sum of squares due to error as

$$SSE = SST - SSTR$$

$$= 2{,}872 - 90 = 2{,}782$$

Now that all the sums of squares are computed we can compute the mean squares as

$$MSTR = \frac{SSTR}{k - 1} = \frac{90}{3 - 1} = 45$$

$$MSE = \frac{SSE}{n_T - k} = \frac{2782}{15 - 3} = 231.83$$

Then we determine the F value as

$$F = \frac{MSTR}{MSE} = \frac{45}{231.83} = 0.194$$

From Table 4 of Appendix B of your text, we read the critical value of F_α with $k - 1 = 3 - 1 = 2$ degrees of freedom for the numerator and $n_T - k = 15 - 3 = 12$ degrees of freedom for the denominator as $F_{.01} = 6.93$. Since 0.194 is less than 6.93, the null hypothesis is not rejected. Thus, we conclude that there is not sufficient evidence to conclude that the average scores of the students from the three universities are significantly different.

7. The three major automobile manufacturers have entered their cars in the Indianapolis 500 race. The speeds of the tested cars are given below.

G	F	C
180	177	162
175	180	174
169	160	187
174	172	
190		

At $\alpha = 0.05$, test to see if there is a significant difference in the average speeds of the cars of the three auto manufacturers. (*Use the procedure of Appendix 13.1 of your textbook.*)

***8.** Refer to exercise 6. Part of the ANOVA table for exercise 6 is shown below.

Source of Variation	Sum of Squares	Degrees of Freedom	Mean Square	F
Between Treatments	90	2	MSTR	
Error (Within Treatments)	2782	12	MSE	

Compute the missing values and use $\alpha = 0.01$ to determine if there is any significant difference among the means.

Answer: As you can see, the above results are the same as those calculated in exercise 1. The MSTR and MSE can be calculated as

$$\text{MSTR} = \frac{\text{SSR}}{\text{DF(Between)}} = \frac{90}{2} = 45$$

$$\text{MSE} = \frac{\text{SSE}}{\text{DF(Within)}} = \frac{2{,}782}{12} = 231.83$$

Thus, we can calculate the F value as

$$F = \frac{\text{MSTR}}{\text{MSE}} = \frac{45}{231.83} = 0.194$$

Since $0.194 < 6.93$, the null hypothesis is not rejected; therefore, we conclude that there is no significant difference between the average scores of the students from the three universities.

9. Refer to exercise 7. Part of the ANOVA table for exercise 7 is shown below.

Source of Variation	Sum of Squares	Degrees of Freedom	Mean Square	F
Between Treatments	65.375			
Error (Within Treatments)	798.625			

Complete all the missing values in the above table and use $\alpha = 0.01$ to determine if there is any significant difference among the means.

***10.** A test of general knowledge was administered to students in the fields of (1) engineering, (2) education, and (3) business (treatments).

In each of the three treatments, 11 students took the test. An analysis of variance was performed on the test scores. Part of the ANOVA table for this study is shown below.

Source of Variation	Sum of Squares	Degrees of Freedom	Mean Square	F
Between Treatments	491.7			
Error (Within Treatments)	1165.3			
Total	1657			

(a) Complete the ANOVA table; and at $\alpha = 0.05$, test to determine if there is a significant difference in the means of the three populations (treatments).

Answer: The degrees of freedom associated with *Between Treatments* is $K - 1 = 3 - 1 = 2$, where K represents the number of treatments. The degrees of freedom associated with *Error (Within Treatments)* is given by $n_T - K$. Since there were 11 observations in each treatment, $n_T = 11 \times 3 = 33$. Therefore, the degrees of freedom is $n_T - K = 33 - 3 = 30$.

Next, we compute MSTR as

$$MSTR = \frac{SSTR}{K-1} = \frac{491.7}{3-1} = 245.85$$

and MSE is computed as

$$MSE = \frac{SSE}{n_T - K} = \frac{1165.3}{33-3} = 38.84$$

Finally, the F statistic is calculated as

$$F = \frac{MSTR}{MSE} = \frac{245.85}{38.84} = 6.33$$

Thus, the complete ANOVA table will be as follows.

Source of Variation	Sum of Squares	Degrees of Freedom	Mean Square	F
Between Treatments	491.7	2	245.85	6.33
Error (Within Treatments)	1165.3	30	38.84	
Total	1657	32		

The hypotheses to be tested are

H_0: $\mu_1 = \mu_2 - \mu_3$

H_a: Not all the population means are equal

From the F table with 2 numerator and 30 denominator degrees of freedom, we read $F_{.05} = 3.32$.

Since $F = 6.33 > 3.32$, the null hypothesis is rejected, and we conclude that at least one mean is different from the others.

(b) Now that we have determined that at least one mean is different from the others, determine which mean(s) is (are) different. The sample means for this study are shown below.

Treatment	Sample Mean
(1) Engineering	84.0
(2) Education	87.5
(3) Business	93.4

Answer: Probably the most widely used method for making pair wise comparison of population means is Fisher's Least-significant-difference (LSD). First, let us test to determine if the means of Population 1 (Engineering) and Population 2 (Education) are different. The hypotheses to be tested are

H_o: $\mu_1 = \mu_2$

H_a: $\mu_1 \neq \mu_2$

The null hypothesis will be rejected if $|\bar{x}_1 - \bar{x}_2| >$ LSD, where LSD is

$$LSD = t_{\alpha/2}\sqrt{MSW\left(\frac{1}{n_1}+\frac{1}{n_2}\right)}$$

The value of MSW (which is also referred to as MSE) was computed in Part a and is shown in the ANOVA table as 38.84. The value of $t_{\alpha/2}$ is read from the t table at 30 degrees of freedom as $t_{.025} = 2.042$. Thus, LSD is computed as

$$LSD = t_{\alpha/2}\sqrt{MSW\left(\frac{1}{n_1}+\frac{1}{n_2}\right)} = (2.042)\sqrt{38.84\left(\frac{1}{11}+\frac{1}{11}\right)} = 5.42$$

Now, we find the absolute value of the difference between the means of Sample 1 and Sample 2 as

$$|\bar{x}_1 - \bar{x}_2| = |84.0 - 87.5| = 3.5$$

Since 3.5 is not greater than the LSD value of 5.42, the null hypothesis is not rejected. Hence, we cannot conclude that there is a significant difference between the means of Population 1 (Engineering) and Population 2 (Education). Once the LSD is computed, we simply find the absolute value of the difference between any pair of sample means; and if this difference is greater than LSD, the null hypothesis will be rejected.

For instance, to test for the significant difference between the means of Population 1 (Engineering) and Population 3 (Business), simply compute the following

$$|\bar{x}_1 - \bar{x}_3| = |84.0 - 93.4| = 9.4$$

Since 9.4 > 5.42, the null hypothesis (i.e., H_0: $\mu_1 = \mu_3$) is rejected. We can then conclude that there is a significant difference between the means of Population 1 (Engineering) and Population 3 (Business).

We can also test to see if there is a significant difference between the means of Population 2 (Education) and Population 3 (Business) by computing the following.

$$|\bar{x}_2 - \bar{x}_3| = |87.5 - 93.4| = 5.9$$

Since 5.9 > 5.42, the null hypothesis (i.e., H_0: $\mu_2 = \mu_3$) is rejected; therefore, we conclude that there is a significant difference between the means of population 2 (Education) and population 3 (Business).

11. Eight observations were selected from each of 3 populations, and an analysis of variance was performed on the data. The following are part of the results.

Source of Variation	Sum of Squares	Degrees of Freedom	Mean Square	F
Between Treatments			34.67	
Error (Within Treatments)				
Total	189.33			

(a) Using $\alpha = .05$, test to see if there is a significant difference among the means of the three populations. Show the complete ANOVA table.

(b) If in Part a you concluded that at least one mean is different from the others, determine which mean is different. The three sample means are $\bar{x}_1 = 28$, $\bar{x}_2 = 27$, and $\bar{x}_3 = 31$. Use Fisher's LSD procedure and let $\alpha = .05$.

12. Ten observations were selected from each of 3 populations, and an analysis of variance was performed on the data. The following information was obtained.

$\bar{x}_1 = 147.1$ $\quad \bar{x}_2 = 180.0$ $\quad \bar{x}_3 = 196.8$

MSB = 614.83 (Also referred to as MSE)

At the $\alpha = 0.05$ level of significance, use Fisher's LSD procedure and determine which mean(s) is (are) different from the others (if any).

***13.** Mary Beth is an instructor in the statistics laboratory. She has noted that some statistics professors give objective examinations (i.e., true or false and multiple choice questions), while other professors give subjective examinations (i.e., questions and problems). In order to evaluate the two types of examinations, she has randomly selected 4 students who have just finished the statistics course and has asked them to take two types of examinations, one objective and another subjective. Table 13.1 shows the scores of the 4 students on the two types of examinations.

Type of Examination Treatment	Student (Blocks) 1	2	3	4	Row or Treatment Totals	Treatment Means $(\bar{x}_{.j})$
Objective	90	70	60	80	300	$\bar{x}_{.1} = \frac{300}{4} = 75$
Subjective	80	90	80	70	320	$\bar{x}_{.2} = \frac{320}{4} = 80$
Column or Block Totals	170	160	140	150	$\sum_i \sum_j x_{ij}$ 620 = Overall sum	
Block Means $(\bar{x}_{i.})$	$\bar{x}_{1.} = \frac{170}{2} = 85$	$\bar{x}_{2.} = \frac{160}{2} = 80$	$\bar{x}_{3.} = \frac{140}{2} = 70$	$\bar{x}_{4.} = \frac{150}{2} = 75$	$\bar{\bar{x}} = \frac{620}{8} = 77.5$	

Table 13.1

(a) Treating the students as *blocks,* at $\alpha = 0.05$, test to see if there is any difference in the scores of the two types of examinations. Use the procedure as explained in section 13.6 of your textbook.

Answer: The ANOVA procedure for the *randomized block design* partitions the sum of squares total (SST) into three sums of squares as follows.

SST	=	SSTR	+	SSBL	+	SSE
↑		↑		↑		↑
Sum of Squares Total		Treatment		Block		Error

Thus, we need to compute each of the sums of squares as shown in the following step-by-step procedure.

STEP 1: Compute the total sum of squares (SST) as follows.

$$SST = \sum_{i=1}^{b}\sum_{j=1}^{k}\left(x_{ij} - \bar{\bar{x}}\right)^2 = (90 - 77.5)^2 + (70 - 77.5)^2 + \ldots + (70 - 77.5)^2 = 750$$

STEP 2: Compute SSTR as

$$SSTR = b\sum_{j=1}^{k}(\bar{x}_{.j} - \bar{\bar{x}})^2 = 4[(75 - 77.5)^2 + (80 - 77.5)^2] = 50$$

STEP 3: Compute SSBL as

$$SSBL = k\sum_{i=1}^{b}\left(\bar{x}_{i.} - \bar{\bar{x}}\right)^2$$
$$= 2[(85 - 77.5)^2 + (80 - 77.5)^2 + (70 - 77.5)^2 + (75 - 77.5)^2] = 250$$

STEP 4: Compute SSE as

$$SSE = SST - SSTR - SSB = 750 - 50 - 250 = 450$$

The above sums of squares divided by their corresponding degrees of freedom provide the mean square values. The mean square values are computed as follows.

$$MSTR = \frac{SSTR}{k - 1} = \frac{50}{2 - 1} = 50$$

$$MSB = \frac{SSB}{b - 1} = \frac{250}{4 - 1} = 83.33$$

$$MSE = \frac{SSE}{(k - 1)(b - 1)} = \frac{450}{(2 - 1)(4 - 1)} = 150$$

Now we can complete the ANOVA table as shown in Table 13.2.

Source of Variation	Sum of Squares	Degrees of Freedom	Mean Square	F
Between Treatments	50	1	50.00	0.33
Between Blocks	250	3	83.33	
Error	450	3	150.00	
Total	750	7		

Table 13.2

The F value is computed as MSTR/MSE = 50/150 = 0.33. The critical F value in Table 4 of Appendix B at $\alpha = 0.05$ is 10.3 (1 numerator degrees of freedom and 3 denominator degrees of freedom). Since 0.33 < 10.3, we do not reject the null hypothesis. Therefore, we do not have sufficient evidence to conclude that the means of the two types of examinations are different.

(b) Test to see if there is any difference in the scores of the two types of examinations. (*Use the procedure as explained in Appendix 13.2 of your textbook.*)

Answer: This is an alternate form of computing each of the sums of squares. Each sum of squares can be computed by the following step-by-step procedure.

STEP 1: Compute the SST as follows.

$$SST = \sum_{i=1}^{b}\sum_{j=1}^{k} x_{ij}^2 - \frac{T^2}{n_T}$$

where x_{ij} = the value of the observation under treatment j in block i.
T = the total of all observations
n_T = the total sample size

First let us compute the following.

$$\sum_{i=1}^{b}\sum_{j=1}^{k} x_{ij}^2 = (90)^2 + (70)^2 + \ldots + (80)^2 + (70)^2 = 48{,}800$$

Then, T = 620 (shown in Table 13.1)
n_T = 8 (the total sample size)

Therefore, $$\frac{T^2}{n_T} = \frac{(620)^2}{8} = 48{,}050$$

Now we can compute SST as

$$SST = 48{,}800 - 48{,}050 = 750$$

STEP 2: Compute the SSTR as follows.

$$SSTR = \frac{\sum_{j=1}^{k} T_{.j}^2}{b} - \frac{T^2}{n_T}$$

where $T_{.j}$ = the total of all observations in treatment i (shown in Table 13.1 under row totals)
b = the number of blocks = 4

Thus, $$SSTR = \frac{(300)^2 + (320)^2}{4} - \frac{(620)^2}{8} = 50$$

STEP 3: Compute the SSBL as follows.

$$SSBL = \frac{\sum_{i=1}^{b} T_{i.}^2}{k} - \frac{T^2}{n_T}$$

where k = the number of treatments = 2

$T_{i.}$ = the total of all observations in block i (shown in Table 13.1 opposite block totals)

Therefore, $$\sum T_{i.}^2 = (170)^2 + (160)^2 + (140)^2 (150)^2 = 96{,}600$$

Now we can compute the SSBL as follows.

$$SSBL = \frac{\sum_{i=1}^{b} T_{i.}^2}{k} - \frac{T^2}{n_T} = \frac{96600}{2} - \frac{(620)^2}{8} = 250$$

STEP 4: Now that the SST, SSTR and SSBL are computed, we can simply compute SSE:

$$SSE = SST - SSTR - SSB$$
$$= 750 - 50 - 250 = 450$$

The above sums of squares divided by their corresponding degrees of freedom provide the mean square values. The mean square values are computed as follows.

$$MSTR = \frac{SSTR}{k - 1} = \frac{50}{2 - 1} = 50$$

$$MSB = \frac{SSBL}{b - 1} = \frac{250}{4 - 1} = 83.33$$

$$MSE = \frac{SSE}{(k - 1)(b - 1)} = \frac{450}{(2 - 1)(4 - 1)} = 150$$

Now we can complete the ANOVA table as shown in Table 13.3.

Source of Variation	Sum of Squares	Degrees of Freedom	Mean Square	F
Between Treatments	50	1	50.00	0.33
Between Blocks	250	3	83.33	
Error	450	3	150.00	
Total	750	7		

Table 13.3

The F value is computed as MSTR/MSE = 50/150 = 0.33. The critical F value in Table 4 of Appendix B at $\alpha = 0.05$ is 10.3 (1 numerator degrees of freedom and 3 denominator degrees of freedom). Since $0.33 < 10.3$, we do not reject the null hypothesis. Thus, we conclude that there is not sufficient evidence to conclude that the means of the two types of examinations are different.

14. Five drivers were selected to test drive 2 makes of automobiles. The following table shows the number of miles per gallon for each driver driving both cars.

Automobile	Drivers (Blocks)				
A	30	31	30	27	32
B	36	35	28	31	30

At $\alpha = 0.05$, test to see if there is any difference in the miles per gallon of the two makes of automobiles.

***15.** In exercise 6 you were introduced to a situation where only one factor (the universities) existed. Let us expand that situation to a two factor problem. Assume that as a second factor we are considering the sex of the students taking the examination. The first factor has 3 treatments (the preparation program at the 3 universities), and the second factor has two treatments (male or female). Thus, there are a total of 3 x 2 = 6 treatment combinations. A sample of two students is selected corresponding to each of the 6 treatments. The scores of the students on the examination are shown in Table 13.4.

		Factor B: Sex of the Students	
		Male	Female
Factor A: University	Northern	79	83
		81	91
	Central	92	84
		86	94
	Southern	90	86
		86	94

Table 13.4

Use the procedure as explained in section 13.7 of your textbook to answer the following questions: (Let $\alpha = 0.05$.)

1. Do the universities' preparation programs differ in terms of effect on the examination scores (Main effect, Factor A)?

2. Do the male and female students differ in terms of their ability to perform on the examination (Main effect, Factor B)?

3. Do students of one sex do better in one university while students of the other sex do better in a different university (Interaction effect, Factors A and B)?

Answer: The analysis of variance for the two-factor factorial experiment partitions the SST as follows.

	SST	=	SSA	+	SSB	+	SSAB	+	SSE
Sum of	↑		↑		↑		↑		↑
Squares	Total		Factor A		Factor B		Interaction Of Factors A and B		Error

First, let us compute a summary of the data as presented in Table 13.5.

	FACTOR B: SEX OF THE STUDENTS			
Treatment Combination Totals	Male	Female	Row Totals	Factor A Means
	79 80	83 91		
Northern	160	174	334	$\bar{x}_{1.} = \frac{334}{4} = 83.5$
	$\bar{x}_{11} = \frac{160}{2} = 80$	$\bar{x}_{12} = \frac{174}{2} = 87$		
	92 86	84 94		
FACTOR A: Central	178	178	356	$\bar{x}_{2.} = \frac{356}{4} = 89$
UNIVERSITY	$\bar{x}_{21} = \frac{178}{2} = 89$	$\bar{x}_{22} = \frac{178}{2} = 89$		
	90 86	86 94		
Southern	176	180	356	$\bar{x}_{3.} = \frac{356}{4} = 89$
	$\bar{x}_{31} = \frac{176}{2} = 88$	$\bar{x}_{32} = \frac{180}{2} = 90$		
Column Totals	514	532	1046 =	Overall Total
			$\bar{x} = \frac{1046}{12} = 87.17$	
Factor B Means	$\bar{x}_{.1} = \frac{514}{6} = 85.67$	$\bar{x}_{.2} = \frac{532}{6} = 88.67$		

Table 13.5

Then, we can compute each of the sums of squares by the following step-by-step procedure.

STEP 1: Compute SST as

$$SST = \sum_{i=1}^{a}\sum_{j=1}^{b}\sum_{k=1}^{r}(x_{ijk} - \bar{\bar{x}})^2$$

$$= (79 - 87.17)^2 + (81 - 87.17)^2 + \ldots + (86 - 87.17)^2 + (94 - 87.17)^2 = 275.67$$

STEP 2: Compute SSA as

$$SSA = br\sum_{i=1}^{a}(\bar{x}_{i.} - \bar{\bar{x}})^2$$
$$= (2)\,(2)\,[(83.5 - 87.17)^2 + (89 - 87.17)^2 + (89 - 87.17)^2] = 80.67$$

STEP 3: Compute SSB as

$$SSB = ar\sum_{j=1}^{b}(\bar{x}_{.j} - \bar{\bar{x}})^2$$
$$= (3)\,(2)\,[(85.67 - 87.17)^2 + (88.67 - 87.17)^2] = 27$$

STEP 4: Compute SSAB as

$$SSAB = r\sum_{i=1}^{a}\sum_{j=1}^{b}\left(\bar{x}_{ij} - \bar{x}_{i.} - \bar{x}_{.j} + \bar{\bar{x}}\right)^2$$
$$= 2\,[(80 - 83.5 - 85.67 + 87.17)^2 + \ldots + (90 - 89 - 88.67 + 87.17)^2] = 26$$

STEP 5: Compute SSE as

$$SSE = SST - SSA - SSB - SSAB$$

$$= 275.67 - 80.67 - 27 - 26 = 142$$

Now that we have computed all the sums of squares, we can compute the mean squares by dividing each sum of square by its corresponding degrees of freedom, where degrees of freedom are as follows.

Source of Variation	Degrees of Freedom
Factor A treatment	a - 1 = 3 - 1 = 2
Factor B treatment	b - 1 = 2 - 1 = 1
Interaction	(a - 1) (b - 1) = (2) (1) = 2
Error	a b (r - 1) = (3) (2) (2 - 1) = 6

With this information, we can set up the ANOVA table as shown in Table 13.6.

Source of Variation	Sum of Squares	Degrees of Freedom	Mean Square	F
Factor A Treatment	80.67	2	40.33	$\frac{40.33}{23.67} = 1.70$
Factor B Treatment	27.00	1	27.00	$\frac{27.00}{23.67} = 1.14$
Interaction (AB)	26.00	2	13.00	$\frac{13.00}{23.67} = 0.55$
Error	142.00	6	23.67	
Total	275.67			

Table 13.6

Now we are in a position to answer the questions set forth at the beginning of the exercise. The F ratio used to test for a difference among the universities' preparation programs is 1.70. The critical F value at $\alpha = 0.05$ (with 2 numerator degrees of freedom and 6 denominator degrees of freedom) is 5.14. Since $1.70 < 5.14$, we cannot reject the null hypothesis. Therefore, we conclude that there is no difference in the preparation provided by the three universities.

The F ratio for the sex of the students is 1.14. The critical F value with 1 numerator and 6 denominator degrees of freedom is 5.99. Since $1.14 < 5.99$, we cannot reject the null hypothesis. Therefore, we conclude that there is no difference in the performance of male and female students.

Finally, the interaction F value is 0.55. The critical F value with 2 numerator and 6 denominator degrees of freedom is 5.14. Since $0.55 < 5.14$, no significant interaction can be identified. Thus, we conclude that there is no reason to believe that the three universities differ in their ability to prepare male and female students.

(b) Use the procedure as explained in Appendix 13.3 and rework Part a of this exercise.

Answer: Once again SST is given by

	SST	=	SSA	+	SSB	+	SSAB	+	SSE
Sum of	↑		↑		↑		↑		↑
Squares	Total		Factor A		Factor B		Interaction Of Factors A and B		Error

Then we can compute each of the sums of the squares by the following step-by-step procedure.

STEP 1: Compute the sum of squares total as

$$SST = \sum_{i=1}^{a}\sum_{j=1}^{b}\sum_{k=1}^{r} x_{ijk}^2 - \frac{T^2}{n_T}$$

where x_{ijk} = the observation corresponding to the kth replicate taken from treatment i of Factor A and treatment j of Factor B.

First let us compute the sum of squares as

$$\sum_{i=1}^{a}\sum_{j=1}^{b}\sum_{k=1}^{r} x_{ijk}^2 = (79)^2 + (81)^2 + \ldots + (86)^2 + (94)^2 = 91{,}452$$

Then we compute T, which is the total of all observations as

$$T = 79 + 83 + \ldots + 86 + 94 = 1046$$

and $n_T = 12$, which is the total number of observations taken in the experiment.

Thus, we can compute SST as

$$SST = SST = \sum_{i=1}^{a}\sum_{j=1}^{b}\sum_{k=1}^{r} x_{ijk}^2 - \frac{T^2}{n_T} = 91{,}452 - \frac{(1046)^2}{12} = 275.67$$

STEP 2: Compute the sum of squares for Factor A as

$$SSA = \frac{\sum_{i=1}^{a} T_{i.}^2}{br} - \frac{T^2}{n_T}$$

where $T_{i.}$ = the total of all observations in treatment i, Factor A (i.e., row total)

Therefore,

$$T_{1.} = 79 + 81 + 83 + 91 = 334$$
$$T_{2.} = 92 + 86 + 84 + 94 = 356$$
$$T_{3.} = 90 + 86 + 86 + 94 = 356$$

and b = the number of levels of Factor B = 2
r = the number of replications = 2

Now we can compute SSA as

$$SSA = \frac{\sum_{i=1}^{a} T_{i.}^2}{br} - \frac{T^2}{n_T} = \frac{(334)^2 + (356)^2 + (356)^2}{(2)\,(2)} - \frac{(1046)^2}{12} = 80.67$$

STEP 3: Compute the sum of squares for Factor B as

$$SSB = \frac{\sum_{j=1}^{b} T_{.j}^2}{ar} - \frac{T^2}{n_T}$$

where $T_{.j}$ = the total of all observations in treatment j, Factor B (i.e., column totals)

Therefore,

$$T_{.1} = 79 + 81 + 92 + 86 + 90 + 86 = 514$$
$$T_{.2} = 83 + 91 + 84 + 94 + 86 + 94 = 532$$

and a = the number of levels of Factor A = 3

Then SSB can be computed as

$$SSB = \frac{\sum_{j=1}^{b} T_{.j}^2}{ar} - \frac{T^2}{n_T} = \frac{(514)^2 + (532)^2}{(3)\,(2)} - \frac{(1046)^2}{12} = 27$$

STEP 4: Compute the sum of squares for the interaction as

$$SSAB = \frac{\sum_{i=1}^{a}\sum_{j=1}^{b} T_{ij}^2}{r} - \frac{T^2}{n_T} - SSA - SSB$$

where T_{ij} = the total of all observations in the combination of treatment i (Factor A) and treatment j (Factor B).

Therefore,

$$\sum_{i=1}^{a}\sum_{j=1}^{b} T_{ij}^2 = (79 + 81)^2 + (83 + 91)^2 + \ldots + (86 + 94)^2 = 182{,}620$$

Then SSAB can be computed as

$$SSAB = \frac{\sum_{i=1}^{a}\sum_{j=1}^{b} T_{ij}^2}{r} - \frac{T^2}{n_T} - SSA - SSB$$

$$= \frac{182620}{2} - \frac{(1046)^2}{12} - 80.67 - 27 = 26$$

STEP 5: Compute the sum of squares due to error as

$$SSE = SST - SSA - SSB - SSAB$$

$$= 275.67 - 80.67 - 27 - 26 = 142$$

Now that we have computed all the sums of squares, we can compute the mean squares by dividing each sum of square by its corresponding degrees of freedom, where degrees of freedom are calculated as follows.

Source of Variation	Degrees of Freedom
Factor A treatment	a - 1 = 3 - 1 = 2
Factor B treatment	b - 1 = 2 - 1 = 1
Interaction	(a - 1) (b - 1) = (2) (1) = 2
Error	a b (r - 1) = (3) (2) (2 - 1) = 6

With the above information, we can set up the ANOVA table as shown in Table 13.7.

Source of Variation	Sum of Squares	Degrees of Freedom	Mean Square	F
Factor A Treatment	80.67	2	40.33	$\frac{40.33}{23.67} = 1.70$
Factor B Treatment	27.00	1	27.00	$\frac{27.00}{23.67} = 1.14$
Interaction (AB)	26.00	2	13.00	$\frac{13.00}{23.67} = 0.55$
Error	142.00	6	23.67	
Total	275.67			

Table 13.7

Now we are in a position to answer the questions set forth at the beginning of the exercise. The F ratio used to test for a difference among the universities' preparation programs is 1.70. The critical F value at $\alpha = 0.05$ (with 2 numerator degrees of freedom and 6 denominator degrees of freedom) is 5.14. Since $1.70 < 5.14$, we cannot reject the null hypothesis; and we conclude that there is no difference in the preparation provided by the three universities.

The F ratio for the sex of the students is 1.14. The critical F value with 1 numerator and 6 denominator degrees of freedom is 5.99. Since $1.14 < 5.99$, we cannot reject the null hypothesis. Therefore, we conclude that there is no difference in the performance of male and female students.

Finally, the interaction F value is 0.55. The critical F value with 2 numerator and 6 denominator degrees of freedom is 5.14. Since $0.55 < 5.14$, no significant interaction can be identified. Thus, we conclude that there is no reason to believe that the three universities differ in their ability to prepare male and female students.

16. A factorial experiment involving 2 levels of Factor A and 2 levels of Factor B resulted in the following.

		Factor B: Level 1	Factor B: Level 2
Factor A	Level 1	14	18
		16	12
	Level 2	18	16
		20	14

Test for any significant main effect and any interaction effect. Use $\alpha = 0.05$.

SELF-TESTING QUESTIONS

In the following multiple choice questions, circle the correct answer. An answer key is provided following the questions.

1. The F ratio in a completely randomized ANOVA is the ratio of

a) MST/MSE
b) MSE/MSTR
c) MSE/MST
d) MSTR/MSE

2. The critical F value with 8 numerator and 6 denominator degrees of freedom at $\alpha = .05$ is

a) 3.58
b) 4.88
c) 4.15
d) none of the above

3. The ANOVA procedure is a statistical approach for determining whether or not

a) the means of two samples are equal
b) the means of more than two samples are equal
c) the means of two or more populations are equal
d) none of the above

4. The variable of interest in an ANOVA procedure is called

a) a factor
b) a treatment
c) either a or b
d) none of the above

5. An ANOVA procedure is applied to data obtained from 5 samples, where each sample contains 9 observations. The degrees of freedom for the critical value of F are

a) 5 numerator and 9 denominator degrees of freedom
b) 4 numerator and 8 denominator degrees of freedom
c) 45 degrees of freedom
d) 4 numerator and 40 denominator degrees of freedom

6. In the ANOVA, treatment refers to

a) experimental units
b) different levels of a factor
c) a factor
d) none of the above

7. The mean square is the sum of squares divided by

a) the total number of observations
b) its corresponding degrees of freedom - 1
c) its corresponding degrees of freedom
d) none of the above

8. In the factorial designs, the response produced when the treatments of one factor interact with the treatments of another in influencing the response variable is known as

a) the main effect
b) interaction
c) replication
d) none of the above

9. An experimental design where the experimental units are randomly assigned to the treatments is known as

a) factor block design
b) random factor design
c) completely randomized design
d) none of the above

10. The number of times each experimental condition is observed in a factorial design is known as

a) replication
b) the experimental condition
c) a factor
d) none of the above

11. When analysis of variance is performed on samples drawn from K populations, the mean square between treatments (MSTR) is SSTR divided by

a) n_T
b) $n_T - 1$
c) K
d) K - 1
e) none of the above

12. In analysis of variance, where the total sample size for the experiment is n_T and the number of populations is K, the mean square within treatments is computed by dividing SSE by

a) n_T
b) $n_T - K$
c) $n_T - 1$
d) $n_T + K$
e) none of the above

Use the following information to answer questions 13 through 15.

In a completely randomized experimental design involving four treatments, a total of 10 observations were recorded for **each** of the four treatments. The following information is provided.

SSTR = 300 (Sum Square Between Treatments)
SST = 1200 (Total Sum Square)

13. The sum of squares within treatments (SSE) is

a) 1500
b) 900
c) 300
d) 1200
e) none of the above

14. The mean square between treatments (MSTR) is

a) 75
b) 25
c) 100
d) 36
e) none of the above

15. The computed F statistic is

a) 3
b) 4
c) 5
d) 6
e) none of the above

Use the following information to answer questions 16 through 20.

Fourteen observations were selected from each of four populations, and analysis of variance was performed on the data. Part of the ANOVA table is shown below.

Source of Variation	Sum of Squares	Degrees of Freedom	Mean Square	F
Between Treatments	60			
Error (Within Treatments)	260			

16. The number of degrees of freedom corresponding to between treatments is

a) 1
b) 2
c) 3
d) 4
e) none of the above

17. The number of degrees of freedom corresponding to within treatments is

a) 51
b) 52
c) 53
d) 54
e) none of the above

18. The mean square between treatments (MSTR) is

a) 15
b) 20
c) 30
d) 40

19. The mean square within treatments ,or mean square error (MSE) is

a) 2
b) 3
c) 4
d) 5
e) none of the above

20. The computed F statistic is

a) 4
b) 3
c) 2
d) 1
e) none of the above

ANSWERS TO THE SELF-TESTING QUESTIONS

1. d
2. c
3. c
4. a
5. d
6. b
7. c
8. b
9. c
10. a
11. d
12. b
13. b
14. c
15. b
16. c
17. b
18. b
19. d
20. a

ANSWERS TO CHAPTER THIRTEEN EXERCISES

2. $F = 1.8 < 3.68$ Thus, do not reject H_0
   There is no significant difference in the average lives of the three brands of tires.

4. $F = 0.66 < 3.89$ Thus, do not reject H_0
   We cannot conclude that there is a significant difference among the average heating bills.

5.

Source of Variation	Sum of Squares	Degrees of Freedom	Mean Square	F
Between Treatments	173.33	2	86.67	0.66
Error (Within Treatments)	1586.00	12	132.17	
Total	1759.33	14		

7. $F = 0.3683 < 4.26$ Thus, do not reject H_0
   We cannot conclude that there is a significant difference among the average speeds of the cars.

9.

Source of Variation	Sum of Squares	Degrees of Freedom	Mean Square	F
Between Treatments	65.375	2	32.687	0.3863
Error (Within Treatments)	798.625	9	88.694	

Since $F = 0.3683 < 8.02$ do not reject H_0
We cannot conclude that there is a significant difference among the average speeds of the cars.

11. a.

Source of Variation	Sum of Squares	Degrees of Freedom	Mean Square	F
Between Treatments	69.33	2	34.665	6.07
Error (Within Treatments)	120.00	21	5.714	
Total	189.33	23		

Since F = 6.07 > 3.47, reject H_0. At least one mean is different from the others.

(b) LSD = 2.47

$|\bar{x}_1 - \bar{x}_2| = 1$; $|\bar{x}_1 - \bar{x}_3| = 3$; $|\bar{x}_2 - \bar{x}_3| = 4$

Therefore, the mean of population 3 is different from the others.

12. LSD = 22.75

$|\bar{x}_1 - \bar{x}_2| = 32.9$; $|\bar{x}_1 - \bar{x}_3| = 49.7$; $|\bar{x}_2 - \bar{x}_3| = 16.8$

Therefore, the mean of population 1 is different from the others.

14. F = 1.428 < 7.71 Thus, do not reject H_0

16. Factor A treatment F = 1.33 < 7.71 Do not reject H_0, not significant
Factor B treatment F = 1.33 < 7.71 Do not reject H_0, not significant
Interaction (AB) F = 1.33 < 7.71 Do not reject H_0, not significant

CHAPTER FOURTEEN

SIMPLE LINEAR REGRESSION

CHAPTER OUTLINE AND REVIEW

In this chapter, you have been introduced to two statistical techniques, namely, regression and correlation. Regression and correlation are used for analyzing the relationship between variables by determining the best mathematical expression which describes their relationship and by measuring the strength of this relationship. The terms and concepts which you have learned include

A. **Independent Variable:** The variable that is doing the predicting or explaining. It is denoted by x in the regression equation.

B. **Dependent Variable:** The variable that is being predicted or explained. It is denoted by y in the regression equation.

C. **Simple Linear Regression:** The simplest kind of regression, involving only two variables (one independent and one dependent variable). The relationship between variables is approximated by a straight line.

D. **Regression Model:** The probability model describing how the dependent variable (y) is related to the independent variable (x) in simple linear regression. The regression model has the form $y = \beta_0 + \beta_1 x + \varepsilon$.

E. **Regression Equation:** The mathematical equation relating the independent variable to the expected value of the dependent variable; that is, $E(y) = \beta_0 + \beta_1 x$.

F. **Estimated Regression Equation:** The estimate of the regression equation developed from sample data using the least squares method; that is, $\hat{y} = b_o + b_1 x$.

G. **Scatter Diagram:** A graph of the data in which the independent variable appears on the horizontal axis and the dependent variable appears on the vertical axis.

H. **Least Squares Method:** The approach used to develop the estimated regression equation which minimizes the sum of squares of the vertical distances from the points to the least squares fitted line. That is, it minimizes $\Sigma(y_i - \hat{y}_i)^2$.

I. **Coefficient of Determination (r^2):** A measure of the proportion of the variation in the dependent variable that is explained by the estimated regression equation. It is a measure of how well the estimated regression equation fits the data.

J. **Correlation Coefficient (r):** A statistical measure of the strength of the linear relationship between two variables.

K. **Mean Square Error:** The unbiased estimate of the variance, σ^2, of the error term ε. It is denoted by MSE or s^2.

L. **Standard Error of the Estimate:** The square root of the mean square error, denoted by s. It is the estimate of σ, the standard deviation of the error term ε.

M. **ANOVA Table:** The analysis of variance table used to summarize the computations associated with the F test for significance.

N. **Confidence Interval Estimate:** The interval estimate of the mean value of y for a given value of x.

O.	**Prediction Interval Estimate:**	The interval estimate of an individual value of y for a given value of x.
P.	**Residual:**	The difference between the observed value of the dependent variable and the value predicted using the estimated regression equation; i.e., $y_i - \hat{y}_i$.
Q.	**Residual Analysis:**	The analysis of the residuals used to determine if the assumptions made about the regression model appear to be valid. Residual analysis is also used to identify unusual and influential observations.
R.	**Residual Plots:**	Graphical representations of the residuals that can be used to determine if the assumptions made about the regression model appear to be valid.
S.	**Standardized Residual:**	The value obtained by dividing the residual by its standard deviation.
T.	**Normal Probability Plot:**	A graph of normal scores plotted against values of the standardized residuals. This plot helps determine if the assumption that the error term has a normal probability distribution appears to be valid.
U.	**Outlier:**	A data point or observation that is unusual compared to the remaining data.
V.	**Influential Observation:**	An observation that has a strong influence or effect on the regression results.
W.	**Leverage:**	A measure of the influence an observation has on the regression results. Influential observations have high leverage.

CHAPTER FORMULAS

Simple Linear Regression Model

$$y = \beta_0 + \beta_1 x + \varepsilon \qquad (14.1)$$

Simple Linear Regression Equation

$$E(y) = \beta_0 + \beta_1 x \qquad (14.2)$$

Least Squares Criterion

$$\text{Min} \sum (y_i - \hat{y}_i)^2 \qquad (14.5)$$

Estimated Simple Linear Regression Equation

$$\hat{y} = b_o + b_1 x \qquad (14.3)$$

where $\hat{y}$ = the estimated value of the dependent variable

b_0 = the y-intercept

b_1 = the slope of the line

and b_0 and b_1 are computed as

$$b_1 = \frac{\sum x_i y_i - \dfrac{\sum x_i \sum y_i}{n}}{\sum x_i^2 - \dfrac{(\sum x_i)^2}{n}} \qquad (14.6)$$

$$b_o = \bar{y} - b_1 \bar{x} \qquad (14.7)$$

Sum of Squares Due to Error

$$SSE = \sum (y_i - \hat{y}_i)^2 \qquad (14.8)$$

CHAPTER FORMULAS
(Continued)

Sum of Squares Due to Regression

$$SSR = \sum (\hat{y} - \bar{y})^2 \qquad (14.10)$$

For computational efficiency, use the following to compute SSR.

$$SSR = \frac{\left(\sum x_i y_i - \frac{\sum x_i \sum y_i}{n} \right)^2}{\sum x_i^2 - \frac{(\sum x_i)^2}{n}} \qquad (14.14)$$

Total Sum of Squares

$$SST = \sum (y_i - \bar{y})^2 \qquad (14.9)$$

For computational efficiency, use the following to compute SST.

$$SST = \sum y_i^2 - \frac{(\sum y_i)^2}{n} \qquad (14.13)$$

Relationship Among SST, SSR, and SSE

$$SST = SSR + SSE \qquad (14.11)$$

Coefficient of Determination

$$r^2 = \frac{SSR}{SST} \qquad (14.12)$$

Sample Correlation Coefficient

$$r_{xy} = (\text{the sign of } b_1) \sqrt{\text{Coefficient of Determination}} = \pm \sqrt{r^2} \qquad (14.15)$$

where b_1 = the slope of the regression equation

CHAPTER FORMULAS
(Continued)

Testing for Significance of Correlation Coefficient

$H_o: \rho_{xy} = 0$

$H_a: \rho_{xy} \neq 0$

t statistic: $t = r_{xy}\sqrt{\dfrac{n-2}{1-r_{xy}^2}}$ (Appendix 14.2)

Reject H_O if $t < -t_{\alpha/2}$ or: $t > t_{\alpha/2}$ (degrees of freedom = n - 2)

Mean Square Error (Estimate of σ^2)

$$s^2 = MSE = \frac{SSE}{n-2} \quad (14.17)$$

Standard Error of the Estimate

$$s = MSE = \sqrt{\frac{SSE}{n-2}} \quad (14.18)$$

t Test for significance in Simple Linear Regression (Testing for the Significance of the Slope)

$H_o: \beta_1 = 0$

$H_a: \beta_1 \neq 0$

t statistic: $t = \dfrac{b_1}{s_{b_1}}$ (14.21)

where s_{b_1} (Estimated Standard Deviation of b_1) is

$$s_{b_1} = \frac{s}{\sqrt{\Sigma x_i^2 - (\Sigma x_i)^2/n}} \quad (14.20)$$

Reject H_O if $t < -t_{\alpha/2}$ or: $t > t_{\alpha/2}$ (degrees of freedom = n - 2)

CHAPTER FORMULAS
(Continued)

F Test for Significance in Simple Linear Regression

$H_o: \beta = 0$

$H_a: \beta \neq 0$

$$F = \frac{MSR}{MSE} \tag{14.23}$$

where MSR (Mean Square Due to Regression) is

$$MSR = \frac{SSR}{\text{Number of Independent Variables}} \tag{14.22}$$

Reject H_o if $F > F_\alpha$

degrees of freedom = n - 2

Confidence Interval Estimate of $E(y_p)$ (for the Mean Value of y)

$$\hat{y}_p \pm t_{\alpha/2} s_{\hat{y}_p} \tag{14.26}$$

where Estimated Standard Deviation of $\hat{y}_p$ is

$$s_{\hat{y}_p} = s\sqrt{\frac{1}{n} + \frac{(x_p - \bar{x})^2}{\Sigma x_i^2 - \frac{(\Sigma x_i)^2}{n}}} \tag{14.25}$$

CHAPTER FORMULAS
(Continued)

Confidence Interval Estimate of y_p (for an Individual Value of y)

$$\hat{y}_p \pm t_{\alpha/2} s_{ind}$$

where

$$s_{ind} = s\sqrt{1 + \frac{1}{n} + \frac{(x_p - \bar{x})^2}{\sum x_i^2 - \frac{(\sum x_i)^2}{n}}} \qquad (14.28)$$

Residual for Observation i

$$y_i - \hat{y}_i \qquad (14.30)$$

Standard Deviation of the ith Residual

$$s_{y_i - \hat{y}_i} = s\sqrt{1 - h_i} \qquad (14.32)$$

where h_i is the leverage for Observation i.

$$h_i = \frac{1}{n} + \frac{(x_i - \bar{x})^2}{\sum (x_i - \bar{x})^2} \qquad (14.33)$$

Standardized Residual for Observation i

$$\frac{y_i - \hat{y}_i}{s_{y_i - \hat{y}_i}} \qquad (14.34)$$

EXERCISES

***1.** Shultz, Inc. is a large carpet manufacturing firm. The following data represent Shultz's yearly sales volume and its advertising expenditures over a period of 10 years. **Note: Since the topics in this chapter are interrelated, the remaining exercises in this chapter that are marked with an "*" are based on this exercise and have been worked for you.**

Year	Sales Volume (y) ($ Millions)	Advertising Expenditure (x) ($ Millions)
1986	26	1.8
1987	31	2.3
1988	28	2.6
1989	30	2.4
1990	34	2.8
1991	38	3.0
1992	41	3.4
1993	44	3.2
1994	40	3.6
1995	43	3.8

(a) Develop a scatter diagram for the above data.

Answer: Plotting the advertising expenditures on the horizontal axis, the scatter diagram is shown in Figure 14.1.

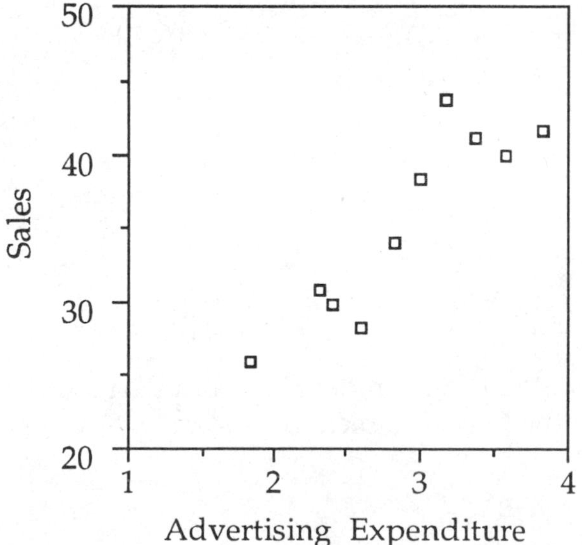

Figure 14.1

(b) What does the scatter diagram developed in Part a indicate about the relationship between the two variables?

Answer: The scatter diagram shows that there is a positive relationship between the variables. The overview of the data shows that as advertising expenditures increase so does the sales volume.

2. Assume you have noted the following prices for books and the number of pages that each book contains. **Note: Exercises 5, 8, 11, 14, 16, and 18 are a continuation of this exercise and use the same data set.**

Book	Pages (x)	Price (y)
A	500	$7.00
B	700	7.50
C	750	9.00
D	590	6.50
E	540	7.50
F	650	7.00
G	480	4.50

(a) Develop a scatter diagram for the above data with the number of pages on the horizontal axis.

(b) What does the scatter diagram developed in Part a indicate about the relationship between the two variables?

3. The following data represent the number of weed-eaters sold per month at a local garden shop and their prices. **Note: Exercises 6, 9, 12, and 19 are a continuation of this exercise and use the same data set.**

Price (x)	Units Sold (y)
$34	3
36	4
32	6
35	5
30	9
38	2
40	1

(a) Develop a scatter diagram for the above data with the prices on the horizontal axis.

(b) What does the scatter diagram developed in Part a indicate about the relationship between the two variables?

***4. Refer to exercise 1.**

(a) What kind of model would be appropriate for representing the relationship between the two variables?

Answer: Since one cannot guarantee a single value of y for each value of x, the relationship between the variables is explained by a probabilistic model. Noting the relationship between sales volume (y) and advertising expenditure (x), one can assume that this relationship can be approximated by a straight line in the form of

$$y = \beta_0 + \beta_1 x + \varepsilon$$

where β_0 = y intercept.
β_1 = the slope of the line.
ε = the error, or deviation, of the actual y value from the line given by $\beta_0 + \beta_1 x$.

The above probabilistic model is known as a regression model which provides a good approximation of the y value at each x. β_0 and β_1 are known as the parameters of the model.

(b) In the regression model shown in Part a, what assumptions are made about the error term ε in the regression model $y = \beta_0 + \beta_1 x + \varepsilon$?

Answer: The 4 major assumptions about the regression model are as follows.

1. The error term ε is a random variable with a mean or expected value of zero. That is, $E(\varepsilon) = 0$. Since $E(\varepsilon) = 0$, then $E(y) = \beta_0 + \beta_1 x$.

2. The variance of ε is the same for all values of x.

3. The values of the error are independent. That is, the size of error for a particular value of x is not related to the size of error for any other value of x.

4. The error term ε is a normally distributed random variable which can take on negative or positive values. Thus, the error term represents the deviation between the y value and the value given by $\beta_0 + \beta_1 x$. Since y is a

linear function of ε, the above implies that y is also a normally distributed random variable.

(c) Develop a least-squares estimated regression line.

Answer: In Part a, it was assumed that the regression model $y = \beta_0 + \beta_1 x + \varepsilon$ related x and y. Furthermore, it was assumed $E(\varepsilon) = 0$. Thus, the regression model became $E(y) = \beta_0 + \beta_1 x$. Now we can use the least squares method to estimate β_0 and β_1, thereby determining the following estimated regression equation:

$$\hat{y} = b_0 + b_1 x$$

where $\hat{y}$ = estimate of E(y)
b_0 = estimate of β_0
b_1 = estimate of β_1

The least squares method determines the estimated regression equation which minimizes the sum of squares of the differences between the observed values of the dependent variable (y_i) and the estimated values of the dependent variable ($\hat{y}_i$). That is, the values of b_0 and b_1 are determined in such a way that

$$\sum_{i=1}^{n} (y_i - \hat{y}_i)^2 \text{ is minimized.}$$

Using differential calculus, it is determined that the following b_0 and b_1 minimizes the above sum of squares of the differences.

$$b_1 = \frac{\sum x_i y_i - \frac{\sum x_i \sum y_i}{n}}{\sum x_i^2 - \frac{(\sum x_i)^2}{n}}$$

$$b_o = \bar{y} - b_1 \bar{x}$$

where x_i = the value of the independent variable for the ith observation
y_i = the value of the dependent variable for the ith observation
$\bar{x}$ = the mean value for the independent variable
$\bar{y}$ = the mean value for the dependent variable
n = the total number of observations

Thus, we need the following calculations for estimating the regression line:

	x_i	y_i	$x_i\ y_i$	x_i^2
	1.8	26	46.8	3.24
	2.3	31	71.3	5.29
	2.6	28	72.8	6.76
	2.4	30	72.0	5.76
	2.8	34	95.2	7.84
	3.0	38	114.0	9.00
	3.4	41	139.4	11.56
	3.2	44	140.8	10.24
	3.6	40	144.0	12.96
	3.8	43	163.4	14.44
Totals	28.9	355	1,059.7	87.09

The values of $\bar{x}$ and $\bar{y}$ are determined to be

$$\bar{x} = \frac{\Sigma x}{n} = \frac{28.9}{10} = 2.89$$

$$\bar{y} = \frac{\Sigma y}{n} = \frac{355}{10} = 35.5$$

Now using the above totals and the means we can compute the slope (b_1) as

$$b_1 = \frac{\Sigma x_i y_i - \dfrac{\Sigma x_i \Sigma y_i}{n}}{\Sigma x_i^2 - \dfrac{(\Sigma x_i)^2}{n}} = \frac{(1059.7) - \dfrac{(28.9)\ (355)}{10}}{(87.09) - \dfrac{(28.9)^2}{10}} = 9.4564$$

Hence, the value of b_o is computed as

$$b_o = \bar{y} - b_1 \bar{x} = 35.5 - (9.4564)\,(2.89) = 8.171$$

Therefore, the estimated regression function is

$$\hat{y} = 8.171 + 9.4564\ x_1$$

The slope of the estimated regression line is 9.4564. Since the slope is positive, it implies that as advertising expenditure increases, sales volume is expected to increase. Since both advertising expenditure and sales volume were measured in millions of dollars, it can be concluded that as advertising expenditure increases by 1 million dollars, sales volume is expected to increase by 9.4565 million dollars.

5. Refer to exercise 2. Develop a least-squares estimated regression line.

6. Refer to exercise 3. Develop a least-squares regression line and explain what the slope of the line indicates.

***7. Refer to exercise 4.**

(a) Compute the coefficient of determination, and comment on the strength of the relationship between advertising and sales.

Answer: The coefficient of determination (r^2) is a measure indicating how well the regression equation fits the observed data. The coefficient of determination is computed as

$$r^2 = \frac{SSR}{SST}$$

where SSR = the sum of squares explained by regression
SST = the total sum of squares

That is

$$SSR = \Sigma\left(\hat{y} - \bar{y}\right)^2$$

$$SST = \Sigma\left(y_i - \bar{y}\right)^2$$

The values of SSR and SST can be computed more efficiently by the following equations.

$$SSR = \frac{\left(\Sigma x_i y_i - \dfrac{\Sigma x_i \Sigma y_i}{n}\right)^2}{\Sigma x_i^2 - \dfrac{(\Sigma x_i)^2}{n}}$$

Note: The numerator of the equation for SSR is the square of the numerator of the equation for b_1, and the denominator of SSR is the same denominator as that of b_1.

$$SST = \Sigma y_i^2 - \frac{(\Sigma y_i)^2}{n}$$

In Part c of exercise 4, the following values were computed.

$\Sigma x_i = 28.9$ $\Sigma y_i = 355$

$\Sigma x_i y_i = 1059.7$ $\Sigma x_i^2 = 87.09$

Now SSR can be computed as

$$SSR = \frac{\left(\Sigma x_i y_i - \frac{\Sigma x_i \Sigma y_i}{n}\right)^2}{\Sigma x_i^2 - \frac{(\Sigma x_i)^2}{n}} = \frac{\left(1059.7 - \frac{(28.9)(355)}{10}\right)^2}{87.09 - \frac{(28.9)^2}{10}} = \frac{(33.750)^2}{3.569}$$
$$= 319.15$$

To calculate SST the following computations are needed.

y_i	y_i^2
26	676
31	961
28	784
30	900
34	1156
38	1444
41	1681
44	1936
40	1600
43	1849
$\Sigma y_i = 355$	$\Sigma y_i^2 = 12987$

Thus, SST is computed as

$$SST = \Sigma y_i^2 - \frac{(\Sigma y_i)^2}{n} = 12{,}987 - \frac{(355)^2}{10} = 384.5$$

Now the coefficient of determination is computed as

$$r^2 = \frac{SSR}{SST} = \frac{319.5}{384.5} = 0.83$$

Therefore, it can be concluded that the estimated regression function has accounted for 83% of the total sum of the squares, which indicates that it provides a good fit for the data.

(b) Compute the sample correlation coefficient between the sales volumes and the advertising expenditures.

Answer: The sample correlation coefficient is simply the square root of the coefficient of determination. In Part a of this exercise the coefficient of determination was computed to be 0.83. Therefore the coefficient of correlation is

$$r_{xy} = (\text{the sign of } b_1)\ \sqrt{\text{Coefficient of Determination}} = +\ \sqrt{0.83} = 0.91$$

(c) At 95% confidence test the following hypotheses. **Note: This topic is covered in Appendix 14.2 of you text.**

$$H_o: \rho_{xy} = 0$$
$$H_a: \rho_{xy} \neq 0$$

Answer: The test statistic is

$$t = r_{xy}\sqrt{\frac{n-2}{1-r^2}} = (0.91)\sqrt{\frac{10 - 2}{1 - (0.91)^2}} = 6.2$$

From Table 2 in Appendix B, we read the t value with n - 2 = 10 - 2 = 8 degrees of freedom as $t_{.025} = \pm 2.306$. Since 6.2 > 2.306, the null hypothesis is rejected; and we conclude that x and y are related.

8. Refer to exercise 5.

(a) Compute the coefficient of determination and explain its meaning.

(b) Compute the correlation coefficient between the price and the number of pages. Test to see if x and y are related. Use $\alpha = 0.10$.

Note: This topic is covered in Appendix 14.2 of your text and the solution procedure is shown to you in Parts b and c of exercise 7.

9. Refer to exercise 6.

(a) Compute the coefficient of determination and comment on the strength of relationship between x and y.

(b) Compute the sample correlation coefficient between the price and the number of weed-eaters sold. Use a = 0.01 to test the relationship between x and y. **Note: This topic is covered in Appendix 14.2 of you text and the solution procedure is shown to you in Parts b and c of exercise 7.**

***10. Refer to exercise 7.**

(a) Using an F test, determine if the advertising expenditures and the sales volumes are related. Let $\alpha = 0.05$.

Answer: If a relationship between x and y of the form $E(y) = \beta_0 + \beta_1 x$ really exists, then β_1 must be different from zero. Thus, the hypotheses to be tested are

H_0: $\beta_1 = 0$ (variables are not related)

H_a: $\beta_1 \neq 0$ (variables are related)

First we need to compute an F value as

$$F = \frac{MSR}{MSE}$$

Then, if this F value is less than or equal to the critical F value (F_a), the null hypothesis will not be rejected. In exercise 7 we computed

SSR = 319.15

SST = 384.5

Now we can compute SSE as

SSE = SST - SSR= 384.5 - 319.15 = 65.35

Then, the mean squares are computed as

$$MSR = \frac{SSR}{\text{Number of Independent Variables}}$$

Since there is one independent variable, MSR will be

$$MSR = \frac{319.15}{1} = 319.15$$

and

$$MSE = \frac{SSE}{n - 2} = \frac{65.35}{10 - 2} = 8.1687$$

Now the F statistic is computed as

$$F = \frac{319.15}{8.1687} = 39.07$$

From Table 4 of Appendix B, the critical value of F_α with 1 numerator degree of freedom (the number of independent variables) and n - 2 = 10 - 2 = 8 denominator degrees of freedom is determined to be $F_{.05} = 5.32$. Since 39.07 > 5.32, the null hypothesis is rejected, and we conclude that there is a significant statistical relationship between sales volume and advertising expenditure.

(b) Use a t test to answer the question asked in Part a.

Answer: The hypotheses to be tested are the same as shown in Part a, which are

H_0: $\beta_1 = 0$ (variables are not related)

H_a: $\beta_1 \neq 0$ (variables are related)

The test statistic is

$$t = \frac{b_1}{s_{b_1}}$$

where s_{b_1} (estimated standard deviation of b_1) is

$$s_{b_1} = \frac{s}{\sqrt{\Sigma x_i^2 - (\Sigma x_i)^2/n}}$$

In Part a of this exercise, MSE was computed to be 8.1687. Using MSE, first we can compute the numerator of the above (s) as

$$s = \sqrt{MSE} = \sqrt{8.1687} = 2.8581$$

In Part c of exercise 4, the following values were computed.

$$\Sigma x_i = 28.9$$

$$\Sigma x_i^2 = 87.09$$

Now we can compute s_{b_1} as

$$s_{b_1} = \frac{s}{\sqrt{\Sigma x_i^2 - (\Sigma x_i)^2/n}} = \frac{2.8581}{\sqrt{87.09 - (28.9)^2/10}} = 1.5128$$

Now we can compute the t statistic. (Recall that b_1 was calculated in Part c of exercise 4 as 9.4564.)

$$t = \frac{b_1}{s_{b_1}} = \frac{9.4564}{1.5128} = 6.25$$

The decision rule for this t test is

Reject H_0 if $t < -t_{\alpha/2}$ or $t > t_{\alpha/2}$

From Table 2 of Appendix B, we find that $t_{\alpha/2}$ with 8 degrees of freedom (n - 2) is $t_{.025} = 2.306$. Since 6.25 > 2.306, the null hypothesis is rejected, and we conclude that there is a significant statistical relationship between sales volume and advertising expenditure. Note that this is the same conclusion we had in Part a.

11. Refer to exercise 8.

(a) Perform an F test and determine if the price and the number of pages of the books are related. Let $\alpha = 0.01$.

(b) Perform a t test and determine if the price and the number of pages of the books are related. Let $\alpha = 0.01$.

12. Refer to exercise 9.

(a) Perform an F test and determine if the price and the number of weed-eaters sold are related. Let $\alpha = 0.01$

(b) Perform a t test and determine if the price and the number of weed-eaters sold are related. Let $\alpha = 0.01$

***13. Refer to exercises 4 and 10.**

(a) Determine the expected sales volume when the advertising expenditures are 3.5 million dollars.

Answer: With advertising expenditures of 3.5 million dollars, the expected sales volume is $\hat{y}_p$.

$$\hat{y}_p = 8.171 + 9.4564\,(3.5) = 41.2684$$

Thus, we conclude that the expected sales volume is $41,268,400.

(b) Develop a 95% confidence interval for estimating the mean sales volume with advertising expenditures of 3.5 million dollars.

Answer: The estimated standard deviation of $\hat{y}_p$ is

$$s_{\hat{y}_p} = s\sqrt{\frac{1}{n} + \frac{\left(x_p - \bar{x}\right)^2}{\sum x_i^2 - \frac{(\sum x_i)^2}{n}}}$$

Recall that in Part c of exercise 4 we have calculated the following values.

$$\sum x_i = 28.9$$

$$\sum x_i^2 = 87.09$$

And in Part b of exercise 10, we determined s = 2.8581. Thus, we can calculate the standard deviation of $\hat{y}_p$ as

$$s_{\hat{y}_p} = s\sqrt{\frac{1}{n} + \frac{\left(x_p - \bar{x}\right)^2}{\sum x_i^2 - \frac{(\sum x_i)^2}{n}}} = 2.8581\sqrt{\frac{1}{10} + \frac{(3.5 - 2.89)^2}{87.09 - (28.9)^2/10}}$$

$$= 1.29 \text{ (rounded)}$$

Therefore, we can find an interval estimate for $E(y_p)$ as

$$\hat{y}_p \pm t_{a/2}\, s_{\hat{y}_p}$$

From Table 2 of Appendix B, we can read the t value with n - 2 = 10 - 2 = 8 degrees of freedom as $t_{.025} = 2.306$. Hence, the confidence interval is as follows.

(Note: $\hat{y}_p$ was determined in Part a of this exercise.)

$$\hat{y}_p \pm t_{\alpha/2}\, s_{\hat{y}_p}$$

$$41.2684 \pm (2.306)\,(1.29) =$$

$$41.2684 \pm 2.9747$$

Thus, the 95% confidence interval estimate in terms of dollars is from \$38,293,700 to \$44,243,100.

14. Refer to exercises 5 and 11. Develop a 90% confidence interval for estimating the average price of books which contain 800 pages.

***15. Refer to exercise 13.** Develop a 95% confidence interval to estimate the sales volume for a specific advertising expenditure of 3.5 million dollars.

Answer: In exercise 13, we determined an interval for the average sales volume. In exercise 15, we are interested in determining an interval for the sales volume corresponding to a specific advertising expenditure of 3.5 million dollars. The procedure is basically the same as that which was shown in exercise 13. However, the standard deviation is

$$s_{ind} = s\sqrt{1+\frac{1}{n}+\frac{(x_p-\bar{x})^2}{\sum x_i^2-\frac{(\sum x_i)^2}{n}}} = 2.8581\sqrt{1+\frac{1}{10}+\frac{(3.5-2.89)^2}{87.09-(28.9)^2/10}}$$

$$= 3.14 \text{ (rounded)}$$

Thus, the interval estimate is

$$\hat{y}_p \pm t_{a/2}\, s_{ind}$$

$$41.2684 \pm (2.306)(3.14) = 41.2684 \pm 7.2408$$

Therefore, the 95% confidence interval estimate in terms of dollars is from \$34,027,600 to \$48,509,200.

16. As an extension of exercise 14, develop a 90% confidence interval to estimate the price of a specific book which has 800 pages.

***17. Refer to exercise 4.**

(a) Prepare a plot of the residuals against the independent variable x and comment on it.

Answer: The residuals are computed as shown in Table 14.1.

(Note: $\hat{y} = 8.171 + 9.4564\,x$, and $\hat{y}$ values are computed by substituting x values in this equation.)

x_i	y_i	$\hat{y}_i$	$(y_i - \hat{y}_i)$
1.8	26	25.1925	0.8075
2.3	31	29.9207	1.0793
2.6	28	32.7576	-4.7576
2.4	30	30.8664	-0.8664
2.8	34	34.6489	-0.6489
3.0	38	36.5402	1.4598
3.4	41	40.3228	0.6772
3.2	44	38.4315	5.5685
3.6	40	42.2140	-2.2140
3.8	43	44.1053	-1.1053

Table 14.1

Now we plot the x_i on the horizontal and $(y_i - \hat{y}_i)$ on the vertical axis, thus producing the residual plot as shown in Figure 14.2

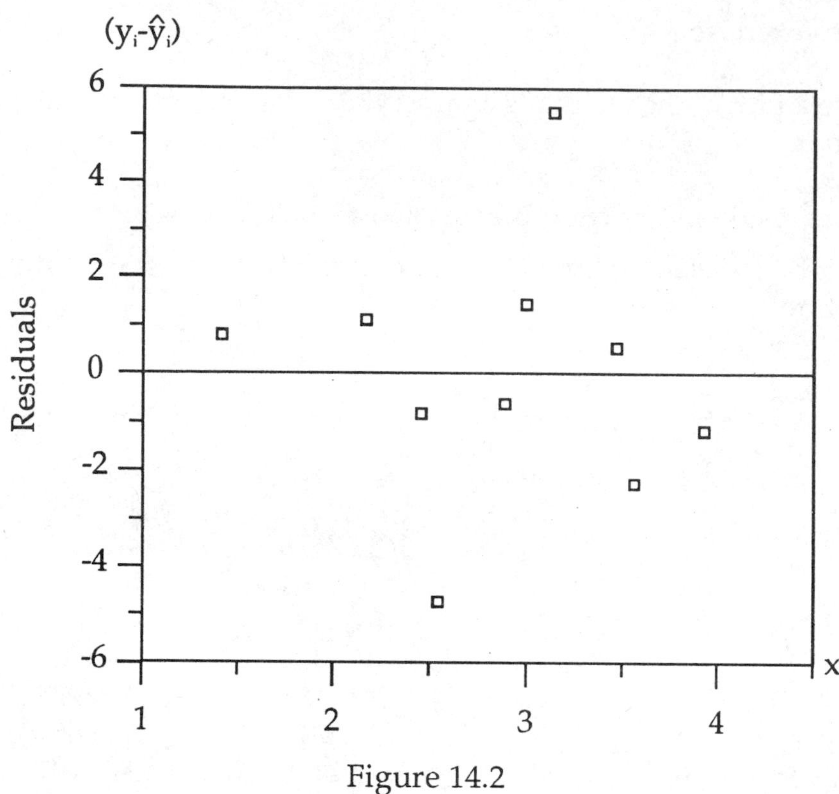

Figure 14.2

Note that the largest residual is 5.5685 and the smallest is - 4.7576. Also, residuals appear to be randomly distributed with no particular pattern. Therefore, it can be concluded that the assumptions made in Part b of exercise 4 are satisfied and the proposed linear relationship between x and y is an appropriate model.

(d) Prepare a residual plot of $\hat{y}$ versus $(y - \hat{y})$ and comment on it.

Answer: The residuals are computed as shown in Table 14.1. Now we plot the $\hat{y}_i$ on the horizontal and $(y_i - \hat{y}_i)$ on the vertical axis. This residual plot is shown in Figure 14.3.

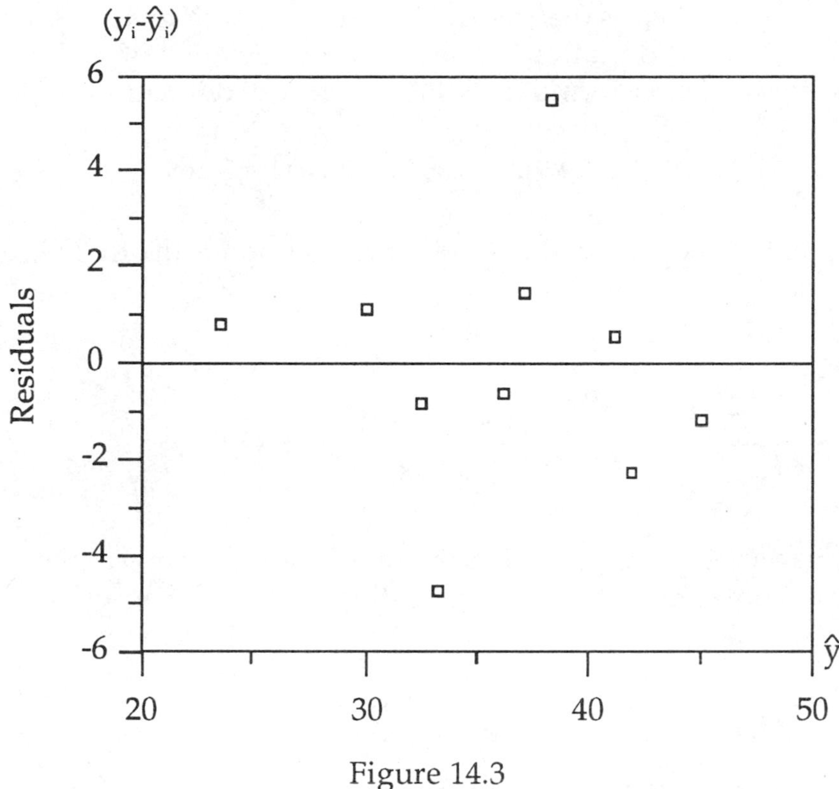

Figure 14.3

As you note, the pattern of this residual plot is the same as the residual plot shown in Figure 14.2. In fact, either residual plot can be constructed, and the same conclusions can be drawn.

(e) Construct a residual plot of the standardized residuals against the independent variable.

Answer: Any random variable is standardized by subtracting its mean from it and dividing the results by its standard deviation. Since the mean of the residuals is zero, we can simply divide each residual by its standard deviation and arrive at the standardized values. The standard deviation of the ith residual is given by

$$S_{y_i - \hat{y}_i} = S\sqrt{1 - h_i}$$

where $h_i = \frac{1}{n} + \frac{\left(x_i - \bar{x}\right)^2}{\Sigma\left(x_i - \bar{x}\right)^2}$

The above formula indicates that residuals corresponding to different values of x have different standard deviations. The standard deviation of each residual is computed using the above formula. Note that the denominator of h_i is simply

$$\Sigma(x_i - \bar{x})^2 = (1.8\text{-}2.89)2 + (2.3\text{-}2.89)2 + \ldots + (3.8\text{-}2.89)2 = 3.569$$

As an example, let us compute the standard deviation for the first observation (i.e., x = 1.8). First, we can compute h_1 as

$$h_1 = \frac{1}{n} + \frac{(x_1 - \bar{x})^2}{\Sigma(x_i - \bar{x})^2} = \frac{1}{10} + \frac{(1.8 - 2.89)^2}{3.569} = 0.4329$$

All other values of h_i are computed in the same manner and are shown in Table 14.2. Next, we compute the standard deviation for the first observation as

$$S_{y_i - \hat{y}_i} = S\sqrt{1 - h_i} = 2.858\sqrt{(1 - 0.4329)} = 2.1523$$

For the above computations, the value of S = 2.8581, which was computed in Part b of exercise 10. All other variances and ultimately the standard deviations are computed in the same manner and their values are shown in Table 14.2.

x_i	y_i	$\hat{y}$	$y_i - \hat{y}$	h_i	Standard Deviation	Standardized Residual
1.8	26	25.1925	0.8075	0.4329	2.1523	0.3752
2.3	31	29.9207	1.0793	0.1975	2.5603	0.4215
2.6	28	32.7576	-4.7576	0.1236	2.6757	-1.7781
2.4	30	30.8664	-0.8664	0.1673	2.6081	-0.3322
2.8	34	34.6489	-0.6489	0.1023	2.7080	-0.2396
3.0	38	36.5402	1.4598	0.1034	2.7063	0.5394
3.4	41	40.3228	0.6772	0.1729	2.5993	0.2605
3.2	44	38.4315	5.5685	0.1269	2.6706	2.0852
3.6	40	42.2140	-2.2140	0.2412	2.4896	-0.8893
3.8	43	44.1053	-1.1053	0.3320	2.3359	-0.4732

Table 14.2

Finally, the standardized residuals are computed by dividing each residual by its corresponding standard deviation. These values are also shown in Table 14.2. Now we can construct the plot of the standardized residuals against x values as shown in Figure 14.4.

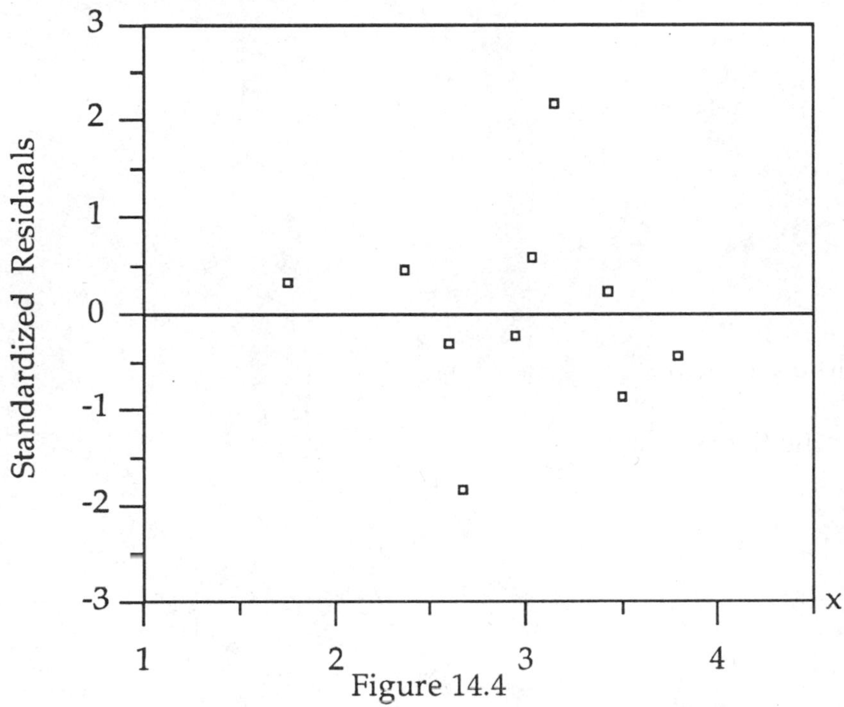

Figure 14.4

As you note, the standardized residual plot has the same general pattern as the original residual plot shown in Figure 14.2.

18. Refer to Exercise 5.

(a) Determine the residuals.

(b) Prepare a residual plot of x versus $(y - \hat{y})$.

19. Refer to Exercise 6.

(a) Determine the residuals.

(b) Prepare a residual plot of x versus $(y - \hat{y})$.

SELF-TESTING QUESTIONS

In the following multiple choice questions, circle the correct answer. An answer key is provided following the questions.

1. In regression analysis, the variable that is being predicted is

a) the independent variable
b) the dependent variable
c) usually denoted by x
d) none of the above

2. In the regression equation $y = b_0 + b_1x$, b_0 is

a) the slope of the line
b) the independent variable
c) the y intercept
d) none of the above

3. In the regression equation (given in question 2), b_1 is

a) the slope of the line
b) an independent variable
c) the y intercept
d) none of the above

4. In regression analysis, the variable that is doing the predicting or explaining is

a) the independent variable
b) usually denoted by y
c) the dependent variable
d) none of the above

5. The coefficient of determination is

a) the square root of the correlation coefficient
b) usually less than zero
c) the correlation coefficient squared
d) none of the above

6. The value of the coefficient of determination ranges between

a) -1 to +1
b) -1 to 0
c) 1 to infinity
d) 0 to +1

7. The coefficient of correlation

a) is the coefficient of determination squared
b) is the square root of the coefficient of determination
c) can never be negative
d) none of the above

8. The range of the correlation coefficient is

a) 0 to +1
b) -1 to 0
c) -1 to infinity
d) -1 to +1

9. If the slope of the regression equation $\hat{y} = b_0 + b_1x$ is positive, then

a) as x increases y decreases
b) as x increases so does y
c) Either a or b is correct.
d) none of the above

10. The residual refers to

a) $\bar{y}_i - \hat{y}_i$
b) $y_i - \hat{y}_i$
c) $\hat{y}_i - \bar{y}_i$
d) none of the above

ANSWERS TO THE SELF-TESTING QUESTIONS

1. b
2. c
3. a
4. a
5. c
6. d
7. b
8. d
9. b
10. b

ANSWERS TO CHAPTER FOURTEEN EXERCISES

2. (b) There appears to be a positive relationship between x and y.

3. (b) There appears to be a negative relationship between x and y.

5. $\hat{y} = 1.0416 + 0.0099x$

6. $\hat{y} = 29.7857 - 0.7286x$
   The slope indicates that as the price goes up by $1, the number of units sold goes down by 0.7286 units.

8. (a) $r^2 = .5629$; the regression equation has accounted for 56.29% of the total sum of squares.
   (b) $r_{xy} = 0.75$
   $t = 2.53 > 2.015$ Reject H_0, and conclude x and y are related.

9. (a) $r^2 = .8556$; the regression equation has accounted for 85.56% of the total sum of squares.
   (b) $r_{xy} = -0.92$
   $t = -5.41 < -4.032$; Reject H_0, and conclude x and y are related.

11. (a) $F = 6.439 < 16.26$; Do not reject H_0, x and y are not related.
    (b) $t = 2.5376 < 4.032$; Do not reject H_0, x and y are not related.

12. (a) $F = 29.624 > 16.26$; Reject H_0, x and y are related.
    (b) $t = -5.4428 < -4.032$; Reject H_0, x and y are related.

14. $7.29 to $10.63 (rounded)

16. $5.62 to $12.31 (rounded)

18. (a) Residuals

1.00487
-0.476563
0.528078
-0.386777
1.10858
-0.481206
-1.29699

(b)

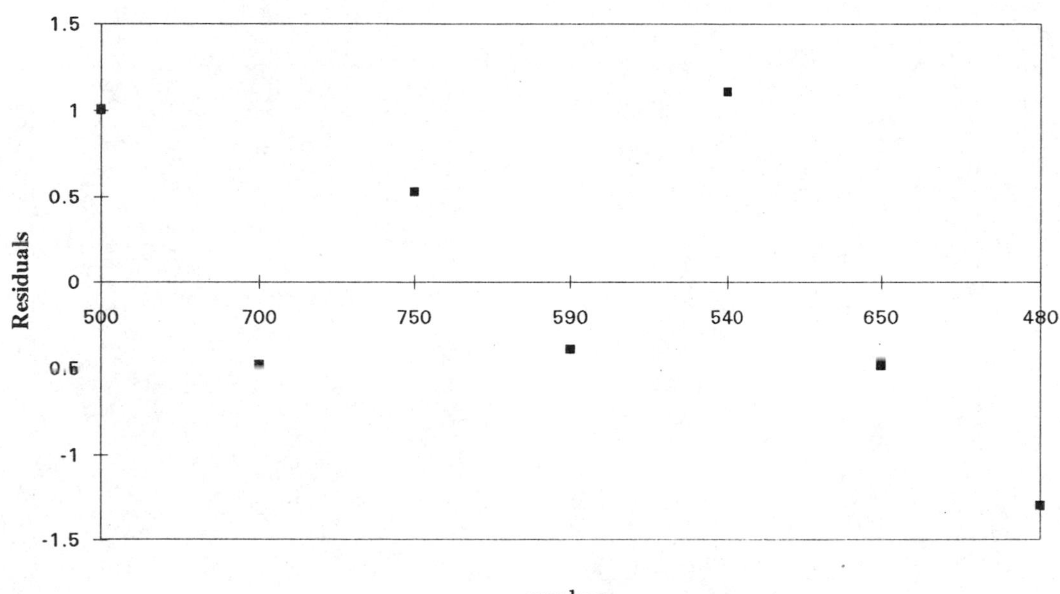

19. (a)

Residuals
-2.01429
0.44285
-0.47144
0.71428
1.07142
-0.10000
0.35714

(b)

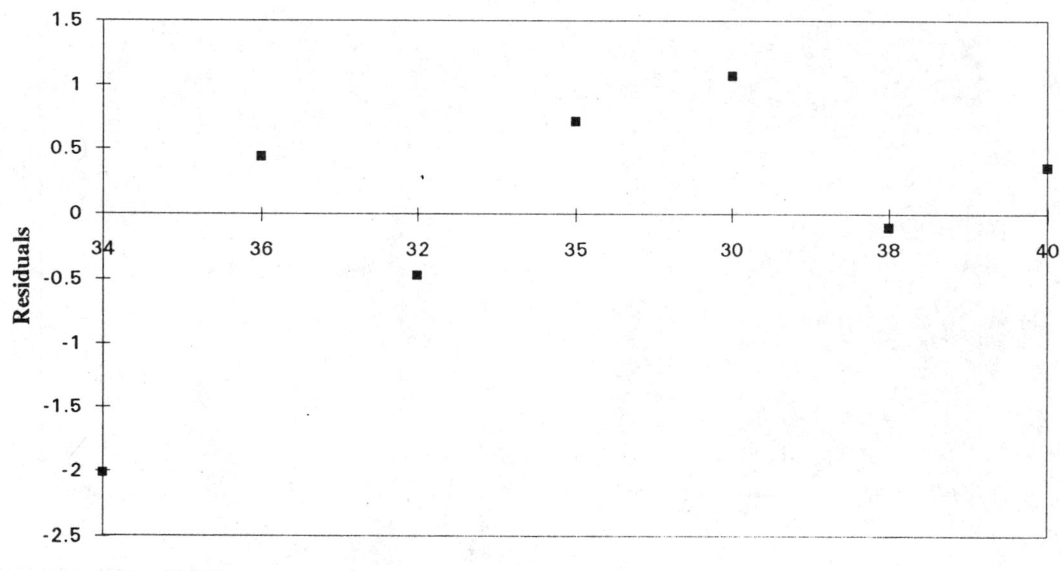

CHAPTER FIFTEEN

MULTIPLE REGRESSION

CHAPTER OUTLINE AND REVIEW

In Chapter 14, you were introduced to the concepts of regression and correlation. You learned how to determine a linear functional relationship between two variables, one dependent and one independent. However, there are many situations in which merely one independent variable cannot explain or predict the dependent variable (y). Thus, we must rely on multiple regression analysis, where the dependent variable is explained or predicted by more than one independent variable. For instance, the estimated regression function, in which there are two independent variables (x_1 and x_2) that predict the dependent variable (y), has the form of

$$\hat{y} = b_0 + b_1x_1 + b_2x_2$$

The coefficients b_0, b_1, and b_2 are calculated from the available data. The procedure for their calculation is based on the least-squares method (as explained in Chapter 14). The mathematics involved in the estimation of b_0, b_1, and b_2 are very tedious. Thus, I recommend that you use a computer package available on your school's computer system for solving problems dealing with multiple regression. In the section of exercises, I will give you the output from a computer package which I used and interpret the results. Keep in mind that no matter what statistical software you use, the results are the same, even though the format of the output may differ among various software packages. In this chapter, you were also shown how qualitative variables can be incorporated into a regression model.

Some of the main terms which you have learned in this chapter include

A. **Multiple Regression:** Regression analysis involving two or more independent variables.

B. **Multiple Regression Equation:** The mathematical equation relating the expected value or mean value of the dependent variable to the value of the independent variables; that is,

$$E(y) = \beta_0 + \beta_1 x_1 + \beta_2 x_2 + \cdots + \beta_p x_p.$$

C. **Estimated Multiple Regression Equation:** The estimate of the multiple regression equation based on sample data and the least squares method; it is $\hat{y} = b_0 + b_1 x_1 + b_2 x_2 + \cdots + b_p x_p$.

D. **Least Squares Method:** The method used to develop the estimated regression equation. It minimizes the sum of squared residuals (the deviations between the observed values of the dependent variable, y_i, and the estimated values of the dependent variable, $\hat{y}_i$.

E. **Multicollinearity:** A term used to describe the case when the independent variables in a multiple regression model are correlated.

F. **Multiple Coefficient of Determination (R^2):** A measure of the goodness of fit for the estimated regression equation.

G. **Adjusted Multiple Coefficient of Determination (R_a^2):** A measure of the goodness of fit for the estimated regression equation which accounts for the number of independent variables in the model.

H. **Qualitative Variable:** A variable with qualitative data. A qualitative variable is not measured in terms of how much or how many.

I. **Dummy Variable:** A variable used to incorporate the effect of the qualitative variable in a regression model. The dummy variable takes on the values of 0 or 1.

J.	**Outlier:**	An observation that does not fit the pattern of the rest of the data.
K.	**Influential Observation:**	An observation which has a great influence in determining the estimated regression equation.
L.	**Studentized Deleted Residuals:**	Standardized residuals that are based on deleting observation **i** from the data set and then performing the regression analysis and computing the standardized residual for observation **i** using the remaining **n-1** observations.
M.	**Leverage:**	A measure of the effect of an unusual x value on the regression results.
N.	**Cook's D:**	A measure of the influence of an observation based on the residual and leverage.

CHAPTER FORMULAS

Multiple Regression Model

$$y = \beta_0 + \beta_1 x_1 + \beta_2 x_2 + \ldots \beta_p x_p + \varepsilon \qquad (15.1)$$

Multiple Regression Equation

$$E(y) = \beta_0 + \beta_1 x_1 + \beta_2 x_2 + \ldots \beta_p x_p \qquad (15.2)$$

Estimated Multiple Regression Equation

$$\hat{y} = b_0 + b_1 x_1 + b_2 x_2 + \ldots + b_p x_p \qquad (15.3)$$

Least Squares Criterion

$$\text{Min } \Sigma(y_i - \hat{y}_i)^2 \qquad (15.4)$$

where

y_i = observed value of the dependent variable for the ith observation.
$\hat{y}_i$ = estimated value of the dependent variable for the ith observation.

Relationship among SST, SSR, and SSE

$$SST = SSR + SSE \qquad (15.7)$$

Multiple Coefficient of Determination

$$R^2 = \frac{SSR}{SST} \qquad (15.8)$$

Adjusted Multiple Coefficient of Determination

$$R_a^2 = 1 - (1 - R^2)\left(\frac{n-1}{n-p-1}\right) \qquad (15.9)$$

CHAPTER FORMULAS
(Continued)

Mean Square Due to Regression

$$MSR = \frac{SSR}{p} \tag{15.12}$$

Where p is the number of independent variables.

Mean Square Due to Error

$$MSE = \frac{SSE}{n - p - 1} \tag{15.13}$$

F Test for Overall Significance in Multiple Regression

H_0: $\beta_1 = \beta_2 = \ldots = \beta_p = 0$

H_a: One or more of the coefficients is not equal to zero

Test Statistic

$$F = \frac{MSR}{MSE} \tag{15.14}$$

Reject H_0 if $F > F_\alpha$

where F_α is based on an F distribution with p numerator degrees of freedom and n-p-1 denominator degrees of freedom

CHAPTER FORMULAS
(Continued)

t Test for Individual Significance in Multiple Regression

$H_0\colon \ \beta_i = 0$

$H_a\colon \ \beta_i \neq 0$ for any parameter β_i

Test Statistic

$$t = \frac{b_i}{s_{b_i}} \tag{15.15}$$

Reject H_0 if $t > t_{\alpha/2}$ or if $t < -t_{\alpha/2}$

Where $t_{\alpha/2}$ is based on a t distribution with n - p - 1 degrees of freedom

Residual for Observation i

$$y_i - \hat{y}_i \tag{15.23}$$

Standardized Residual for Observation i

$$\frac{y_i - \hat{y}_i}{s_{y_i - \hat{y}_i}} \tag{15.24}$$

Cook's Distance Measure

$$D_i = \frac{(y_i - \hat{y}_i)^2}{(p-1)s^2}\left[\frac{h_i}{(1-h_i)^2}\right] \tag{15.26}$$

EXERCISES

***1.** In exercise 1 of Chapter 14, we looked at the sales volume of Shultz, Inc., and then determined a regression function relating sales volume to advertising expenditure. Obviously, advertising expenditure is not the only variable which affects sales volume. Assume we believe that besides the advertising expenditure, the number of salespeople also plays an important role in the sales volume. The following data show the sales volumes, the advertising expenditures (in millions of dollars), and the number of salespeople for the years 1986-1995.

Year	Sales Volume (y_i)	Advertising Expense (x_1)	Number of Salespeople (x_2)
1986	26	1.8	35
1987	31	2.3	38
1988	28	2.6	33
1989	30	2.4	40
1990	34	2.8	38
1991	38	3.0	32
1992	41	3.4	42
1993	44	3.2	49
1994	40	3.6	53
1995	43	3.8	55

(a) What is the regression model for the above?

Answer: If it is believed that the sales volume is related to the advertising expenditure (x_1) and the number of salespeople (x_2), then the regression model involving two independent variables will be

$$y = \beta_0 + \beta_1 x_1 + \beta_2 x_2 + \varepsilon$$

where ε is the error term. The same assumptions that were made in Chapter 14 will also apply here. In multiple regression analysis (similar to simple linear regression analysis), the least squares method is used to estimate the parameters β_0, β_1 and β_2. These estimates are denoted by b_0, b_1 and b_2. Thus, the estimated regression equation has the form of

$$\hat{y} = b_0 + b_1 x_1 + b_2 x_2$$

(b) Use a computer package available at your computer center and determine a regression function.

Answer: I have used the computer package **Minitab** and have named my variables Advertising as (ADV) and the Number of Salespeople as (PEOPLE) and am presenting part of the output below.

Predictor	Coef	Stdev
Constant	7.0174	5.3146
ADV	8.6233	2.3968
PEOPLE	0.0858	0.1845

s = 3.0092 R-sq =83.51% R-sq (adj) = 78.79%

Analysis of Variance

SOURCE	DF	SS	MS	F
Regression	2	321.11	160.555	17.73
Error	7	63.39	9.055	
Total	9	384.50		

From the above results, we note that

$b_0 = 7.0174$
$b_1 = 8.6233$
$b_2 = 0.0858$

Thus, the least-squares estimated regression function is

$$\hat{y} = 7.0174 + 8.6233\, x_1 + 0.0858\, x_2$$

which can be written with the variables' names:

$$\hat{y} = 7.0174 + 8.6233 \text{ ADV} + 0.0858 \text{ PEOPLE}$$

With the above equation, we can estimate the sales volume for a given value of advertising expenditure and the number of salespeople.

(c) Interpret the coefficients of the estimated regression equation which were found in Part a.

Answer: In this equation $b_1 = 8.6233$, which indicates that for each 1 million dollar increase in advertising expenditure, the sales volume is expected to increase by 8.6233 million dollars when the number of salespeople is held constant. Similarly, $b_2 = 0.0858$, which indicates that as the number of salespeople is increased by 1, the sales volume is expected to increase by 0.0858 million dollars (that is $85,800), when advertising expenditure is held constant.

(d) Estimate the sales volume for an advertising expenditure of 3.5 million dollars and 45 salespeople.

Answer: Using the given data and the regression equation as determined in Part a, we can estimate the sales volume:

$$\hat{y} = 7.0174 + 8.6233(3.5) + 0.0858(45) = 41.05995$$

Hence, the expected sales volume in terms of dollars is $41,059,950.

2. The following data represent the number of automobiles sold per month by J. J. Macomber Autos, Inc., their prices, and the number of advertising spots they used on a local television station.

Units Sold (y)	Price (In $1,000s) (PRICE)	Advertising Spots (ADV)
10	8.2	10
8	8.7	6
12	7.9	14
13	7.8	18
9	8.1	10
14	8.8	19
15	8.9	20

(a) Using the above data, determine the least-squares regression function relating units sold (y) to PRICE and ADV. Use a computer software package available to you.

(b) A computer software was used to determine the least-squares regression equation. Part of the output is shown below.

Predictor	Coef	Stdev
Constant	0.8051	0.0387
PRICE	0.4977	0.4617
ADV	0.4773	0.0387

Analysis of Variance

Source	DF	SS	MS	F
Regression	2	40.700	20.350	80.1
Error	4	1.016	0.254	

Use the output shown above and write an equation that can be used to predict the monthly sales of automobiles.

(c) Interpret the coefficients of the estimated regression equation which were determined in Part b.

(d) If the company charges $8,000 for each car and uses 15 advertising spots, how many cars would you expect them to sell?

3. The prices of ABC, Inc. stock (y) over a period of 12 days, the number of shares (in 100s) of company's stocks sold (x_1), and the volume of exchange (in millions) on the New York Stock Exchange (x_2) are shown below.

Day	(y)	(x_1)	(x_2)
1	87.50	950	11.00
2	86.00	945	11.25
3	84.00	940	11.75
4	83.00	930	11.75
5	84.50	935	12.00
6	84.00	935	13.00
7	82.00	932	13.25
8	80.00	938	14.50
9	78.50	925	15.00
10	79.00	900	16.50
11	77.00	875	17.00
12	77.50	870	17.50

A computer software was used to determine the least-squares regression equation. Part of the computer output is shown below:

Predictor	Coef	Stdev	t-ratio
Constant	118.5055		
x_1	-0.0163	0.0315	-0.52
x_2	-1.5726	0.3590	-4.38

Analysis of Variance

Source	DF	SS	MS	F
Regression	2	118.85	59.425	40.92
Error	9	13.07	1.452	

(a) Use the output shown above and write an equation that can be used to predict the price of the stock.

(b) Interpret the coefficients of the estimated regression equation which you found in Part a.

(c) If in a given day, the number of shares of the company which were sold was 94,500 and the volume of exchange on the New York Stock Exchange was 16 million, what would you expect the price of the stock to be?

***4.** At $\alpha = 0.01$ level of significance, test to determine if the fitted equation developed in exercise 1 represents a significant relationship between the independent variables and the dependent variable.

Answer: To determine whether or not the regression function is significant, we want to test the following hypotheses:

H_0: $\beta_1 = \beta_2 = 0$

H_a: At least one of the two coefficients is not equal to zero.

To test the above, we must determine an F value from the data and compare its value with F_α. The null hypothesis will be rejected if the test statistic $F > F_\alpha$. The F value is given in Part b of exercise 1, and its value is computed as follows.

$$F = \frac{MSR}{MSE}$$

where

$$MSR = \frac{SSR}{p}$$ (p represents the number of independent variables)

$$MSE = \frac{SSE}{(n - p - 1)}$$

In this exercise, there are 2 independent variables (p = 2). Therefore, SSR has 2 degrees of freedom, and SSE has n - p - 1 = 10 - 2 - 1 = 7 degrees of freedom. Now we can compute MSR and MSE:

$$MSR = \frac{321.11}{2} = 160.555$$

$$MSE = \frac{63.39}{7} = 9.055$$

Thus, the F value is calculated as

$$F = \frac{160.555}{9.055} = 17.73$$

Now, from Table 4 of Appendix B in your text, we read the F value with 2 degrees of freedom for the numerator and 7 degrees of freedom for the denominator as $F_{.01} = 9.55$. Since the computed F = 17.73 is greater than $F_{.01}$, the null hypothesis is rejected, and we conclude that there is a significant linear relationship between the sales volume and the two independent variables.

Now, refer to the computer output given in the solution to exercise 1. You can identify the following values:

SSR = 321.11	p = 2	MSR = 160.555	F = 17.73
SSE = 63.39	n - p - 1 = 7	MSE = 9.055	

5. At $\alpha = 0.05$ level of significance, test to determine if the fitted equation which you developed in exercise 2 represents a significant relationship between the independent variables and the dependent variable.

6. At $\alpha = 0.01$ level of significance, test to determine if the fitted equation which you developed in exercise 3 represents a significant relationship between the independent variables and the dependent variable.

***7.** Refer to exercise 1. Use a level of significance of 0.05 to test the significance of β_1.

Answer: After using the F test, it was concluded that the multiple regression was significant, indicating that at least one of the two coefficients (β_1 or β_2) was not zero (see exercise 4). Now a t test can be applied for testing the significance of the individual parameters, in this case β_1. The hypotheses to be tested are

$$H_o: \beta_1 = 0$$

$$H_a: \beta_1 \neq 0$$

The null hypothesis will not be rejected if $-t_{\alpha/2} \leq t \leq t_{\alpha/2}$. Now referring to the Minitab printout (exercise 1, Part b), the t statistic can be computed as 8.6233/2.3968 = 3.5978. From Table 2 of Appendix B, the critical t values with 7 degrees of freedom are

$$t_{.025} = \pm 2.365$$

Since 3.5978 > 2.365, the null hypothesis is rejected, thus, concluding that β_1 is significantly different from zero.

8. Refer to exercise 1. Use a level of significance of 0.05 to test the significance of β_2.

***9.** Refer to exercise 1. Compute the multiple coefficient of determination.

Answer: The multiple coefficient of determination (R^2) is provided on the output of the computer, but we could compute its value as follows:

$$R^2 = \frac{SSR}{SST}$$

Thus, the R^2 for this situation is

$$R^2 = \frac{321.11}{384.5} = 0.8351$$

Therefore, 83.51% of the variability in sales volume is explained by the variability in both advertising expenditure and the number of salespeople.

10. Refer to exercise 2. Compute the multiple coefficient of determination.

***11.** Refer to exercise 1. Compute the adjusted R^2 (i.e., R_a^2).

Answer: To avoid overestimating the impact of adding an independent variable on the amount of explained variability, it is recommended to adjust the coefficient of determination:

$$R_a^2 = 1 - (1 - R^2)\left(\frac{n - 1}{n - p - 1}\right)$$

For this example, the adjusted R^2 is provided for you on the computer output. But we can compute its value as

$$R_a^2 = 1 - (1 - 0.8351)\left(\frac{10 - 1}{10 - 2 - 1}\right) = 0.7879$$

12. Refer to exercise 2. Determine the adjusted R_a^2.

***13.** Refer to exercise 1 of this chapter. Assume we also have data available on whether only TV advertising was used or whether multimedia advertising was used. To incorporate the effect of the type of advertising on the sales volume, let us define a new variable x_3, where

$$x_3 = \begin{cases} 0 & \text{if multimedia advertising was used} \\ 1 & \text{if only TV advertising was used} \end{cases}$$

Thus, our available data appear as follows:

Year	Sales Volume (y)	Advertising Expenditure (ADV)	Number of Salespeople (PEOPLE)	Type of Advertising (ADTYPE)
1986	26	1.8	35	0
1987	31	2.3	38	0
1988	28	2.6	33	1
1989	30	2.4	40	0
1990	34	2.8	38	1
1991	38	3.0	32	0
1992	41	3.4	42	0
1993	44	3.2	49	0
1994	40	3.6	53	1
1995	43	3.8	55	1

(a) Use the above data to determine a regression function relating y to ADV, PEOPLE, and ADTYPE.

Answer: The estimated regression function using the least-squares method with incorporation of the type of advertising will have the form of

$$\hat{y} = b_0 + b_1x_1 + b_2x_2 + b_3x_3$$

Using the variables' names, it will be in the form of

$$\hat{y} = b_0 + b_1\text{ADV} + b_2\text{PEOPLE} + b_3\text{ADTYPE}$$

Remember that ADTYPE represents the type of advertising used, and it can only take values of 0 or 1. Using the Minitab computer package, we obtain the following output.

Predictor	Coef	Stdev
Constant	4.0928	1.4400
ADV	10.0230	1.6512
PEOPLE	0.1020	0.1225
ADTYPE	-4.4811	1.4400

s = 1.9964 R-sq = 0.9378

Analysis of Variance

Source	DF	SS	MS	F
Regression	3	360.59	120.197	30.16
Error	6	23.91	3.985	

Thus, the regression function will be as follows:

$$\hat{y} = 4.0928 + 10.0230\ \text{ADV} + 0.102\ \text{PEOPLE} - 4.4811\ \text{ADTYPE}$$

(b) Interpret the meaning of the coefficient of ADTYPE (i.e., -4.4811).

Answer: Recall that ADTYPE is a dummy variable indicating the type of advertising used. ADTYPE = 0 if multimedia advertising was used and ADTYPE = 1 if only TV advertising was used. Thus, -4.4811 indicates that in those years where only TV advertising was used (ADTYPE = 1), sales were lower by 4.4811 (in thousands) than those years in which multimedia advertising was used.

14. In exercise 2 of this chapter , data on units sold, price, and the number of television spots the company used per month were given. During some months the company also used a radio promotion campaign. This qualitative variable is shown in the last column where 0 indicates that a radio promotion campaign was not used and 1 indicates that one was.

Units Sold (y)	Price (In $1000) ($x_1$)	Television Advertising Spots (x_2)	Radio Campaign (x_3)
10	8.2	10	1
8	8.7	6	0
12	7.9	14	0
13	7.8	18	1
9	8.1	10	0
14	8.8	19	1
15	8.9	20	1

(a) Use the above data to determine a regression function relating y to x_1, x_2, and x_3. Use a computer software available to you.

(b) Interpret the meaning of the coefficient of x_3.

15. In exercise 3 of this chapter , the prices of ABC, Inc. stock (y) over a period of 12 days, the number of shares (in 100s) of the company's stocks sold (x_1), and the volume of exchange (in millions) on the New York Stock Exchange (x_2) were given. As an extension of that exercise a dummy variable is also used for predicting the price of the stock. This dummy variable represents the volume of exchange on the NYSE on the previous day. If the volume of exchange on the previous day was high, $x_3 = 1$, otherwise $x_3 = 0$. These data are presented below.

Day	(y_i)	(x_1)	(x_2)	(x_3)
1	87.50	950	11.00	1
2	86.00	945	11.25	1
3	84.00	940	11.75	0
4	83.00	930	11.75	0
5	84.50	935	12.00	1
6	84.00	935	13.00	0
7	82.00	932	13.25	1
8	80.00	938	14.50	1
9	78.50	925	15.00	0
10	79.00	900	16.50	1
11	77.00	875	17.00	1
12	77.50	870	17.50	0

(a) Use a computer software available to you to determine a regression function relating y to x_1, x_2, and x_3.

***16.** Refer to exercise 1 of this chapter.

(a) Compute the estimated values of sales volume ($\hat{y}_i$) and the residuals.

Answer: In Part b of exercise 1, the computer package Minitab was used and the Least-Squares estimated regression function was determined to be

$$\hat{y} = 7.0174 + 8.6233\ x_1 + 0.0858\ x_2$$

Thus, the $\hat{y}_i$ values can be computed by substituting x_1 and x_2 values in the above equation. Then the residuals are computed by subtracting the estimated values from the observed values. The results are shown below.

Year	Observed Sales Volume (y_i)	Predicted Values ($\hat{y}_i$)	Residuals ($y_i - \hat{y}_i$)
1986	26	25.54281	0.45719
1987	31	30.11190	0.88810
1988	28	32.26983	-4.26983
1989	30	31.14586	-1.14586
1990	34	34.42356	-0.42356
1991	38	35.63334	2.36666
1992	41	39.94079	1.05921
1993	44	38.81682	5.18318
1994	40	42.60940	-2.60940
1995	43	44.50569	-1.50569

(b) Determine the standardized residuals and plot them versus $\hat{y}_i$. Does the standardized residual plot support the assumptions involving ε?

Answer: The computation of standardized residuals is too complex to be done manually. The computer package Minitab was used and the values of $\hat{y}_i$ and the standardized residuals were determined to be as follows.

Year	Predicted Values ($\hat{y}_i$)	Standardized Residuals
1986	25.54281	0.15193
1987	30.11190	0.29513
1988	32.26983	-1.41894
1989	31.14586	-0.38079
1990	34.42356	-0.14075
1991	35.63334	0.78648
1992	39.94079	0.35199
1993	38.81682	1.72246
1994	42.60940	-0.86715
1995	44.50569	-0.50037

Now we can plot the predicted values($\hat{y}_i$) versus the standardized residuals as shown below.

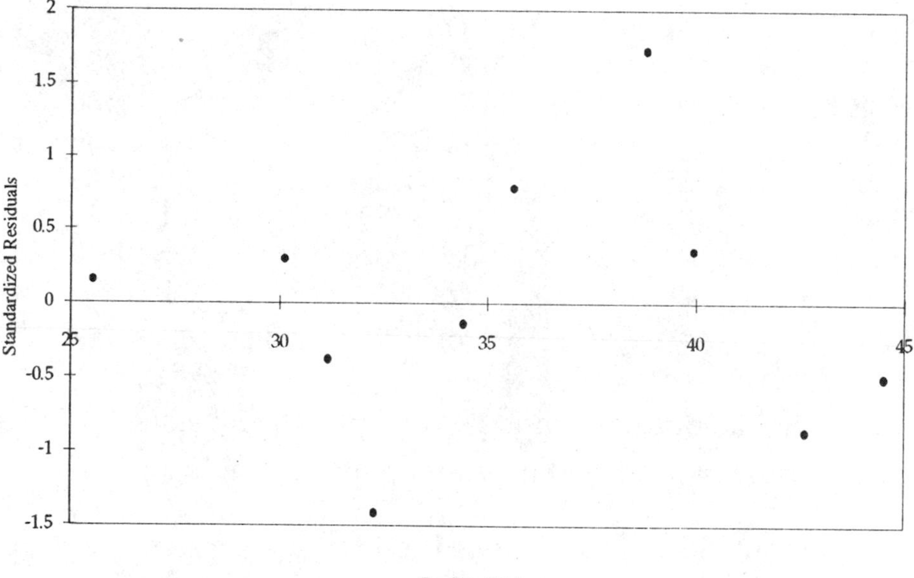

This standardized residuals plot does not show any abnormalities. The standardized residuals appear to be randomly distributed with no particular pattern and their values range between -2 an +2. Thus we conclude that the standardized residual plot support the assumptions involving ε.

17. Refer to exercise 2. Use a computer package available to you and determine the predicted values ($\hat{y}_i$) and the standardized residuals. Plot the standardized residuals against the predicted values. Does the standardized residual plot support the assumptions involving ε?

***18.** In exercise 1 you were given the following information regarding the sales volumes, the advertising expenditures (in millions of dollars), and the number of salespeople for the years 1986-1995.

Year	Sales Volume (y)	Advertising Expense (x_1)	Number of Salespeople (x_2)
1986	26	1.8	35
1987	31	2.3	38
1988	28	2.6	33
1989	30	2.4	40
1990	34	2.8	38
1991	38	3.0	32
1992	41	3.4	42
1993	44	3.2	49
1994	40	3.6	53
1995	43	3.8	55

Write the normal equations as explained in the Appendix of your textbook (chapter 15) and determine b_0, b_1 and b_2, and then solve for b_0, b_1 and b_2.

Answer: For the case of two independent variables, the normal equations are

$$nb_0 + (\Sigma x_{1i})b_1 + (\Sigma x_{2i})b_2 = \Sigma y_i \qquad (15.A5)$$

$$(\Sigma x_{1i})b_0 + (\Sigma x_{1i}^2)b_1 + (\Sigma x_{1i}x_{2i})b_2 = \Sigma x_{1i}y_i \qquad (15.A6)$$

$$(\Sigma x_{2i})b_0 + (\Sigma x_{1i}x_{2i})b_1 + (\Sigma x_{2i}^2)b_2 = \Sigma x_{2i}y_i \qquad (15.A7)$$

The coefficients for the above normal equations are computed as shown in Table 15.1.

y_i	x_{1i}	x_{2i}	x_{1i}^2	x_{2i}^2	$x_{1i}x_{2i}$	$x_{1i}y_i$	$x_{2i}y_i$
26	1.8	35	3.24	1225	63.0	46.8	910
31	2.3	38	5.29	1444	87.4	71.3	1178
28	2.6	33	6.76	1089	85.8	72.8	924
30	2.4	40	5.76	1600	96.0	72.0	1200
34	2.8	38	7.84	1444	106.4	95.2	1292
38	3.0	32	9.00	1024	96.0	114.0	1216
41	3.4	42	11.56	1764	142.8	139.4	1722
44	3.2	49	10.24	2401	156.8	140.8	2156
40	3.6	53	12.96	2809	190.8	144.0	2120
43	3.8	55	14.44	3025	209.0	163.4	2365
355	28.9	415	87.09	17825	1234.0	1059.7	15083

Table 15.1

Now the information provided in Table 15.1 can be substituted into equations 15.A5 to 15.A7 in order to arrive at the following normal equations:

$$(10)b_0 + (28.9)b_1 + (415)b_2 = 355$$

$$(28.9)b_0 + (87.09)b_1 + (1234)b_2 = 1059.7$$

$$(415)b_0 + (1234)b_1 + (17825)b_2 = 15083$$

Then the values of b_0, b_1 and b_2 are determined by solving the above system of three simultaneous linear equations. The solution yields

$$b_0 = 7.0174$$
$$b_1 = 8.6233$$
$$b_2 = 0.0858$$

Thus, the estimated regression equation is

$$\hat{y} = 7.0174 - 8.6233\, x_1 + 0.0858\, x_2$$

19. Refer to exercise 2 of this chapter. Write the normal equations as explained in the Appendix of this chapter (in your textbook) and find the regression equation.

SELF-TESTING QUESTIONS

In the following multiple choice questions, circle the correct answer. Answers are provided following the questions.

1. A regression model in which more than one independent variable is used to predict the dependent variable is called

a) a simple linear regression model
b) a multiple regression model
c) an independent model
d) none of the above

2. A term used to describe the case when the independent variables in a multiple regression model are correlated is

a) regression
b) correlation
c) multicollinearity
d) none of the above

3. A multiple regression model has the form: $y = 2 + 3x_1 + 4x_2$.
As x_1 increases by 1 unit (holding x_2 constant), y will

a) increase by 3 units
b) decrease by 3 units
c) increase by 4 units
d) decrease by 4 units

4. A multiple regression model has

a) only one independent variable
b) more than one dependent variable
c) more than one independent variable
d) none of the above

5. A measure of goodness of fit for the estimated regression equation is the

a) multiple coefficient of determination
b) mean square due to error
c) mean square due to regression
d) none of the above

6. The adjusted multiple coefficient of determination accounts for

a) the number of dependent variables in the model
b) the number of independent variables in the model
c) unusually large predictors
d) none of the above

7. The multiple coefficient of determination is computed by

a) dividing SSR by SST
b) dividing SST by SSR
c) dividing SST by SSE
d) none of the above

8. For a multiple regression model, SST = 200 and SSE = 50. The multiple coefficient of determination is

a) 0.25
b) 4.00
c) 0.75
d) none of the above

9. The correct relationship between SST, SSR, and SSE is given by

a) SSR = SST + SSE
b) SST = SSR + SSE
c) SSE = SSR - SST
d) all of the above
e) none of the above

10. The ratio of MSR/MSE yields

a) the t statistic
b) SST
c) the F statistic
d) none of the above

11. A variable that cannot be measured in terms of how much or how many is called

a) an interaction.
b) a constant variable.
c) a dependent variable.
d) a qualitative variable.

12. A variable that takes on the values of 0 or 1 and is used to incorporate the effect of qualitative variables in a regression model is called:

a) an interaction
b) a constant variable
c) a dummy variable
d) none of the above

ANSWERS TO THE SELF-TESTING QUESTIONS

1. b
2. c
3. a
4. c
5. a
6. b
7. a
8. c
9. b
10. c
11. d
12. c

ANSWERS TO CHAPTER FIFTEEN EXERCISES

2. (a) $b_0 = 0.8051$
   $b_1 = 0.4977$
   $b_2 = 0.4773$

   (b) $\hat{y} = 0.8051 + 0.4977x_1 + 0.4773x_2$
   (c) As price increases by 1 unit ($1,000s), units sold will increase by 0.4977 (holding advertising spots constant). As advertising spots increase by 1 unit, units sold will increase by 0.4773 (holding the price constant).
   (d) 12 (rounded)

3. (a) $\hat{y} = 118.5055 - 0.0163x_1 - 1.5726x_2$
   (b) As the number of shares of the stock sold goes up by 1 unit, the stock price goes down by $0.0163 (holding the volume of exchange on the NYSE constant). As the volume of exchange on the NYSE goes up by 1 unit, the stock price goes down by $1.5726 (holding the number of shares of the stock sold constant).
   (c) $77.95

5. $F = 80.0 > 6.94$; Reject H_0, there is a significant relationship.

6. $F = 40.92 > 8.02$; Reject H_0, there is a significant relationship.

8. Since $0.46 < 2.365$; Do not reject H_0, there is no significant relationship.

10. $R^2 = 0.9756$

12. $R_a^2 = 0.9632$

14. (a) $\hat{y} = 0.855 + 0.493\, x_1 + 0.475\, x_2 + 0.03\, x_3$
    (b) When radio promotions were used, sales were higher by 0.03.

15. $\hat{y} = 123.294 - 0.021\, x_1 - 1.618\, x_2 + 0.738\, x_3$

17. (a)

Predicted Values ($\hat{y}$)	Standardized Residuals
9.65931	0.67633
7.99896	0.00207
11.4192	1.15299
13.27863	-0.55314
9.60954	-1.21004
14.25364	-0.50353
14.78072	0.43532

(b)

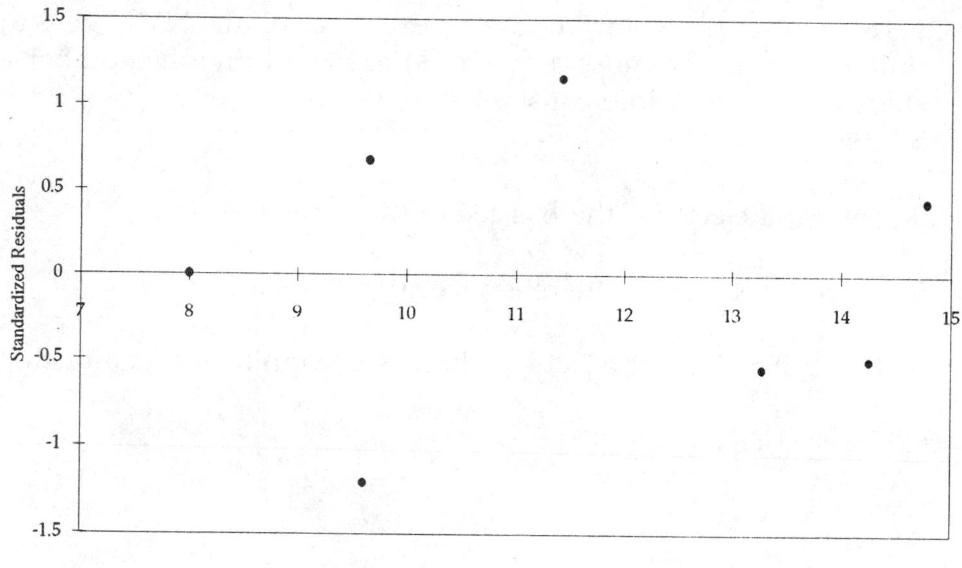

This standardized residuals plot does not show any abnormalities. The standardized residuals appear to be randomly distributed with no particular pattern and their values range between -1.5 an +1.5. Thus we conclude that the standardized residual plot supports the assumptions involving ε.

19. $\hat{y} = 0.8051 + 0.4977x_1 + 0.4773x_2$

CHAPTER SIXTEEN

REGRESSION ANALYSIS: MODEL BUILDING

CHAPTER OUTLINE AND REVIEW

This chapter is an extension of Chapters 14 and 15. In Chapter 16, you were introduced to the concept of *general linear models* and were shown how the general linear model can be used for situations where a curvilinear relationship exists between the independent variables and the dependent variable. Finally, you were shown how to perform an F test in order to determine when to add or delete variables to or from the model.

Some of the main terms which you have learned in this chapter include

A.	**General Linear Model:**	A model in the form of $y = b_0 + b_1z_1 + b_2z_2 + \ldots + b_pz_p + e$ where each independent variable z_j (for $j = 1, 2, \ldots, p$) is a function of x_j. x_j is the variable for which data has been collected.
B.	**Interaction:**	The joint effect of two variables acting together.
C.	**Variable Selection Procedures:**	Computer based methods for selecting subsets of the potential independent variables for a regression model.
D.	**Leverage:**	A measure which indicates how far an observation is from the others in terms of the values of the independent variables.

E. **Autocorrelation (Serial Correlation):** The correlation in the error terms at successive points in time. First order autocorrelation is when e_t and e_{t-1} are related. Second order is when e_t and e_{t-2} are related, and so on.

F. **Durbin-Watson Test:** A test used to determine whether or not first order autocorrelation is present.

CHAPTER FORMULAS

General Linear Statistical Model

$$y = \beta_0 + \beta_1 z_1 + \beta_2 z_2 + \ldots + \beta_p z_p + \varepsilon \tag{16.1}$$

where z_j (for j = 1, 2, . . ., p) is a function of x_j

First Order Model with One Predictor Variable

$$y = \beta_0 + \beta_1 x_1 + \varepsilon \tag{16.2}$$

Second Order Model with One Predictor Variable

$$y = \beta_0 + \beta_1 x_1 + \beta_1 x_1^2 + \varepsilon \tag{16.3}$$

Second Order Model with Interaction

$$y = \beta_0 + \beta_1 x_1 + \beta_2 x_2 + \beta_3 x_1^2 + \beta_4 x_2^2 + \beta_5 x_1 x_2 + \varepsilon \tag{16.4}$$

Exponential Model

$$E(y) = \beta_0 \beta_1^x \tag{16.7}$$

Exponential model can be transformed to linear logarithm model as

$$\text{Log } E(y) = \text{Log } \beta_0 + x \text{ Log } \beta_1 \tag{16.8}$$

F Statistic for Determining When to Add or Delete x_2

$$F = \frac{\dfrac{SSE(x_1) - SSE(x_1, x_2)}{1}}{\dfrac{SSE(x_1, x_2)}{n - p - 1}} \tag{16.10}$$

CHAPTER FORMULAS
(Continued)

General F Test for Adding or Deleting Variables

$$F = \frac{\dfrac{SSE(x_1, x_2, \ldots, x_q) - SSE(x_1, x_2 + \ldots + x_q, x_{q+1} + \ldots + x_p)}{p - q}}{\dfrac{SSE(x_1, x_2, \ldots, x_q, x_{q+1}, \ldots, x_p)}{n - p - 1}} \qquad (16.13)$$

which can be written in the form

$$F = \frac{\dfrac{SSE(\text{reduced}) - SSE(\text{full})}{\text{number of extra items}}}{MSE(\text{full})} \qquad (16.15)$$

or

$$F = \frac{\dfrac{SSR(\text{full}) - SSR(\text{reduced})}{\text{number of extra items}}}{MSE(\text{full})}$$

Autocorrelated Error Terms

$$e_t = \rho e_{t-1} + z_t \qquad (16.16)$$

Durbin-Watson Statistic

$$d = \frac{\sum_{t=2}^{n} (e_t - e_{t-1})^2}{\sum_{t=1}^{n} e_t^2} \qquad (16.17)$$

EXERCISES

***1.** Monthly total production costs and the number of units produced at a local company over a period of 10 months are shown below.

Month	Production Costs (y_i) (in $ millions)	Units Produced (x_i) (in millions)
1	1	2
2	1	3
3	1	4
4	2	5
5	2	6
6	4	7
7	5	8
8	7	9
9	9	10
10	12	10

(a) Draw a scatter diagram for the above data. Does the relationship between production costs and units produced appear to be linear?

Answer: The scatter diagram is shown in Figure 16.1.

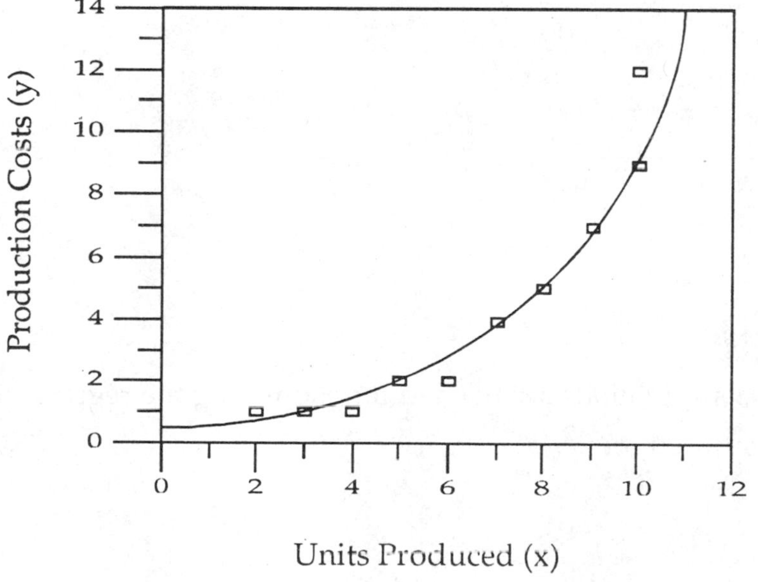

Figure 16.1

As you can see, the relationship between production costs and units produced appears to be curvilinear.

(b) From the scatter diagram, it appears that a model in the form of

$$y = \beta_0 + \beta_1 x^2 + \varepsilon$$

best describes the relationship between x and y. Estimate the parameters of this curvilinear regression equation.

Answer: Let a new variable $z = x^2$. Then we have a simple linear regression model in the form of

$$y = \beta_0 + \beta_1 z + \varepsilon$$

Then the estimated regression model will be in the form of

$$\hat{y} = b_0 + b_1 z$$

The parameters of this model can simply be computed by substituting z_i for x_i^2 in the formulas which were used to compute b_0 and b_1 as explained in Chapter 14. Thus, b_0 and b_1 are computed as

$$b_1 = \frac{\sum z_i y_i - \dfrac{\sum z_i \sum y_i}{n}}{\sum z_i^2 - \dfrac{(\sum z_i)^2}{n}}$$

$$b_0 = \bar{y} - b_1 \bar{z}$$

The following computations are needed for estimating the regression function, where z is substituted for x^2.

Month	y_i	x_i	$z_i = x_i^2$	$z_i\ y_i$	z_i^2
1	1	2	4	4	16
2	1	3	9	9	81
3	1	4	16	16	256
4	2	5	25	50	625
5	2	6	36	72	1296
6	4	7	49	196	2401
7	5	8	64	320	4096
8	7	9	81	567	6561
9	9	10	100	900	10000
10	12	10	100	1200	10000
Totals	44	64	484	3334	35332

First let us compute $\bar{z}$ and $\bar{y}$ as

$$\bar{z} = \frac{\Sigma z_i}{n} = \frac{484}{10} = 48.4$$

$$\bar{y} = \frac{\Sigma y_i}{n} = \frac{44}{10} = 4.4$$

Now we can compute b_1 as

$$b_1 = \frac{\Sigma z_i y_i - \dfrac{\Sigma z_i \Sigma y_i}{n}}{\Sigma z_i^2 - \dfrac{(\Sigma z_i)^2}{n}} = \frac{3334 - \dfrac{(484)(44)}{10}}{35332 - \dfrac{(484)^2}{10}} = 0.10116 \text{ (rounded)}$$

and finally, b_0 is computed:

$$b_0 = \bar{y} - b_1 \bar{z} = 4.4 - (0.10116)(48.4) = -0.496$$

Thus, the regression function in terms of z_i is

$$\hat{y} = -0.496 + 0.10116\ z$$

Since, originally we had substituted z for x^2, now we can substitute x^2 back for z and arrive at the following curvilinear regression function.

$$\hat{y} = -0.496 + 0.10116\,x^2$$

2. For the following data,

y_i	x_i
2	1
3	4
5	6
8	7
10	8

(a) Draw a scatter diagram. Does the relationship between x and y appear to be linear?

(b) Develop an estimated regression model for the above data. Assume the relationship between x and y can best be given by $\hat{y} = b_0 + b_1x^2$.

3. Consider the following data for two variables x and y. Use a computer software package available to you to solve this problem.

y	x
1	1
3	2
5	3
6	4
8	5
7	6
5	7
4	8

(a) Develop an estimated regression equation for the above data of the form $\hat{y} = b_0 + b_1 x$. Comment on the adequacy of this equation for predicting y. Let $\alpha = .05$.

(b) Develop an estimated regression equation for the above data of the form $\hat{y} = b_0 + b_1 x + b_2 x^2$. Comment on the adequacy of this equation for predicting y. Let $\alpha = .05$.

(c) Use the results of Part b and predict y when x = 4.

4. Consider the following data for two variables x and y. Use a computer software package available to you to solve this problem.

x	y
1	1
4	6
7	9
8	7
9	4
10	3

(a) Develop an estimated regression equation for the above data of the form $\hat{y} = b_0 + b_1 x$. Comment on the adequacy of this equation for predicting y. Let $\alpha = .05$.

(b) Develop an estimated regression equation for the above data of the form $\hat{y} = b_0 + b_1x + b_2x^2$. Comment on the adequacy of this equation for predicting y. Let $\alpha = .05$.

(c) Predict the value of y when x = 5.

5. The following estimated regression equation has been developed for the relationship between y, the dependent variable, and x, the independent variable.

$$\hat{y} = 60 + 200x - 6x^2$$

The sample size for this regression model was 23, and SSR = 600 and SSE = 400.

(a) Compute the coefficient of determination.

(b) Using $\alpha = .05$, test for a significant relationship.

***6.** The following information was given in exercise 1 of Chapter 15.

Year	Sales Volume (y)	Advertising Expense (x_1)	Number of Salespeople (x_2)
1986	26	1.8	35
1987	31	2.3	38
1988	28	2.6	33
1989	30	2.4	40
1990	34	2.8	38
1991	38	3.0	32
1992	41	3.4	42
1993	44	3.2	49
1994	40	3.6	53
1995	43	3.8	55

The Minitab computer package was first used to determine a regression function relating sales volume (y) and advertising expenditure (x_1). The regression function is

$$\hat{y} = 8.171 + 9.4564\, x_1$$

and the ANOVA table associated with this function is

Source	DF	SS	MS	F
Regression	1	319.15	319.1500	39.07
Error	8	65.35	8.1687	

Then the estimated regression function relating sales volume (y) to both advertising expenditure (x_1) and number of salespeople (x_2) was determined to be

$$\hat{y} = 7.0174 + 8.6233\, x_1 + 0.0858\, x_2$$

The ANOVA table associated with this latter function is as follows.

Source	DF	SS	MS	F
Regression	2	321.11	160.555	17.73
Error	7	63.39	9.055	

Use the above information and determine if x_2 contributes significantly to the model. Let $\alpha = 0.05$.

Answer: The second variable (x_2) should be added to the model if the reduction in SSE is significant. To test whether or not the reduction is significant, the following F statistic is used.

$$F = \frac{\dfrac{SSE(x_1) - SSE(x_1, x_2)}{1}}{\dfrac{SSE(x_1, x_2)}{n - p - 1}}$$

In this equation, SSE (x_1) represents the error sum of squares when x_1 is the only independent variable, and SSE (x_1, x_2) represents the error sum of squares when both x_1 and x_2 are independent variables. Note p is the number of independent variables.

Our hypotheses to be tested are

H_0: Adding x_2 to the model does not reduce the error sum of squares significantly.

H_a: Adding x_2 to the model does reduce the error sum of squares significantly.

If the null hypothesis is rejected, then we can conclude that x_2 contributes significantly to the model.

From the computer solutions (provided earlier in this problem), we note SSE (x_1) = 65.35 and SSE (x_1, x_2) = 63.39. Now we can compute the F statistic:

$$F = \frac{\dfrac{SSE(x_1) - SSE(x_1, x_2)}{1}}{\dfrac{SSE(x_1, x_2)}{n - p - 1}} = \frac{\dfrac{65.35 - 63.39}{1}}{\dfrac{63.39}{10 - 2 - 1}} = 0.216$$

From Table 4 of Appendix B (in your textbook), the F value is 5.59. Note that the numerator degrees of freedom is 1 and the denominator degrees of freedom is 7. Since the computed F statistic of 0.216 is less than 5.59, the null hypothesis is not rejected, and we conclude that the addition of x_2 to the model does not reduce the error sum of squares significantly. Thus, x_2 does not contribute to the model significantly.

7. In exercise 3 of Chapter 15, you were given the following information.

Day	Stock Price (y)	No. of Shares Sold (x_1)	Volume of Exchange on NYSE (x_2)
1	87.50	950	11.00
2	86.00	945	11.25
3	84.00	940	11.75
4	83.00	930	11.75
5	84.50	935	12.00
6	84.00	935	13.00
7	82.00	932	13.25
8	80.00	938	14.50
9	78.50	925	15.00
10	79.00	900	16.50
11	77.00	875	17.00
12	77.50	870	17.50

If the number of shares sold (x_1) is used as the sole independent variable, the following function is provided.

$$\hat{y} = 0.108 - 17.817x_1$$

The SSE for the above model is 40.936.

On the other hand, when both the number of shares sold (x_1) and the volume of exchange on the NYSE (x_2) are used, the following function is provided.

$$\hat{y} = 118.506 - 0.0163x_1 - 1.5726x_2$$

This latter model's SSE is 13.069.

(a) Use a computer software package available to you and verify that the above functions and error sums of squares are correct.

(b) Use an F test and determine if x_2 contributes significantly to the model. Let $\alpha = 0.05$.

8. In exercise 2 of Chapter 15, the following information was provided.

Units Sold (y)	Price (In $1,000s) ($x_1$)	Advertising Spots (x_2)
10	8.2	10
8	8.7	6
12	7.9	14
13	7.8	18
9	8.1	10
14	8.8	19
15	8.9	20

Using price (x_1) as the only independent variable, the following function is provided.

$$\hat{y} = 0.408 + 1.338x_1$$

The SSE for the above model is 39.535.

Using both x_1 and x_2 as independent variables yields the following function.

$$\hat{y} = 0.805 + 0.498x_1 - 0.477x_2$$

The SSE for this latter function is 1.015.

(a) Use a computer software package available to you and verify that the above functions and error sums of squares are correct.

(b) Use an F test and determine if x_2 contributes significantly to the model. Let $\alpha = 0.05$.

9. A regression analysis was applied in order to determine the relationship between a dependent variable and 6 independent variables. The following information was obtained from the regression analysis.

$r^2 = 0.6$
SSR = 1,200
Total number of observations n = 32

(a) Fill in the blanks in the following ANOVA table.

Source of Variation	DF	SS	MS	F
Regression	______?	______?	______?	______?
Error (Residual)	______?	______?	______?	
Total	______?	______?		

(b) Is the model significant? Let $\alpha = 0.05$.

10. In a regression analysis involving 21 observations and 4 independent variables, the following information was obtained.

$r^2 = 0.80$
$S = 5.0$

Based on the above information, fill in all the blanks in the following ANOVA.

Hint: $r^2 = \frac{SSR}{SST}$, but also $r^2 = 1 - \frac{SSE}{SST}$.

Source	DF	SS	MS	F
Regression	_____?	_____?	_____?	_____?
Error (Residual)	_____?	_____?	_____?	
Total	_____?	_____?		

*11. We are interested in determining an estimated regression equation relating the yearly incomes of a sample of 20 professors and the following variables.

1. AGE = Age of the professor
2. EXP = Years of teaching experience
3. CONSULT = Years of consulting experience
4. GENDER = Gender of the professor (male = 1, female = 0)

The Minitab program was used for the analysis of the data, and the results are shown below.

The regression equation is

INCOME = 19,855.8 + 642.3 AGE + 273.4 EXP + 3.4 CONSULT - 14,906.9 SEX

Predictor	coef	Stdv	t-ratio	p-value
Constant	19,855.8	6,581.2	3.02	0.009
AGE	642.3	158.4	4.05	0.001
EXP	273.4	528.5	0.52	0.618
CONSULT	3.4	583.2	0.01	0.991
GENDER	-14,906.9	3,190.9	-4.67	0.001

s = 6,498.821 R-sq = 82.3% R-sq (adj) = 77.5%

Analysis of Variance

Source	DF	SS	MS	F	p-value
Regression	4	2,937,268,224	734,317,056	17.387	0.000
Error	15	633,520,128	42,234,676		
Total	19	3,570,788,352			

(a) Is the relationship between y and the independent variables significant? Let $\alpha = .05$.

Answer: In this case, we need to perform an F test in order to test the following hypotheses.

H_o: $\beta_1 = \beta_2 = \beta_3 = \beta_4 = \beta_5 = 0$

H_a: One or more of the coefficients is not equal to zero

If H_o is rejected, we can conclude that there is a significant relationship, which means the estimated regression equation is useful for predicting the yearly salaries. From the computer output provided above, we note the F statistic is 17.387. Then we read F_α with 5 numerator and 14 denominator degrees of freedom as $F_{.05} = 2.96$. Since $17.387 > 2.96$, H_o is rejected, and we conclude that the relationship is significant.

Note that we could have used the p-value for our test. The p-value for the analysis of variance is almost zero (0.000), which is less than $\alpha = .05$. Thus, H_o is rejected.

(b) Is AGE a significant variable? Let $\alpha = .05$.

Answer: Now we are interested in testing for the significance of an individual variable where the hypotheses are

H_o: $\beta_1 = 0$

H_a: $\beta_1 \neq 0$

First, let us read $t_{\alpha/2}$ with 15 degrees of freedom (from the table) as $t_{0.025} = 2.131$. Then from the computer printout, we note that the t-ratio has a value of 4.05. Since $4.05 > 2.131$, H_0 is rejected; and we conclude that AGE is a significant variable. Note, we could have considered the p-value for the variable AGE whose value is 0.001, which is less than .05, and rejected H_0.

12. Refer to exercise 11.

(a) At 95% confidence, test to see which other coefficient is significant.

(b) Interpret the meaning of the coefficient of GENDER (i.e., -14,906.90).

SELF-TESTING QUESTIONS

In the following multiple choice questions, circle the correct answer. Answers are provided following the questions.

1. A regression model in the form of

   $y = \beta_0 + \beta_1 x_1 + \varepsilon$

   is referred to as

a) a simple first-order model with two predictor variables
b) a simple second-order model with one predictor variable
c) a simple second-order model with two predictor variables
d) a simple first-order model with one predictor variables
e) none of the above

2. A regression model in the form of

   $y = \beta_0 + \beta_1 x_1 + \beta_2 x_1^2 + \varepsilon$

   is referred to as a

a) second-order model with three predictor variables
b) second-order model with two predictor variables
c) second-order model with one predictor variable
d) first-order model with one predictor variables
e) none of the above

3. Serial correlation is

a) the correlation between serial numbers of products
b) the same as autocorrelation
c) the same as leverage
d) none of the above

4. The joint effect of two variables acting together is called

a) autocorrelation
b) interaction
c) serial correlation
d) none of the above

5. A test to determine whether or not first order autocorrelation is present is

a) a t test
b) an F test
c) a test of interaction
d) none of the above

6. Which of the following tests is used to determine whether an additional variable makes a significant contribution to a multiple regression model?

a) A t test
b) A Z test
c) An F test
e) none of the above

7. In multiple regression analysis, the general linear model

a) cannot be used to accommodate curvilinear relationships between dependent variables and independent variables
b) can be used to accommodate curvilinear relationships between independent variables and dependent variables
c) must contain more than two independent variables
d) none of the above

8. The range of the Durbin-Watson statistic is between

a) -1 and 1
b) 0 and 1
c) - infinity and + infinity
d) 0 and 4
e) none of the above

9. If SSR = 200 and SSE = 50, then r^2 will be

a) 0.25
b) 4.00
c) 0.80
d) 0.20

10. If $r^2 = 0.25$ and SST = 2,000, then SSE will have a value of

a) 2,666.6
b) 500.0
c) 8,000.0
d) 1,500.0

ANSWERS TO SELF-TESTING QUESTIONS

1. d
2. c
3. b
4. b
5. d
6. c
7. b
8. d
9. c
10. d

ANSWERS TO CHAPTER SIXTEEN EXERCISES

2. (a) The relationship appears to be curvilinear.
   (b) $\hat{y} = 1.253 + 0.131\ x^2$

3. (a) $\hat{y} = 2.786 + 0.464\ x$
   $r^2 = 26\%$ Only 26% of variation is explained.
   P-value = 0.196; no significant relationship exists.
   The model is not adequate for predicting y.
   (b) $\hat{y} = -2.839 + 3.839\ x - 0.375\ x^2$
   $r^2 = 93.7\%$, which means 93.7% of variation in y is explained by both x and x^2. Both x and x^2 are significant. (Both p-values are 0.001.) The p-value for the analysis of variance is 0.002, which is less than 0.05. Therefore, the model is adequate for predicting y.
   (c) 6.517

4. (a) $\hat{y} = 3.304 + 0.261\ x$
   $r^2 = 9.3\%$ Only 9.3% of variation is explained.
   P-value = 0.56, no significant relationship exists.
   The model is not adequate for predicting y.
   (b) $\hat{y} = -2.681 + 3.68\ x - 0.313\ x^2$
   $r^2 = 90.4\%$, which indicates 90.4% of variation in y is explained by both x and x^2. The p-value for x is 0.012 and for x^2 is 0.013. Both are significant. The p-value for the analysis of variance is 0.029 which is less than 0.05. Therefore, the model is adequate for predicting y.
   (c) 7.894

5. (a) 0.60
   (b) $F = 15 > 3.49$; reject H_0, the relationship is significant.

7. $F = 19.19$ and $F_{.05} = 5.12$
   Since $19.19 > 5.12$, reject H_0 and conclude x_2 contributes significantly to the model.

8. $F = 151.8$ and $F_{.05} = 7.71$
   Since $151.8 > 7.71$, reject H_0 and conclude x_2 contributes significantly to the model.

9. (a)

Source of Variation	DF	SS	MS	F
Regression	6	1,200	200	6.25
Error (Residual)	25	800	32	
Total	31	2,000		

(b) Since $F = 6.25 > 2.49$, reject H_0. The model is significant.

10.

Source of Variation	DF	SS	MS	F
Regression	4	1,600	400	16
Error (Residual)	16	400	25	
Total	20	2,000		

12. (a)

Variable	P-value	
EXP	0.618 > .05	not significant
CONSULT	0.991 > .05	not significant
GENDER	0.001 < .05	significant

(b) A value of 1 was used for male and 0 for female. Therefore, this coefficient indicates that in this data set, females' incomes were higher than males by $14,906.90.

CHAPTER SEVENTEEN

INDEX NUMBERS

CHAPTER OUTLINE AND REVIEW

In this chapter, you have studied a variety of index numbers. More specifically, you have learned the following.

A.	**Price Relative:**	The price relative is a price index which is determined by dividing the current unit price by a base period unit price and multiplying the result by 100.
B.	**Aggregate Price Index:**	A composite price index based on the prices of a group of items is known as an aggregate price index.
C.	**Weighted Aggregate Price Index:**	A composite price index where the prices are weighted by their relative importance is called a weighted aggregate price index.
D.	**Laspeyres Index:**	A weighted aggregate price index where the weight for each item is its base period quantity is known as the Laspeyres index.
E.	**Paasche Index:**	The Paasche index is a weighted aggregate price index where the weight for each item is its current period quantity.

F.	**Consumer Price Index:**	The consumer price index is a monthly price index that uses the price changes in consumer goods and services for measuring the changes in consumer price over time.
G.	**Producer Price: Index:**	A monthly price index which measures the changes in the prices of goods sold in a primary market is called the producer price index.
H.	**Dow Jones Averages:**	The Dow Jones Averages are aggregate price indexes showing the prices of stocks listed on the New York Stock Exchange.
I.	**Quantity Index:**	A quantity index measures the changes in quantity over time.
J.	**Index of Industrial Production:**	The Index of Industrial Production is a quantity index which measures the changes in the volume of the production of industrial goods over time.

CHAPTER FORMULAS

Price Relative

$$\text{Price Relative in Period t} = \frac{\text{Price in Period t}}{\text{Base Period Price}}(100) \qquad (17.1)$$

Unweighted Aggregate Price Index

$$I_t = \frac{\sum P_{it}}{\sum P_{io}}(100) \qquad (17.2)$$

where

P_{it} = the unit price of item "i" in period "t"

P_{io} = the unit price of item "i" in the base period

Weighted Aggregate Price Index

$$I_t = \frac{\sum P_{it} Q_i}{\sum P_{io} Q_i}(100) \qquad (17.3)$$

where

Q_i = the quantity of item "i"

Laspeyres Index

$$I_t = \frac{\sum P_{it} Q_{io}}{\sum P_{io} Q_{io}}(100) \qquad (17.4)$$

where

Q_{io} = the quantity of item "i" in the base period

CHAPTER FORMULAS

CONTINUED

Paasche Index

$$I_t = \frac{\sum P_{it} Q_{it}}{\sum P_{io} Q_{it}}(100) \qquad (17.5)$$

where

Q_{it} = the quantity of item "i" in period "t"

Weighted Average of Price Relative

$$I_t = \frac{\sum \frac{P_{it}}{P_{io}} W_i}{\sum W_i}(100) \qquad (17.6)$$

where

$$W_i = P_{io} Q_i \qquad (17.7)$$

Weighted Aggregate Quantity Index

$$I_t = \frac{\sum Q_{it} W_i}{\sum Q_{io} W_i}(100) \qquad (17.9)$$

EXERCISES

***1.** The prices for a kilowatt hour of electricity for the years 1992 through 1995 are given below.

Year	Price/KWH
1992	$.035
1993	.039
1994	.040
1995	.041

Determine the price relatives for the years 1992 through 1995. Let 1992 be the base.

Answer: The price relative in period t is calculated as

$$\text{Price Relative in Period t} = \frac{\text{Price in Period t}}{\text{Base Period Price}}(100)$$

Thus, the price relatives can be calculated as

Year	Price Relatives (Base 1992 = 100)
1992	(.035/.035)(100) = 100.00
1993	(.039/.035)(100) = 111.43
1994	(.040/.035)(100) = 114.28
1995	(.041/.035)(100) = 117.14

2. The price of a pound of coffee for the years 1990 through 1995 are given below.

Year	Price/Pound
1990	$1.60
1991	1.85
1992	2.10
1993	2.50
1994	2.60
1995	2.90

Let 1990 be the base and compute the price relatives for the years 1990 through 1995.

***3.** An automobile dealership sells 3 different types of automobiles. Unit prices for the years 1990 and 1995 are shown below.

Automobile Type	Unit Price ($) 1990 ($P_o$)	1995 (P_t)
Small	9,000	12,000
Medium	14,000	17,000
Large	18,000	22,000

Compute an unweighted aggregate price index for the automobiles.

Answer: An unweighted aggregate price index in period t can be determined as

$$I_t = \frac{\Sigma P_{it}}{\Sigma P_{io}} (100)$$

$$I_{1995} = \frac{12000 + 17000 + 22000}{9000 + 14000 + 18000} (100) = 124.39$$

Thus, we can say that the prices of the automobiles have increased 24.39% over the period from 1990 to 1995.

4. The prices of 3 products for the years 1994 and 1995 are shown below.

Product	Unit Price ($) 1994	1995
A	10.00	11.00
B	8.00	10.00
C	7.00	9.00

Compute an unweighted aggregate price index for the products.

***5.** Refer to exercise 3. Assume the dealer has indicated that his 1990 sales for the three different types of automobiles were as follows.

Automobile Type	Units Sold (1990)
Small	400
Medium	310
Large	150

Compute a weighted aggregate price index (Laspeyres index) for 1995 with 1990 as the base period.

Answer: The weighted aggregate price index is determined by

$$I_t = \frac{\sum P_{it}\, Q_{io}}{\sum P_{io}\, Q_{io}}(100)$$

$$I_{1995} = \frac{(12000)(400) + (17000)(310) + (22000)(150)}{(9000)(400) + (14000)(310) + (18000)(150)}(100)$$

$$= 125.66$$

Thus, we can conclude that the prices of the automobiles have increased by 25.66% over the period from 1990 to 1995.

6. Refer to exercise 4. Assume the number of units of each product sold in 1994 are as follows:

Product	Units Sold (1994)
A	1000
B	800
C	400

Compute a weighted aggregate price index (Laspeyres index) for 1995 with 1994 as the base period.

***7.** Refer to exercise 3. Assume the dealer's records show the following sales volume in 1995.

Automobile Type	Units Sold (1995)
Small	800
Medium	340
Large	100

Compute a weighted aggregate price index using the sales values of 1995.

Answer: When the current period quantities are used, the resulting price index is called the Paasche index; and the value is

$$I_t = \frac{\Sigma P_{it}\ Q_{it}}{\Sigma P_{io}\ Q_{it}}(100)$$

Thus,

$$I_{1995} = \frac{(12000)(800) + (17000)(340) + (22000)(100)}{(9000)(800) + (14000)(340) + (18000)}(100)$$

Therefore, the prices (using the quantities of 1995) show a 27.76% increase.

8. Refer to exercise 4. Assume the sales quantities in 1995 were

Product	Units Sold (1995)
A	1500
B	1100
C	100

Compute the Paasche price index.

***9.** The yearly incomes of a college graduate for the years 1992 through 1995 and the Consumer Price Index are shown below.

Year	Yearly Income ($)	CPI (Base 1992)
1992	17,000	100
1993	20,000	108
1994	24,000	116
1995	26,000	131

Use the Consumer Price Index to deflate the yearly income series.

Answer: The above series can be deflated as follows:

Year	Deflated Yearly Income
1992	(17000/100)(100) = 17,000.00
1993	(20000/108)(100) = 18,518.52
1994	(24000/116)(100) = 20,689.65
1995	(26000/131)(100) = 19,847.33

As you will note, even though the individual's income for the years 1994 through 1995 has increased from $24,000 to $26,000, his/her "real income" has decreased (from $20,689.65 to $19,847.33).

10. The following shows an individual's weekly income for the years 1992 to 1995 and the Consumer Price Index for those years.

Year	Weekly Income ($)	CPI (1992 Base)
1992	400	100
1993	440	108
1994	480	116
1995	600	131

Use the Consumer Price Index to deflate the weekly income series.

11. The average weekly grocery expenditures of a particular family for the years 1992 through 1995 and the Consumer Price Index are shown below.

Year	Weekly Expenditure ($)	CPI (1992 Base)
1992	60	100
1993	75	108
1994	87	116
1995	114	131

Use the Consumer Price to deflate the weekly expenditure series.

SELF-TESTING QUESTIONS

In the following multiple choice questions, circle the correct answer. An answer key is provided following the questions.

1. The price relative is a price index which is determined by

a) (price in period t/base period price)(100)
b) (base period price/price in period t)(100)
c) (price in period t + base period price)(100)
d) none of the above

2. A composite price index based on the prices of a group of items is known as the

a) Laspeyres Index
b) Paasche Index
c) aggregate price index
d) Consumer Price Index

3. A weighted aggregate price index where the weight for each item is its base period quantity is known as the

a) Paasche Index
b) Consumer Price Index
c) Producer Price Index
d) Laspeyres Index

4. A monthly price index that uses the price changes in consumer goods and services for measuring the changes in consumer prices over time is known as the

a) Paasche Index
b) Consumer Price Index
c) Producer Price Index
d) Laspeyres Index

5. A monthly price index which measures the changes in the price of goods sold in a primary market is known as the

a) Consumer Price Index
b) quantity index
c) Index of Industrial Production
d) Producer Price Index

6. The Dow Jones Averages are aggregate price indexes showing the prices of stocks listed

a) on the American Stock Exchange
b) over the counter
c) on the New York Stock Exchange
d) none of the above

7. A composite price index where the prices of the items in the composite are weighted by their relative importance is known as the

a) price relative
b) weighted aggregate price index
c) Consumer Price Index
d) none of the above

8. A weighted aggregate price index where the weight for each item is its current-period quantity is called the

a) Laspeyres Index
b) aggregate index
c) Consumer Price Index
d) Index of Industrial Production
e) Paasche Index

9. An index which is designed to measure changes in quantities over time is known as the

a) time index
b) quantity index
c) Paasche index
d) change index
e) none of the above

10. A quantity index which is designed to measure changes in physical volume or production levels of industrial goods over time is known as the

a) physical volume index
b) time index
c) Index of Industrial Production
d) none of the above

ANSWERS TO THE SELF-TESTING QUESTIONS

1. a
2. c
3. d
4. b
5. d
6. c
7. b
8. e
9. b
10. c

ANSWERS TO CHAPTER SEVENTEEN EXERCISES

2.

Year	Price Relatives (1990 = 100)
1990	100.000
1991	115.625
1992	131.250
1993	156.250
1994	162.500
1995	181.250

4. $I_t = 120$

6. $I_t = 117.71$ (rounded)

8. $I_t = 115.92$ (rounded)

10.

Year	Deflated Income
1992	$400.00
1993	407.41
1994	413.79
1995	458.01

11.

Year	Deflated Income
1992	$60.00
1993	69.44
1994	75.00
1995	87.02

CHAPTER EIGHTEEN

FORECASTING

CHAPTER OUTLINE AND REVIEW

In this chapter, you have studied the concept of time series analysis and have learned how to predict the value of a variable in a future time period based on past data. Specifically, you have learned the following:

A.	**Time Series:**	A group of observations measured at successive points in time.
B.	**Forecast:**	A prediction of the future values of a time series.
C.	**Multiplicative Time Series Model:**	A model that assumes that the separate components of trend, cyclical, seasonal, and irregular effects can be multiplied together to identify the actual time series value $Y = T \cdot C \cdot S \cdot I$.
D.	**Trend Component:**	The long run movement in the time series over several periods.
E.	**Cyclical Component:**	The component of the time series model that results in periodic above-trend and below-trend behavior of the time series lasting more than one year.

F.	**Seasonal Component:**	The component of the time series model which shows a periodic pattern over one year or less.
G.	**Irregular Component:**	The component of the time series model which reflects the random variation of the actual time series values beyond what can be explained by the trend, cyclical, and seasonal components.
H.	**Moving Averages:**	A method of smoothing a time series by averaging each successive group of data points. The moving averages method can be used to identify the combined trend/cyclical component of the time series.
I.	**Weighted Moving Averages:**	A method of forecasting or smoothing a time series by computing a weighted average of past data values. The sum of the weights must equal one.
J.	**Exponential Smoothing:**	A forecasting technique that uses a weighted average of past time series values in order to arrive at smoothed time series values which can be used as forecasts.
K.	**Smoothing Constant:**	A parameter of the exponential smoothing model which provides the weight given to the most recent time series value in the calculation of the forecast value.
L.	**Mean Square Error (MSE):**	One approach to measuring the accuracy of a forecasting model. This measure is the average of the sum of the squared differences between the forecast values and the actual time series values.
M.	**Mean Absolute Deviation (MAD):**	A measure of forecast accuracy. MAD is the average of the sum of absolute values of the forecast errors.
N.	**Deseasonalized Time Series:**	A time series that has had the effect of season removed by dividing each original time series observation by the corresponding seasonal factor.

O. **Causal Forecasting Methods:** Forecasting methods that relate a time series value to other variables that are believed to explain or cause its behavior.

P. **Autoregressive Model:** A time series model that uses a regression relationship based on past time series values to predict the future time series values.

Q. **Delphi Approach:** A qualitative forecasting method that obtains forecasts through "group consensus."

R. **Scenario Writing:** A qualitative forecasting method which consists of developing a conceptual scenario of the future based upon a well defined set of assumptions.

CHAPTER FORMULAS

Moving Average

$$\text{Moving Average} = \frac{\Sigma(\text{most recent n data values})}{n} \tag{18.1}$$

Exponential Smoothing Model

$$F_{t+1} = \alpha Y_t + (1 - \alpha)\, F_t \tag{18.2}$$

where F_{t+1} = the forecast of the time series for period t + 1
Y_t = the actual value of the time series in period t
F_t = the forecast of the time series for period t
α = the smoothing constant $(0 \leq \alpha \leq 1)$

The exponential smoothing model can be rewritten as

$$F_{t+1} = F_t + \alpha\,(Y_t - F_t) \tag{18.3}$$

Forecast Error in Period t

Linear Trend Equation

$$T_t = b_0 + b_1 t \tag{18.5}$$

where T_t = the forecast value (based upon trend) of the time series in period t
b_0 = the intercept of the trend line
b_1 = the slope of the trend line
t = a period in time

CHAPTER FORMULAS
(Continued)

Computation of the Trend Equation Coefficients

$$b_1 = \frac{\Sigma t y_t - \dfrac{\Sigma t \Sigma y_t}{n}}{\Sigma t^2 - \dfrac{(\Sigma t)^2}{n}} \tag{18.6}$$

$$b_o = \bar{y} - b_1 \bar{t} \tag{18.7}$$

where y_t = the actual value of the time series in period t

n = the number of periods

$\bar{y}$ = the average value of the time series: That is, $\bar{y} = \dfrac{\Sigma y_t}{n}$

$\bar{t}$ = the average value of t; that is, $\bar{t} = \dfrac{\Sigma t}{n}$

Multiplicative Model

$$Y_t = T_t \cdot C_t \cdot S_t \cdot I_t \tag{18.10}$$

Seasonal-Irregular Effect in the Time Series

$$S_t I_t = \frac{Y_t}{T_t C_t}$$

EXERCISES

***1.** The following time series shows the sales of a clothing store over a 10 week period.

Week	Sales ($,1000s)
1	15
2	16
3	19
4	18
5	19
6	20
7	19
8	22
9	15
10	21

(a) Compute a 4 week moving average for the above time series.

Answer: The moving averages method consists of computing an average of the most recent n data values in the time series. This average is then used as the forecast for the next period. In the above example, the moving average for the first 4 weeks of the time series is computed as

$$\text{Moving average (weeks 1 - 4)} = \frac{15 + 16 + 19 + 18}{4} = 17$$

This moving average value is then used as the forecast for week 5. To compute the next moving average, we drop the oldest sales value (that is, 15 for week 1) and add a new week's sale (that is, 16 for week 5). Thus, we compute the next moving average as

$$\text{Moving Average (weeks 2 - 5)} = \frac{16 + 19 + 18 + 19}{4} = 18$$

Continuing in the same manner, we can compute the rest of the moving averages as

Week	Sales ($1,000s)	Moving Average Forecast	Forecast Error	$(\text{Error})^2$
1	15			
2	16			
3	19			
4	18			
5	19	17	2	4
6	20	18	2	4
7	19	19	0	0
8	22	19	3	9
9	15	20	-5	25
10	21	19	2	4
				46

Table 18.1

(b) Compute the mean square error (MSE) for the 4 week moving average forecast.

Answer: The mean square error is the average of the sum of squared errors. The total sum of squared errors is 46. (The computations are shown in the last two columns of Table 18.1 above.) Therefore, the mean square error is computed as

$$\text{MSE} = \frac{46}{6} = 7.67$$

This measure (MSE) is used as a measure of the accuracy of the method.

(c) Use $\alpha = 0.3$ to compute the exponential smoothing values for the time series.

Answer: Exponential smoothing is a forecasting technique that uses a smoothed value of the time series in one period to forecast the value of the time series in the next period. The exponential smoothing model as given in equation 18.2 is

$$F_{t+1} = \alpha Y_t + (1 - \alpha) F_t$$

where F_{t+1} = the forecast of the time series for period t + 1
Y_t = the actual value of the time series in period t
F_t = the forecast of the time series for period t
α = the smoothing constant ($0 \leq \alpha \leq 1$)

Since no forecast is available for week 1, we let F_1 be the actual value of the time series in period 1. That is, with $Y_1 = 15$, we will assume $F_1 = 15$ in order to get the computations started. Using a smoothing constant of $\alpha = 0.3$, the forecast for week 2 becomes

$$F_2 = 0.3\,Y_1 + (1 - 0.3)\,F_1$$

$$= (0.3)(15) + (0.7)(15) = 15$$

Then continuing our computations, we determine the following forecast for week 3:

$$F_3 = 0.3\,Y_2 + 0.7\,F_2$$

$$= (0.3)(16) + (0.7)(15) = 15.3$$

By continuing the exponential smoothing calculations, we can forecast the weekly sales as shown in Table 18.2.

Week (t)	Time Series Values (Y_t)	Forecast (F_t)	Forecast Error ($Y_t - F_t$)	Squared Error $(Y_t - F_t)^2$
1	15	15.00	—	—
2	16	15.00	1.00	1.00
3	19	15.30	3.70	13.69
4	18	16.41	1.59	2.53
5	19	16.89	2.11	4.45
6	20	17.52	2.48	6.15
7	19	18.26	0.74	0.55
8	22	19.38	2.62	6.86
9	15	18.07	-3.07	9.42
10	21	18.95	2.05	4.20
			Total	48.85

Table 18.2

$$\text{Mean Square Error (MSE)} = \frac{48.85}{9} = 5.43$$

Now we can use the above information to generate a forecast for week 11 before the actual value of the sales in week 11 becomes known.

$$F_{11} = 0.3\,Y_{10} + 0.7\,F_{10}$$

$$= (0.3)(21) + (0.7)(18.95) = 19.56$$

The above figure indicates that the forecasted sales for week 11 is $19,560.

2. The following time series shows the number of units of a particular product sold over the past six months:

Month	Units Sold (Thousands)
1	8
2	3
3	4
4	5
5	12
6	10

(a) Compute a 3 month moving average for the above time series.

(b) Compute the mean square error (MSE) for the 3 month moving average.

(c) Use $\alpha = 0.2$ to compute the exponential smoothing values for the time series.

(d) Forecast the sales volume for month 7.

***3.** The sales volumes of MNM, Inc., a computer firm, for the past 8 years are given below.

Year (t)	Sales (In Millions of Dollars)
1	3.0
2	4.0
3	3.0
4	6.0
5	5.0
6	8.0
7	7.0
8	9.0

(a) Develop a linear trend expression for the above time series.

Answer: The linear trend expression has the form of

$$T_t = b_0 + b_1 t$$

where T_t = the trend value for sales in year t
b_0 = the intercept of the trend line
b_1 = the slope of the trend line
t = a period in time

The values of b_0 and b_1 can be calculated as previously explained by the regression coefficients:

$$b_1 = \frac{\Sigma t y_t - \frac{\Sigma t \Sigma y_t}{n}}{\Sigma t^2 - \frac{(\Sigma t)^2}{n}}$$

and

$$b_o = \bar{y} - b_1 \bar{t}$$

Thus, the following computations are needed.

	t	Y_t	tY_t	t^2
	1	3.0	3	1
	2	4.0	8	4
	3	3.0	9	9
	4	6.0	24	16
	5	5.0	25	25
	6	8.0	48	36
	7	7.0	49	49
	8	9.0	72	64
Totals	36	45.0	238	204

$$\bar{t} = \frac{36}{8} = 4.5$$

$$\bar{Y} = \frac{45}{8} = 5.625$$

Now the slope of the trend line can be computed as

$$b_1 = \frac{\Sigma t y_t - \dfrac{\Sigma t \Sigma y_t}{n}}{\Sigma t^2 - \dfrac{(\Sigma t)^2}{n}} = \frac{(238) - \dfrac{(36)(45)}{8}}{(204) - \dfrac{(36)^2}{8}} = 0.85 \quad \text{(rounded)}$$

Then, the intercept of the trend line is determined as

$$b_o = \bar{y} - b_1 \bar{t} = 5.625 - (0.85)(4.5) = 1.8$$

Therefore, the trend equation can be written as

$$T_t = b_0 + b_1 t = 1.8 + 0.85\ t$$

where T_t is the trend value for sales in period (year) t.

(b) Using the trend equation determined in Part a, project the sales for t = 9.

Answer: Assuming that the past 8 years' data is a good indicator of the future, we can project the sales as

$$T_t = 1.8 + (0.85)(9) = 9.45$$

Since the sales figures are in millions of dollars, we can say that we project the sales in year 9 to be $9,450,000.

4. The sales records of a major auto manufacturer over the past ten years are shown below.

Year (t)	Number of Cars Sold (In Thousands of Units)
1	200
2	300
3	400
4	350
5	600
6	650
7	800
8	700
9	900
10	1000

Develop a linear trend expression and project the sales (the number of cars) for $t = 11$.

***5.** The following data show the quarterly sales of MNM, Inc. (exercise 3) for the years 6 through 8.

Year	Quarter	Sales
6	1	2.5
	2	1.5
	3	2.4
	4	1.6
7	1	2.0
	2	1.4
	3	1.7
	4	1.9
8	1	2.5
	2	2.0
	3	2.4
	4	2.1

(a) Compute the four-quarter moving average values for the above time series.

Answer: The moving average for the first four quarters of year 6 is

$$\text{Moving Average} = \frac{2.5 + 1.5 + 2.4 + 1.6}{4} = 2.0$$

In order to compute the next moving average, we drop the oldest quarterly sales (2.5 of year 6, quarter 1) and add a new quarter's sales (quarter 1 of year 7 which had a scale of 2.0). Thus, we calculate the next moving average:

$$\text{Moving Average} = \frac{1.5 + 2.4 + 1.6 + 2.0}{4} = 1.88$$

Continuing in the same manner, we can compute the moving averages as

Year	Quarter	Sales	Four-Quarter Moving Average	Centered Moving Average
6	1	2.5		
	2	1.5		
			2.00	
	3	2.4		1.94
			1.88	
	4	1.6		1.87
			1.85	
7	1	2.0		1.77
			1.68	
	2	1.4		1.72
			1.75	
	3	1.7		1.82
			1.88	
	4	1.9		1.96
			2.03	
8	1	2.5		2.12
			2.20	
	2	2.0		2.26
			2.25	
	3	2.4		
	4	2.1		

Now note that the moving averages (under the heading "Four-Quarter Moving Average") do not correspond directly to the original quarters of the time series. For instance, the value of 2.0 corresponds to the first half of quarter 3 (of year 6); and 1.88 corresponds to the last half of quarter 3. To resolve this problem, we simply find the average of 2.0 and 1.88 as $(2.0 + 1.88)/2 = 1.94$ and consider this value as the moving average for quarter 3. These averages (called *centered moving averages*) are calculated and shown in the last column above.

(b) Compute the seasonal factors for the four quarters.

Answer: We can determine the seasonal/irregular effect by

$$S_t I_t = \frac{Y_t}{T_t C_t}$$

Thus, by dividing each time series observation (Y_t) by the corresponding moving average value ($T_t C_t$), we can identify the seasonal/irregular effect:

Year	Quarter	Sales (Y_t)	Moving Average Component ($T_t C_t$)	Seasonal-Irregular Component ($S_t I_t = Y_t / T_t C_t$)
6	1	2.5	—	—
	2	1.5	—	—
	3	2.4	1.94	1.24
	4	1.6	1.87	0.86
7	1	2.0	1.77	1.13
	2	1.4	1.72	0.81
	3	1.7	1.82	0.93
	4	1.9	1.96	0.97
8	1	2.5	2.12	1.18
	2	2.0	2.26	0.88
	3	2.4	—	—
	4	2.1	—	—

Now let us consider the seasonal/irregular components for quarter 1. We note $S_t I_t$ had a value of 1.13 in year 7, quarter 1 and a value of 1.18 in year 8, quarter 1. Since year-to-year fluctuations in the seasonal/irregular component can be attributed to an irregular component, we can conclude that the average of $S_t I_t$ values (mentioned above) is an estimate of the first quarter's seasonal influences. Thus, for the first quarter, the seasonal effect = (1.13 + 1.18)/2 = 1.16 (rounded). Therefore, the seasonal components for quarters 2, 3, and 4 can be calculated as shown on the following page.

Quarter	Seasonal/Irregular Components ($S_t I_t$)	Seasonal Factor (S_t)
1	1.13, 1.18	1.16
2	0.81, 0.88	0.85
3	1.24, 0.93	1.09
4	0.86, 0.97	0.92

Considering the seasonal factors, we note that the best sales quarter is quarter 1, where the sales averaged 16.0% above the average quarterly sales; while the worst sales quarter is quarter 2 with the sales averaging 15.0% below the average quarterly sales.

(c) Use the seasonal factors developed in Part b to adjust the forecast for the effect of season for year 6.

Answer: By dividing each time series observations by the corresponding seasonal factor, the seasonal effect is removed from the time series. The following table summarizes the removal of the seasonal factor for year 6.

Quarter	Sales (Y_t)	Seasonal Factor (S_t)	Deseasonalized Sales ($Y_t/S_t = T_t C_t I_t$)
1	2.5	1.16	2.16
2	1.5	0.85	1.76
3	2.4	1.09	2.20
4	1.6	0.92	1.74

6. The following data show the quarterly sales of a major auto manufacturer (introduced in exercise 4) for the years 8 through 10:

Year	Quarter	Sales
8	1	160
	2	180
	3	190
	4	170
9	1	200
	2	210
	3	260
	4	230
10	1	210
	2	240
	3	290
	4	260

(a) Compute the four-quarter moving average values for the above time series.

(b) Compute the seasonal factors for the four quarters.

(c) Use the seasonal factors developed in Part b to adjust the forecast for the effect of season for year 9.

SELF-TESTING QUESTIONS

In the following multiple choice questions, circle the correct answer. An answer key is provided following the questions.

1. A group of observations measured at successive time intervals is known as

a) a trend component
b) a time series
c) a forecast
d) an additive time series model

2. A component of the time series model that results in the multi-period above-trend and below-trend behavior of a time series is

a) a trend component
b) a cyclical component
c) a seasonal component
d) an irregular component

3. The model which assumes that the actual time series value is the product of its components is the

a) forecast time series model
b) multiplicative time series model
c) additive time series model
d) none of the above

4. A method of smoothing a time series which can be used to identify the combined trend/cyclical component is

a) the moving average
b) the percent of trend
c) exponential smoothing
d) the trend/cyclical index

5. A method which uses a weighted average of past values for arriving at smoothed time series values is known as

a) the smoothing average
b) the moving average
c) the exponential average
d) exponential smoothing

6. In the linear trend equation $T = b_0 + b_1 t$, b_1 represents the

a) trend value in period t
b) intercept of the trend line
c) slope of the trend line
d) point in time

7. In the linear trend equation $T = b_0 + b_1 t$, b_0 represents the

a) time
b) slope of the trend line
c) trend value in period 1
d) the Y intercept

8. A parameter of the exponential smoothing model which provides the weight given to the most recent time series value in the calculation of the forecast value is known as the

a) mean square error
b) mean absolute deviation
c) smoothing constant
d) none of the above

9. One measure of the accuracy of a forecasting model is

a) the smoothing constant
b) a deseasonalized time series
c) the mean square error
d) none of the above

10. A qualitative forecasting method that obtains forecasts through "group consensus" is known as the

a) Autoregressive model
b) Delphi approach
c) mean absolute deviation
d) none of the above

ANSWERS TO THE SELF-TESTING QUESTIONS

1. b
2. b
3. b
4. a
5. d
6. c
7. d
8. c
9. c
10. b

ANSWERS TO CHAPTER EIGHTEEN EXERCISES

2. (a)

Month	Moving Average
1	
2	
3	
4	5
5	4
6	7

(b) MSE = 73/3 = 24.33

(c)

Month	F_t
1	8
2	8
3	7
4	6.4
5	6.12
6	7.296

(d) $F_7 = 7.837$

4. $T_t = 113.33 + 86.67\,t$
$T_{11} = 1066.7$ (rounded: 1067)

6.

(a)

Year	Quarter	Centered Moving Average
8	1	
	2	
	3	180.00
	4	188.75
9	1	201.25
	2	217.50
	3	226.25
	4	231.25
10	1	238.75
	2	245.25
	3	
	4	

(b)

Quarter	SI	Seasonal Factor
1	0.99, 0.88	0.935
2	0.96, 0.99	0.975
3	1.05, 1.15	1.1
4	0.90, 0.99	0.945

(c)

Quarter	Sales	Seasonal Factor	Adjusted Sales
1	200	0.935	213.90
2	210	0.975	215.38
3	260	1.1	236.36
4	230	0.945	243.39

CHAPTER NINETEEN

NONPARAMETRIC METHODS

CHAPTER OUTLINE AND REVIEW

In this chapter, you have been introduced to various distribution - free (nonparametric) statistical methods. The main points of this chapter have been

A. Parametric Methods:	Statistical methods which require assumptions about the distribution of a population.
B. Nonparametric Methods:	Statistical methods which require few assumptions about the distribution of a population. These methods are also called distribution-free methods.
C. Sign Test:	A nonparametric method for determining the differences between two populations based on two matched samples, where only preference data (ordinal data) is available.
D. Wilcoxon Signed-Rank Test:	A nonparametric method for analyzing matched sample data whenever interval or ratio scaled data are available for each matched pair.
E. Mann-Whitney-Wilcoxon Test:	A nonparametric method for testing for a difference between two populations based on independent random samples from the two populations.

F. Kruskal-Wallis Test: This test is an extension of the Mann-Whitney-Wilcoxon test dealing with three or more populations. The Kruskal-Wallis test is a nonparametric version of the parametric analysis of variance test for differences among population means.

G. Spearman Rank Correlation Coefficient: A correlation coefficient based on rank-order data for two variables.

CHAPTER FORMULAS

Sign Test (Large Sample Case)

Mean: $\mu = 0.5\,n$ (19.1)

Standard deviation: $s = \sqrt{.25\,n}$ (19.2)

Wilcoxon Signed-Rank Test

Mean: $\mu_T = 0$ (19.3)

Standard deviation: $\sigma_T = \sqrt{\dfrac{n\,(n+1)\,(2n+1)}{6}}$ (19.4)

Distribution form: approximately normal provided $n \geq 10$

Mann-Whitney-Wilcoxon Test (Large Sample)

Mean: $\mu_T = \dfrac{1}{2}\,n_1\,(n_1 + n_2 + 1)$ (19.6)

Standard deviation: $\sigma_T = \sqrt{\dfrac{1}{12}\,n_1\,n_2\,(n_1 + n_2 + 1)}$ (19.7)

Kruskal-Wallis Test Statistic

$$W = \frac{12}{n_T\,(n_T + 1)} \sum_{i=1}^{k} \frac{R_i^2}{n_i} - 3(n_T + 1) \qquad (19.8)$$

where k = the number of populations

n_i = the number of items in sample i

$n_T = \sum n_i$ = total number of items in all samples

R_i = sum of the ranks for sample i

Chapter Formulas
(Continued)

Spearman Rank-Correlation Coefficient

$$r_s = 1 - \frac{6\sum d_i^2}{n\,(n^2 - 1)} \qquad (19.9)$$

where n = the number of items or individuals being ranked

x_i = the rank of item i with respect to one variable

y_i = the rank of item i with respect to a second variable

$d_i = x_i - y_i$

Sampling Distribution of r_s

Mean: $\mu = 0$ (19.10)

Standard deviation: $\sigma_{r_s} = \sqrt{\frac{1}{(n - 1)}}$ (19.11)

Distribution form: approximately normal provided $n \geq 10$

EXERCISES

***1.** Fifteen people were asked to indicate their preference for Cola versus Uncola soft drinks. The following data showed their preferences.

Individual	Cola vs. Uncola
1	+
2	+
3	-
4	+
5	-
6	-
7	-
8	+
9	+
10	+
11	-
12	+
13	+
14	+
15	-

With $\alpha = 0.06$, test for a significant difference in the preferences for the two soft drinks. A "+" indicates a preference for Cola over Uncola.

Answer: We can use the sign test and test the following hypotheses.

H_o: P = 0.5 (There is no preference for one product over the other.)

H_a: P ≠ 0.5 (There is a preference for one product over the other.)

The number of "+'s" in the sample follows a binomial distribution with P = 0.5.

With a confidence level of 0.06, the rejection region will be an area of approximately 0.03 in each tail of the distribution. Now let us refer to the binomial probability tables (Table 5 in Appendix B). With a sample size n = 15 and P = 0.5, we add the probability of having 0, 1, 2, . . . "+" signs until a value close to 0.03 is reached. Adding the probability of having 0,1,2, or 3 "+" signs, we obtain .0000 + .0005 + .0032 + .0139 = .0176. Note that we stopped at 3 because adding the probability of 4 "+" signs, which is .0417, would result in a value of .0176 + .0417 = .0598, exceeding the desired area of 0.03. Similarly, at the other end of the distribution, we find the probability of having 12, 13, 14, or 15 "+" signs to be .0139 + .0032 + .0005 + .0000 = .0176. The sum of the two areas at the two ends of the distribution is .0176 + .0176 = .0352, which is the closest we can come to the level of significance of 0.06 without exceeding it. Therefore, the decision rule can be stated:

> Do not reject H_0 if the number of "+" signs is greater or equal to 4 and less than or equal to 11.
>
> Reject H_0 if the number of "+" signs is less than 4 or greater than 11.

Now we note that there are 9 "+" signs in the sample. Since 9 is not in the rejection range, the null hypothesis is not rejected; hence, we conclude that there is no preference for Cola over Uncola.

2. The following data show the preference of 20 people for two candidates running for public office. A "+" indicates a preference for the Democratic candidate, and a
"-" indicates a preference for the Republican candidate.

Individual	Republican vs. Democrat
1	+
2	-
3	+
4	+
5	+
6	+
7	-
8	+
9	+
10	+
11	+
12	-
13	-
14	+
15	+
16	-
17	-
18	+
19	+
20	+

With $\alpha = 0.05$ test for a significant difference in the preference for the candidates.

***3.** In a sample of 120 people, 50 indicated that they prefer domestic automobiles, while 60 said they prefer foreign made cars, and 10 indicated no preference. At a 0.05 level of significance, determine if there is evidence of a significant difference in the preferences for the two makes of automobiles.

Answer: The null and the alternative hypotheses are stated as

H_o: $P = 0.5$

H_a: $P \neq 0.5$

Since the sample size $n > 20$, the normal approximation of the sampling distribution for the number of "+" signs is

Distribution Form: Approximately normal
Mean: $\mu = 0.5\,n$

Standard Deviation: $\sigma = \sqrt{p\,(1 - p)\,n} = \sqrt{0.25\,n}$

Since 10 people indicated no preference, our sample is actually $n = 120 - 10 = 110$. Now we can compute the mean and the standard deviation:

$$\mu = pn = 0.5\,n = (0.5)(110) = 55$$

$$\sigma = \sqrt{0.25\,n} = \sqrt{(0.25)(110)} = 5.244$$

Since the distribution is approximately normal, we can read the z value from Table 1 of Appendix B as ± 1.96. Therefore, the decision rule will become

Do not reject H_o if $-1.96 \leq z \leq 1.96$
Reject H_o otherwise

Next, we can compute the z statistic as

$$z = \frac{\bar{x} - \mu}{\sigma} = \frac{50 - 55}{5.244} = -0.953$$

(In computing the above z statistic, $\bar{x}$ represents the number of "+" signs; in this case, the number of people who preferred domestic automobiles.)

Since $z = -0.953$ is not in the rejection range, the null hypothesis is not rejected; and we conclude that there is no evidence of preference for one kind of automobile over the other.

4. In a sample of 300 shoppers, 160 indicated they prefer fluoride toothpaste, 120 favored non-fluoride, and 20 were indifferent. At a 0.05 level of significance, test for a difference in the preference for the two kinds of toothpaste.

***5.** Ten secretaries were sent to take a typing efficiency course. The following data show the typing speeds of the secretaries before and after the course:

Secretary	Typing Speed Before the Course	Typing Speed After the Course
1	59	57
2	57	62
3	60	60
4	66	63
5	68	69
6	59	63
7	72	74
8	52	56
9	58	64
10	63	64

At $\alpha = 0.05$, what can be concluded about the effectiveness of the course?

Answer: We can use the Wilcoxon signed-rank test to test the following hypotheses:

H_0: The two populations of typing speeds are identical.

H_a: The two populations of typing speeds are not identical.

To use the Wilcoxon signed-rank test, we must determine the differences between the pair of observations and then rank the absolute values of the differences. (Any difference of zero is discarded.) After the ranks of the absolute differences have been determined, the ranks are given the sign of the original difference in the data. The procedure just mentioned is shown below.

Secretary	Difference	Absolute Value of Difference	Rank of Absolute Difference	Signed Rank
1	2	2	3.5	3.5
2	-5	5	8.0	-8.0
3	0	0	—	—
4	3	3	5.0	5.0
5	-1	1	1.5	-1.5
6	-4	4	6.5	-6.5
7	-2	2	3.5	-3.5
8	-4	4	6.5	-6.5
9	-6	6	9.0	-9.0
10	-1	1	1.5	-1.5
				T = -28.0

The sampling distribution of T is approximately normal with a mean of

$$\mu_T = 0$$

and a standard deviation of

$$\sigma_T = \sqrt{\frac{n\,(n+1)\,(2n+1)}{6}} = \sqrt{\frac{(9)\,(10)\,(19)}{6}} = 16.88$$

At a 0.05 level of significance, the null hypothesis will not be rejected if $-1.96 \le z \le 1.96$, where the z statistic is computed as

$$z = \frac{T - \mu_T}{\sigma_T} = \frac{-28 - 0}{16.88} = -1.659$$

Since -1.659 is not in the rejection range, the null hypothesis is not rejected; and we conclude that the two populations are identical.

6. Ten drivers were asked to drive two models of a car. Each car was given one gallon of gasoline. The distance that each automobile traveled on a gallon of gasoline is shown below:

Driver	Distance Traveled (Miles) Model A	Model B
1	27.7	27.1
2	28.4	28.0
3	28.9	28.7
4	27.9	27.6
5	26.5	26.0
6	29.1	29.0
7	28.9	28.2
8	28.9	28.0
9	28.8	28.0
10	28.0	27.0

At $\alpha = 0.05$, what can be concluded about the performance of the two models?

***7.** The sales records of two branches of a department store over the last 12 months are shown below. (Sales figures are in thousands of dollars.)

Month	Sales of Branch A	Sales of Branch B
1	257	210
2	280	230
3	200	250
4	250	260
5	284	275
6	295	300
7	297	320
8	265	290
9	330	310
10	350	325
11	340	329
12	272	335

Use $\alpha = 0.05$ and test to determine if there is a significant difference in the populations of the sales of the two branches.

Answer: To answer the above question, we can use the Mann-Whitney Wilcoxon test, where the hypotheses to be tested can be written as

H_0: The two populations are identical.

H_a: The two populations are not identical.

The procedure for the Mann-Whitney-Wilcoxon test requires us to rank the combined data (the sales of both stores) from low to high, where the lowest sales value receives a rank of 1. Where ties appear, we assign a rank equal to the average of the ranks associated with the tied items. For instance, in the following table, you will note that the sales of Branch A in month 4 was 250 and the sales of Branch B in month 3 was also 250. They would have received the rank of 4 and 5; but since the sales values are equal, we assign to each a rank of 4.5 (the average of 4 and 5). The sales and the ranks of both branches and the total ranks for each branch are shown on the next page.

Month	Branch A	Rank	Branch B	Rank
1	257	6	210	2
2	280	11	230	3
3	200	1	250	4.5
4	250	4.5	260	7
5	284	12	275	10
6	295	14	300	16
7	297	15	320	18
8	265	8	290	13
9	330	21	310	17
10	350	24	325	19
11	340	23	329	20
12	272	9	335	22
Sum of Ranks		148.5		151.5

The sum of the ranks for Branch A is T = 148.5, and the sampling distribution of T is approximately normal with a mean of

$$\mu_1 = \frac{1}{2} n_1 (n_1 + n_2 + 1)$$

and a standard deviation of

$$\sigma_T = \sqrt{\frac{1}{12} n_1 n_2 (n_1 + n_2 + 1)}$$

where n_1 and n_2 are the sample sizes. Thus, we can calculate the mean and the standard deviation as

$$\mu_T = \frac{1}{2} (12)(12 + 12 + 1) = 150$$

$$\sigma_T = \sqrt{\frac{1}{12} (12)(12)(12 + 12 + 1)} = 17.3$$

Now, we can use the usual hypothesis testing procedure. The null hypothesis will not be rejected if $-1.96 \leq z \leq 1.96$, where the z statistic is

$$z = \frac{T - \mu_T}{\sigma_T} = \frac{148.5 - 150}{17.3} = -0.087$$

Since the z statistic is not in the rejection range, the null hypothesis is not rejected; and we conclude that the distribution of the sales in both branches is identical.

8. Independent random samples of ten day students and ten evening students at a university showed the following age distributions.

Age of Day Students	Age of Evening Students
26	32
18	24
25	23
27	30
19	40
30	41
34	42
21	39
33	45
31	35

Use $\alpha = 0.05$ and test for any significant differences in the age distribution of the two populations.

***9.** A comprehensive statistics examination is given to 16 students in order to determine whether or not there is a significant difference in the performance of students majoring in the various disciplines of Business Administration. The following data show the scores of the sixteen students (five majoring in accounting, six majoring in management, and five majoring in marketing).

Accounting	Management	Marketing
91	63	95
80	92	80
70	86	70
60	75	60
85	70	90
	99	

At a 0.05 level of significance, test to see if there is a significant difference in the performance of the students in the three majors.

Answer: The Kruskal-Wallis test can be applied to this situation. The null and the alternative hypotheses are stated as

H_o: All three populations are identical

H_a: Not all populations are identical

The Kruskal-Wallis test statistic is given as

$$W = \frac{12}{n_T\,(n_T + 1)} \sum_{i=1}^{k} \frac{R_i^2}{n_i} - 3(n_T + 1)$$

where k = the number of the populations
n = the number of the items in sample i

n_T = the total number of items in all the samples
R_i = the sum of the ranks for sample i.

The first step for this test is to rank the scores of the sixteen students, with the lowest score receiving a rank of 1 and assign an average of the ranks to tie data elements. Thus, the ranking of the scores can be shown as follows.

Accounting	Rank	Management	Rank	Marketing	Rank
91	13	63	3	95	15
80	8.5	92	14	80	8.5
70	5	86	11	70	5
60	1.5	75	7	60	1.5
85	10	70	5	90	12
		99	16		
	38		56		42

Now, the W statistic can be computed as

$$W = \frac{12}{n_T\,(n_T + 1)} \sum_{i=1}^{k} \frac{R_i^2}{n_i} - 3(n_T + 1)$$

$$= \frac{12}{16\,(16 + 1)} \left[\frac{(38)^2}{5} + \frac{(56)^2}{6} + \frac{(42)^2}{5} \right] - 3\,(16 + 1) = 0.36$$

The Kruskall-Wallis "W" can be approximated by a chi-square distribution with k - 1 = 3 - 1 = 2 degrees of freedom. From the chi-square tables, we can read the chi-square to be 5.99147. Since the test statistic "W" (0.36) is less than the chi-square, the null hypothesis is not rejected, and we can conclude there is no significant difference in the performance of the students of the three majors.

10. In exercise 6 of Chapter 13, you were given the following information. Three universities in your state have decided to administer the same comprehensive examination to the recipients of MBA degrees from the three institutions. From each institution, a random sample of MBA recipients has been selected and given the test. The following table shows the scores of the students from each university.

Northern University	Central University	Southern University
56	62	94
85	97	72
65	91	93
86	82	78
93		54
		77

Use the Kruskal-Wallis test to determine if there is a significant difference in the average scores of the students from the three universities. Use a 0.01 level of significance.

***11.** A survey of male and female students showed the following ranking of 12 professors in the management department.

Professor	Ranking By Female Students	Ranking By Male Students
1	7	8
2	8	7
3	1	2
4	2	3
5	9	1
6	3	10
7	10	9
8	11	4
9	4	6
10	6	11
11	12	5
12	5	12

Do the rankings given by the female students agree with the rankings given by the male students? Use $\alpha = 0.05$.

Answer: The Spearman rank correlation coefficient is

$$r_s = 1 - \frac{6\sum d_i^2}{n\,(n^2 - 1)}$$

where d_i is the difference between the rankings by male and female students.

Thus, we can calculate

Professor	Difference d_i	d_i^2
1	-1	1
2	1	1
3	-1	1
4	-1	1
5	8	64
6	-7	49
7	1	1
8	7	49
9	-2	4
10	-5	25
11	7	49
12	-7	49
		294

Therefore, the Spearman rank correlation is

$$r_s = 1 - \frac{6\sum d_i^2}{n(n^2 - 1)} = 1 - \frac{6\ (294)}{12\ (144 - 1)} = -\ 0.028$$

Now that we have computed the sample's rank correlation, we can test for the significance of the rank correlation of the population. The hypotheses to be tested are

$$H_0: \rho_s = 0$$

$$H_a: \rho_s \neq 0$$

The sampling distribution of r_s is

$$\mu = 0$$

$$\sigma_{r_s} = \sqrt{\frac{1}{(n - 1)}} = \sqrt{\frac{1}{(12 - 1)}} = 0.3$$

At a 0.05 level of significance, the null hypothesis is not rejected if the z statistic is between -1.96 and 1.96. The z statistic is computed as

$$z = \frac{r_s - \mu_{r_s}}{\sigma_{r_s}} = \frac{-0.028 - 0}{0.3} = -0.093$$

Since -0.093 is not in the rejection range, the null hypothesis is not rejected; and we conclude that the Spearman rank correlation is not significantly different from zero.

12. Two individuals were asked to rank the performances of eight different automobiles. The following show their rankings.

Automobile	Ranking By First Person	Ranking By Second Person
1	3	2
2	5	1
3	1	4
4	6	7
5	2	5
6	4	8
7	7	6
8	8	3

Determine the Spearman rank correlation coefficient and test for a significant correlation with $\alpha = 0.05$.

13. Two groups of students were asked to rank the activities sponsored by the Student Government Association on campus. The following show their rankings.

Activity	Resident Student Ranking	Non-Resident Student Ranking
1	3	6
2	1	2
3	8	5
4	2	1
5	5	7
6	7	8
7	4	3
8	6	4

Determine the Spearman rank correlation coefficient and test for a significant correlation with $\alpha = 0.05$.

SELF-TESTING QUESTIONS

In the following multiple choice questions, circle the correct answer. An answer key is provided following the questions.

1. A collection of statistical methods that generally requires very few, if any assumptions about the population distribution is known as

a) parametric methods
b) nonparametric methods
c) distribution-free methods
d) either b or c is correct
e) none of the above

2. Which of the following tests would be an example of nonparametric method?

a) Mann-Whitney-Wilcoxon test
b) Wilcoxon signed-rank test
c) sign test
d) all of the above
e) only b and c

3. A nonparametric version of the parametric analysis of variance test is

a) Kruskal-Wallis Test
b) Mann-Whitney-Wilcoxon Test
c) sign-test
d) Wilcoxon Signed-rank test
e) none of the above

4. A nonparametric method for determining the differences between two populations based on two matched samples where only preference data is required is the

a) Mann-Whitney-Wilcoxon test
b) Wilcoxon signed-rank test
c) sign test
d) Kruskal-Wallis Test

5. When ranking combined data in a Wilcoxon signed rank test, the data which receives a rank of 1 is the

a) lowest value
b) highest value
c) middle value
d) This can vary according to data

6. The collection of statistical methods that require assumptions about the population is known as

a) distribution free methods
b) nonparametric methods
c) either a or b
d) parametric methods
e) none of the above

7. The Spearman rank-correlation coefficient is

a) a correlation measure based on the average of data items
b) a correlation measure based on rank-ordered data for two variable
c) either a or b
d) none of the above

8. The level of measurement which allows for the rank ordering of data items is

a) nominal measurement
b) ratio measurement
c) interval measurement
d) ordinal measurement
e) none of the above

9. The level of measurement which is simply a label for the purpose of identifying an item is

a) ordinal measurement
b) ratio measurement
c) nominal Measurement
d) none of the above

10. The labeling of parts as "defective" or "nondefective" is an example of

a) ordinal data
b) ratio data
c) interval data
d) nominal data
e) none of the above

ANSWERS TO THE SELF-TESTING QUESTIONS

1. d
2. d
3. a
4. c
5. a
6. d
7. b
8. d
9. c
10. d

ANSWERS TO CHAPTER NINETEEN EXERCISES

2. Probability (0, 1, 2, 3, 4) = 0.0207
   Probability (15, 16, 17, 18, 19, 20) = 0.0207
   Number of "+" signs is 14
   Do not reject H_o, there is no significant difference.

4. Mean = 140
   Standard deviation = 8.367
   $z = 2.39 > 1.96$ Reject H_o, there is a significant difference.

6. T = 55
   Standard deviation = 19.62
   $z = 2.8$ Reject H_o, there is a significant difference.

8. T = 74.5
   Reject H_o, the populations are different.

10. W = 0.2
    Chi-square = 9.21034
    Do not reject H_o, there is no significant difference.

12. $r_s = 0.0714$
    Do not reject H_o, the Spearman rank correlation is not significantly different from zero.

13. $r_s = 0.64$
    Do not reject H_o, the Spearman rank correlation is not significantly different from zero.

CHAPTER TWENTY

STATISTICAL METHODS FOR QUALITY CONTROL

CHAPTER OUTLINE AND REVIEW

In this chapter, the concept of quality control was explained, and two statistical methods which are used in quality control were presented. The two methods are *acceptance sampling* and *statistical process control*. Most of the material in this chapter has relied heavily on the concepts which you studied in Chapters 5, 7, and 9. A quick review of these chapters will help you greatly in understanding the material covered in Chapter 20. The major topics which you should have learned from Chapter 20 include

A. **Quality Control:** Inspection and measurements performed to determine whether or not the quality standards are being met. If the quality standards are not being met, corrective actions should be taken to insure quality.

B. **Lot:** A group of items such as final products, parts, or raw material.

C. **Acceptance Sampling:** A statistical quality control procedure by which one can decide whether to accept or reject a given lot based on the number of defective items in the lot.

D. **Acceptance Criterion:** The maximum number of defective items that can be found in a sample which will still permit the acceptance of the lot.

E. **Producer's Risk:** A Type I error or the risk of rejecting a good quality lot in acceptance sampling.

F. **Consumer's Risk:** A Type II error or the risk of accepting a poor quality lot in acceptance sampling.

G. **Operating Characteristic Curve:** A graph of the probability of accepting the lot versus the percent of defectives in the lot.

H. **Multiple Sampling Plan:** An acceptance sampling where more than one sample is used. The results of the sample indicate whether to accept or reject the lot or continue sampling.

I. **Common Causes:** Natural variations in process outputs which are due purely to chance.

J. **Assignable Causes:** Variations in process output which are due to causes other than common causes.

K. **Control Chart:** A graphical procedure for determining whether or not a process is in control.

L. **$\bar{x}$ Chart:** A control chart showing the *upper* and the *lower control limits* around the *mean* of the output of a production process.

M. **P Chart:** A control chart used when the output of a production process is measured in terms of the percent defective.

CHAPTER FORMULAS

Standard Error of the Mean

$$\sigma_x = \frac{\sigma}{\sqrt{n}} \tag{20.1}$$

Control Limits for an $\bar{x}$ Chart: Process Mean and Standard Deviation

$$UCL = \mu + 3\sigma_{\bar{x}} \tag{20.2}$$

$$UCL = \mu - 3\sigma_{\bar{x}} \tag{20.3}$$

Overall Sample Mean

$$\bar{\bar{x}} = \frac{\bar{x}_1 + \bar{x}_2 + \ldots + \bar{x}_k}{k} \tag{20.4}$$

Average Range

$$\bar{R} = \frac{R_1 + R_2 + \ldots + R_k}{k} \tag{20.5}$$

Control Limits for $\bar{x}$ Chart: Process Mean and Standard Deviation Unknown

$$\bar{\bar{x}} \pm A_2 \bar{R}$$

Control Limits for an R Chart

$$UCL = \bar{R}D_4 \tag{20.14}$$

$$LCL = \bar{R}D_3 \tag{20.15}$$

Standard Error of the Proportion

$$\sigma_p = \sqrt{\frac{p\,(1 - p)}{n}} \tag{20.16}$$

Chapter Formulas
(Continued)

Control Limits for a P Chart

$$\text{UCL} = \text{P} + 3\sigma_{\bar{p}} \tag{20.17}$$

$$\text{UCL} = \text{P} - 3\sigma_{\bar{p}} \tag{20.18}$$

Control Limits for an np chart

$$\text{UCL} = \text{np} + 3\sqrt{\text{np}(1-\text{p})} \tag{20.19}$$

$$\text{UCL} = \text{np} - 3\sqrt{\text{np}(1-\text{p})} \tag{20.20}$$

The Binomial Probability Function for Acceptance Sampling

$$f(x) = \frac{n!}{x!(n - x)!} p^{x} (1 - p)^{n-x} \tag{20.21}$$

where n = the sample size
p = the proportion of defective components in the entire lot
x = the number of defective components found in the sample
f(x) = the probability of finding x defective components in a sample of n components

EXERCISES

***1.** Brakes Shop, Inc. is a franchise that specializes in repairing brake systems of automobiles. The company purchases brake shoes from a national supplier. Currently, lots of 1,000 brake shoes are purchased, and each shoe is inspected before being installed in an automobile. The company has decided, instead of 100% inspection, to adopt an acceptance sampling plan.

(a) Explain what is meant by the acceptance sampling plan.

Answer: The acceptance sampling plan refers to a statistical quality control procedure for accepting or rejecting a given lot based on the results of the sample. The plan requires the establishment of a sample size (n) and an acceptance criterion (c), where c represents the maximum number of defective units that can be found in the sample and still consider the lot as acceptable.

(b) If the company decides to adopt an acceptance sampling plan, what kinds of risks are there?

Answer: Under the acceptance sampling plan, two types of risks are involved. First, the company could reject a good quality lot based on the sample information. This risk is known as the Producer's Risk (Type I error) since the items will be returned to the producer even though they are good. Next, it is possible for the company to accept a bad quality lot based on the sample information. This latter risk is known as the Consumer's Risk (Type II error) since these bad items could eventually be passed on to consumers. Table 20.1 summarizes the outcomes of acceptance sampling.

	State of Nature	
Decision	Good Quality Lot	Bad Quality Lot
Accept the Lot	Correct Decision	Consumer's Risk (Type II Error) Accepting a Poor Quality Lot
Reject the Lot	Producer's Risk (Type I Error) Rejecting a Good Quality Lot	Correct Decision

Table 20.1

(c) The quality control department of the company has decided to select a sample of 10 brake shoes and inspect them for defects. Furthermore, it has been decided that if the sample contains no defective parts, the entire lot will be accepted. If there are 40 defective shoes in a shipment, what is the probability that the entire lot will be accepted?

Answer: We can use the binomial probability function to compute the desired probability. There are 40 defective units in the lot, which indicates that 4% of the units are defective. Recall that the lot will be accepted if no defective parts are found in the sample. Thus, we have

$n = 10$ (sample size)
$c = 0$ (acceptance criterion)
$p = 0.04$ (percent defectives in the lot)

Now we can compute the probability of no defectives in the sample (which leads to the acceptance of the lot):

$$f(x) = \frac{n!}{x!(n - x)!} p^x (1 - p)^{n-x}$$

$$f(0) = \frac{10!}{0!(10)!} (0.04)^0 (1 - 0.04)^{10} = 0.6648$$

Then the probability of rejecting the lot is 1 - 0.6648 = 0.3352.

(d) What is the probability of accepting the lot if there are 50 defective units in the lot?

Answer: Once again we can use the binomial probability function with

$n = 10$
$c = 0$
$p = 0.05$

and compute f(0) as

$$f(0) = \frac{10!}{0!(10)!}(0.05)^0 (1 - 0.05)^{10} = 0.5987$$

Since we have access to the binomial probability table, we could simply read the above value (0.5987) from Table 5 of Appendix B.

(e) Use the binomial table and read the probability of accepting lots which contain 5, 10, 15, 20, 25, 30, 35, 40, 45, and 50% defective units, and construct an operating characteristic curve (OC).

Answer: From Table 5 of Appendix B we can simply read the probabilities for $x = 0$ and $n = 10$ as

Percent Defective in the Lot	Probability of Accepting the Lot
5	0.5987
10	0.3487
15	0.1969
20	0.1074
25	0.0563
30	0.0282
35	0.0135
40	0.0060
45	0.0025
50	0.0010

The operating characteristic curve for the above data is shown in Figure 20.1.

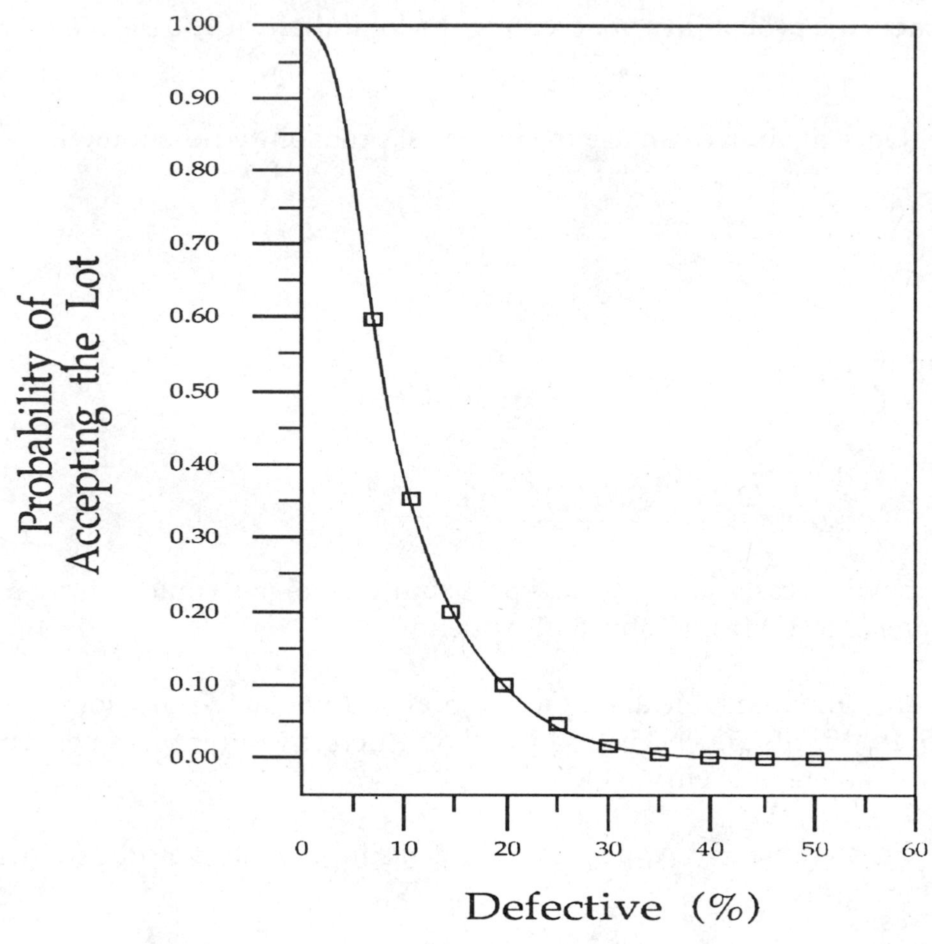

Figure 20.1

2. Refer to exercise 1.

(a) What is the probability of accepting a lot which contains 8% defective units?

(b) What is the probability of rejecting a lot which contains 12% defective units?

3. An acceptance sampling plan uses a sample of 18 with an acceptance criterion of zero. Determine the probability of accepting shipments which contain 5, 10, 15, 20, 25, 30, 35, 40, and 45% defective units, and construct an operating characteristic curve.

***4.** A soft drink filling machine is set up to fill bottles with 12 ounces of soft drink. The standard deviation σ is known to be 0.4 ounces. The quality control department periodically selects samples of 16 bottles and measures their contents. (Assume the distribution of filling volumes is normal.)

(a) Determine the upper and lower control limits and explain what they indicate.

Answer: The process is considered to be in control if the sample mean ($\bar{x}$) is within plus or minus three standard deviations from the mean (μ).

Thus, the upper and lower control limits are given by

$$UCL = \mu + 3\sigma_{\bar{x}}$$

$$LCL = \mu - 3\sigma_{\bar{x}}$$

First, we can compute the standard error of the mean:

$$\sigma_{\bar{x}} = \frac{\sigma}{\sqrt{n}} = \frac{0.4}{\sqrt{16}} = 0.1$$

Then, the upper and lower control limits are computed:

$$UCL = \mu + 3\sigma_{\bar{x}} = 12 + 3\,(0.1) = 12.3$$

$$LCL = \mu - 3\sigma_{\bar{x}} = 12 - 3\,(0.1) = 11.7$$

The above limits indicate that if a sample mean falls between 11.7 to 12.3 ounces, the process is considered to be in control.

(b) The means of six samples were 11.8, 12.2, 11.9, 11.9, 12.1, and 11.8 ounces. Construct an $\bar{x}$ chart and indicate whether or not the process is in control.

Answer: Figure 20.2 shows the $\bar{x}$ chart and the results of the six samples.

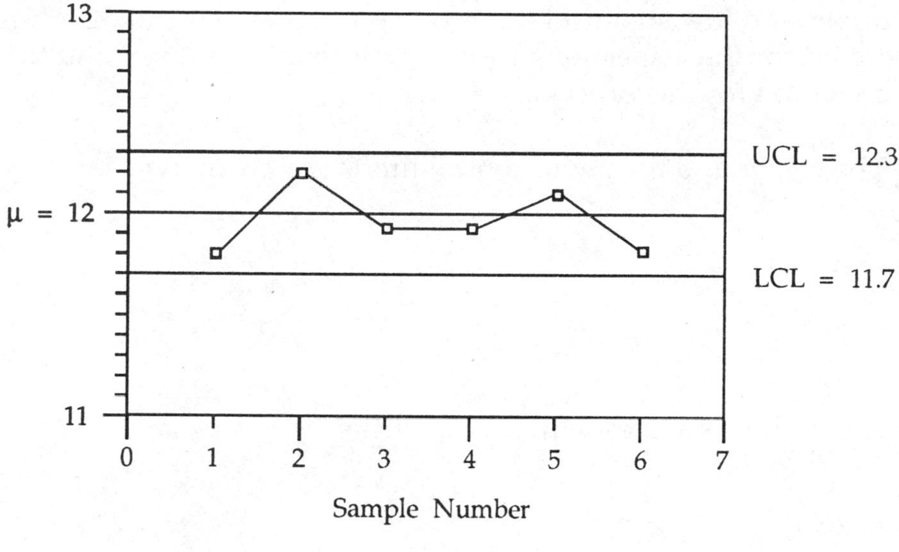

Figure 20.2

As can be seen from Figure 20.2, all the sample means are within the control limits and the process is considered to be in control.

5. A production process which is in control has a mean (μ) of 80 and a standard deviation (σ) of 10.

(a) Determine the upper and the lower control limits for sample sizes of 25.

(b) Five samples had means of 81, 84, 75, 83, and 79. Construct an $\bar{x}$ chart and explain whether or not the process is in control.

***6.** The upper and lower control limits of a process are 66 and 54. Samples of size 16 are used for the inspection process. Determine the mean and the standard deviation for this process.

Answer: The upper and the lower control limits are given by

$$UCL = \mu + 3\sigma_{\bar{x}}$$

$$LCL = \mu - 3\sigma_{\bar{x}}$$

Since the control limits are known, we can write

$$66 = \mu + 3\sigma_{\bar{x}}$$

$$54 = \mu - 3\sigma_{\bar{x}}$$

Adding the above equations together we obtain

$$120 = 2\mu$$

Therefore, the mean is determined as

$$\mu = \frac{120}{2} = 60$$

Now we can substitute the mean in the first equation and get

$$66 = 60 + 3\sigma_{\bar{x}}$$

Solving for $\sigma_{\bar{x}}$ we get

$$3\sigma_{\bar{x}} = 6$$

$$\sigma_{\bar{x}} = \frac{6}{3} = 2$$

Now we can substitute the values of n = 16 and $\sigma_{\bar{x}} = 2$ in the formula $\sigma_{\bar{x}} = \dfrac{\sigma}{\sqrt{n}}$:

$$2 = \frac{\sigma}{\sqrt{16}}$$

Therefore, $\sigma = 8$.

***7.** A production process is considered in control if up to 6% of the items produced are defective. Samples of size 300 are used for the inspection process. Determine the upper and lower control limits for the P chart.

Answer: The upper and lower control limits for the P chart are given by

$$UCL = \mu + 3\sigma_{\bar{p}}$$

$$LCL = \mu - 3\sigma_{\bar{p}}$$

The standard error of the proportion ($\sigma_{\bar{p}}$) is

$$\sigma_{\bar{p}} = \sqrt{\frac{p\,(1 - p)}{n}} \quad \sqrt{\frac{(0.06)(0.94)}{300}} = 0.014 \text{ (rounded)}$$

Thus, the control limits are

$$UCL = 0.06 + (3)\,(.014) = 0.102$$

$$LCL = 0.06 - (3)\,(.014) = 0.108$$

8. A production process is considered to be in control if up to 4% of the items are defective. Samples of size 100 are selected for the inspection process.

(a) Determine the standard error of the proportion.

(b) Determine the upper and the lower control limits for the P chart.

SELF TESTING QUESTIONS

In the following multiple choice questions, circle the correct answer. An answer key is provided following the questions.

1. In acceptance sampling, the risk of rejecting a good quality lot is know as

a) Consumer's risk
b) Producer's risk
c) a Type II error
d) none of the above

2. In acceptance sampling, the risk of accepting a poor quality lot is known as

a) Consumer's risk
b) Producer's risk
c) a Type I error
d) none of the above

3. The maximum number of defective items that can be found in the sample and still lead to acceptance of the lot is

a) the upper control limit
b) the lower control limit
c) the acceptance criterion
d) none of the above

4. Consumer's risk is

a) the same concept as the Producer's risk
b) a Type II error
c) a Type I error
d) none of the above

5. A graph showing the probability of accepting the lot as a function of the percent defective in the lot is

a) a power curve
b) a control chart
c) an operating characteristic curve
d) none of the above

6. A control chart which is used when the output of a production process is measured in terms of the percent defective is

a) a P chart
b) an $\bar{x}$ chart
c) a process chart
d) none of the above

7. If the lower control limit of a P chart is negative,

a) a mistake has been made in the computations
b) use the absolute value of the lower limit
c) it is set to zero
d) none of the above

ANSWERS TO THE SELF TESTING QUESTIONS

1. b
2. a
3. c
4. b
5. c
6. a
7. c

ANSWERS TO CHAPTER TWENTY EXERCISES

2. (a) 0.4344
   (b) 0.7215

3. 0.3972
   0.1501
   0.0536
   0.0180
   0.0056
   0.0016
   0.0004
   0.0001
   0.0000

5. (a) UCL = 86
   LCL = 74
   (b) The process is in control.

8. (a) 0.0196 (rounded)
   (b) UCL = 0.0988
   LCL = 0.0000
   Note: Since the lower control limit is negative, it is set equal to zero.

CHAPTER TWENTY - ONE

SAMPLE SURVEY

CHAPTER OUTLINE AND REVIEW

In this chapter you have been introduced to survey sampling. You have learned that the objective of survey sampling is to make inferences about the population parameters, based on sample information. Furthermore, you have learned that in survey sampling two kinds of errors, i.e., sampling and nonsampling errors, can occur. Four major sample designs: *simple random sampling, stratified simple random sampling, cluster sampling, and systematic sampling* have been presented. The key terms which you should have learned from this chapter (many of which were introduced to you in previous chapters) are

A. **Survey:** A process for collecting data about a situation.

B. **Census:** A survey of the entire population.

C. **Sample Survey:** A survey of a subset of a population.

D. **Element:** The entity on which data are collected.

E. **Population:** The collection of all the elements of interest.

F. **Sample:** A subset of the population.

G. **Target Population:** The population for which inferences are to be made.

H. **Sampled Population:** The population from which the sample is actually selected.

I. **Sampling Unit:** The units selected for sampling. A sampling unit may include several elements.

J. **Frame:** A list of the sampling units for a study. The sample is drawn by selecting units from the frame.

K. **Probabilistic Sampling:** Any method of sampling in which the probability of each possible sample can be computed.

L. **Nonprobabilistic Sampling:** Any method of sampling for which the probability of selecting a sample cannot be computed.

M. **Convenience Sampling:** A nonprobabilistic method of sampling whereby elements are selected on the basis of convenience.

N. **Judgment Sampling:** A nonprobabilistic method of sampling whereby the element selected is based on the judgment of the person doing the study.

0. **Sampling Error:** The error that occurs because a sample, and not the entire population, is used to estimate a population parameter.

P. **Nonsampling Error:** Any error other than sampling error, such as measurement error, interviewer error, processing error, etc.

Q. **Simple Random Sample:** A sample selected in such a manner that each sample of size n has the same probability of being selected.

R. **Bound on Sampling: Error:** A number added to and subtracted from a point estimate to create an approximate 95% confidence interval. It is given by two times the standard error of the point estimator.

S. **Stratified Simple Random Sampling:** A method of selecting a sample in which the population is first divided into strata and a simple random sample is then taken from each stratum.

T. **Cluster Sampling:** A probabilistic method of sampling in which the population is first divided into clusters and then one or more clusters is selected for sampling. In single-stage cluster sampling, every element in each selected cluster is sampled. In two-stage cluster sampling, a sample of the elements in each selected cluster is collected.

U. **Area Sampling:** A version of cluster sampling in which the elements are formed into clusters on the basis of their geographic proximity.

V. **Systematic Sampling:** A method of choosing a sample by randomly selecting the first element and then selecting every k^{th} element thereafter.

CHAPTER FORMULAS

Simple Random Sampling

Interval Estimate of the Population Mean

$$\bar{x} \pm z_{\alpha/2}\sigma_{\bar{x}} \tag{21.1}$$

Estimate of the Standard Error of the Mean

$$s_{\bar{x}} = \sqrt{\frac{N-n}{N}}\left(\frac{s}{\sqrt{n}}\right) \tag{21.2}$$

Interval Estimate of the Population Mean

$$\bar{x} \pm z_{\alpha/2}s_{\bar{x}} \tag{21.3}$$

Approximate 95% Confidence Interval Estimate of the Population Mean

$$\bar{x} \pm 2s_{\bar{x}} \tag{21.4}$$

Point Estimator of a Population Total

$$\hat{x} = N\bar{x} \tag{21.5}$$

Estimate of the Standard Error of $\hat{x}$

$$s_{\hat{x}} = Ns_{\bar{x}} \tag{21.6}$$

Approximate 95% Confidence Interval Estimate of the Population Total

$$N\bar{x} \pm 2s_{\hat{x}} \tag{21.8}$$

CHAPTER FORMULAS
(Continued)

Estimate of the Standard Error of the Proportion

$$s_{\bar{p}} = \sqrt{\left(\frac{N-n}{N}\right)\left(\frac{\bar{p}(1-\bar{p})}{n-1}\right)} \tag{21.9}$$

Approximate 95% Confidence Interval Estimate of the Population Proportion

$$\bar{p} \pm 2s_{\bar{p}} \tag{21.10}$$

Sample Size When Estimating the Population Mean

$$n = \frac{Ns^2}{N\left(\frac{B^2}{4}\right) + s^2} \tag{21.12}$$

Sample Size When Estimating the Population Total

$$n = \frac{Ns^2}{\frac{B^2}{4N} + s^2} \tag{21.13}$$

Sample Size When Estimating the Population Proportion

$$n = \frac{N\bar{p}(1-\bar{p})}{N\left(\frac{B^2}{4}\right) + \bar{p}(1-\bar{p})} \tag{21.14}$$

CHAPTER FORMULAS
(Continued)

Stratified Simple Random Sampling

Point Estimator of the Population Mean

$$\bar{x}_{st} = \sum_{h=1}^{H}\left(\frac{N_h}{N}\right)\bar{x}_h \qquad (21.15)$$

Estimate of the Standard Error of the Mean

$$s_{\bar{x}_{st}} = \sqrt{\left(\frac{1}{N^2}\right)\sum_{h=1}^{H} N_h(N_h - n_h)\frac{s_h^2}{n_h}} \qquad (21.16)$$

Approximate 95% Confidence Interval Estimate of the Population Mean

$$\bar{x}_{st} \pm 2s_{\bar{x}_{st}} \qquad (21.17)$$

Point Estimator of the Population Total

$$\hat{x} = N\bar{x}_{xt} \qquad (21.18)$$

Estimate of the Standard Error of $\hat{x}$

$$s_{\hat{x}} = Ns_{\bar{x}_{st}} \qquad (21.19)$$

Approximate 95% Confidence Interval Estimate of the Population Total

$$N\bar{x}_{st} \pm 2s_{\hat{x}} \qquad (21.20)$$

Point Estimator of the Population Proportion

$$\bar{p}_{st} = \sum_{h=1}^{H}\left(\frac{Nh}{N}\right)\bar{p}_h \qquad (21.20)$$

CHAPTER FORMULAS
(Continued)

Estimate of the Standard Error of $\bar{p}_{st}$

$$s_{\bar{p}_{st}} = \sqrt{\frac{1}{N^2}\sum_{h=1}^{H} N_h\left(N_h - n_h\right)\left[\frac{\bar{p}_h\left(1-\bar{p}_h\right)}{n_h - 1}\right]} \tag{21.22}$$

Approximate 95% Confidence Interval Estimate of the Population Proportion

$$\bar{p}_{st} \pm 2 s_{\bar{p}_{st}} \tag{21.23}$$

Allocating the Total Sample, n, to the Strata: Neyman Allocation

$$n_h = n\left(\frac{N_h s_h}{\sum_{h=1}^{H} N_h s_h}\right) \tag{21.24}$$

Sample Size When Estimating the Population Mean

$$n = \frac{\left(\sum_{h=1}^{H} N_h s_h\right)^2}{N^2\left(\frac{B^2}{4}\right) + \sum_{h=1}^{H} N_h s_h^2} \tag{21.25}$$

CHAPTER FORMULAS
(Continued)

Sample Size When Estimating the Population Total

$$n = \frac{\left(\sum_{h=1}^{H} N_h s_h\right)^2}{\frac{B^2}{4} + \sum_{h=1}^{H} N_h s_h^2} \qquad (21.26)$$

Sample Size When Estimating the Population Proportion

$$n = \frac{\left(\sum_{h=1}^{H} N_h \sqrt{\bar{p}\left(1-\bar{p}_h\right)}\right)^2}{N^2\left(\frac{B^2}{4}\right) + \sum_{h=1}^{H} N_h \bar{p}_h \left(1-\bar{p}\right)} \qquad (21.27)$$

Allocating the Total Sample, n, to the Strata: Proportional Allocation

$$n_h = n\left(\frac{N_h}{N}\right) \qquad (21.28)$$

CHAPTER FORMULAS
(Continued)

Cluster Sampling

Point Estimator for the Population Mean

$$\bar{x}_c = \frac{\sum_{i=1}^{n} x_i}{\sum_{i=1}^{n} M_i} \tag{21.29}$$

Estimate of the Standard Error of the Mean

$$s_{\bar{x}_c} = \sqrt{\left(\frac{N-n}{Nn\bar{M}^2}\right)\frac{\sum_{i=1}^{n}\left(x_i - \bar{x}_c M_i\right)^2}{n-1}} \tag{21.30}$$

Approximate 95% Confidence Interval Estimate of the Population Mean

$$\bar{x}_c \pm 2s_{\bar{x}_c} \tag{21.31}$$

Point Estimator of the Population Total

$$\hat{x} = M\bar{x}_c \tag{21.32}$$

Estimate of the Standard Error of $\hat{x}$

$$s_{\hat{x}} = Ms_{\bar{x}_c} \tag{21.33}$$

Approximate 95% Confidence Interval Estimate of the Population Total

$$M\bar{x}_c \pm 2s_{\bar{x}_c} \tag{21.34}$$

CHAPTER FORMULAS
(Continued)

Point Estimator of the Population Proportion

$$\bar{p}_c = \frac{\sum_{i=1}^{n} a_i}{\sum_{i=1}^{n} M_i} \tag{21.35}$$

Estimate of the Standard Error of $\bar{p}_c$

$$s_{\bar{p}_c} = \sqrt{\left(\frac{N-n}{Nn\bar{M}^2}\right)\frac{\sum_{i=1}^{n}\left(a_i - \bar{p}_c M_i\right)^2}{n-1}} \tag{21.36}$$

Approximate 95% Confidence Interval Estimate of the Population Proportion

$$\bar{p}_c \pm 2 s_{\bar{p}_c} \tag{21.37}$$

EXERCISES

***1.** Nancy June, Inc. has 1500 employees. A simple random sample of 81 employees was selected, and the individuals in the sample were asked how much they contribute (monthly) to their retirement accounts. The sample mean, $\bar{x}$ was $150 with a standard deviation, s, of $45.

(a) Estimate the standard error of the mean.

Answer: When a simple random sample is selected from a finite population, an estimate of the standard error of the mean is

$$s_{\bar{x}} = \sqrt{\frac{N-n}{N}}\left(\frac{s}{\sqrt{n}}\right)$$

In this case, N = 1500, n = 81, and s = 45. Therefore, the standard error of the mean can be estimated as

$$s_{\bar{x}} = \sqrt{\frac{N-n}{N}}\left(\frac{s}{\sqrt{n}}\right) = \sqrt{\frac{1500-81}{1500}}\left(\frac{45}{\sqrt{81}}\right) = 4.86 \text{ (rounded)}$$

(b) Develop an approximate 95% confidence interval for the population mean.

Answer: The interval estimate for the population mean is given by

$$\bar{x} \pm z_{\alpha/2} s_{\bar{x}}$$

When developing an approximate estimate in a sample survey, it is common practice to use a value of z = 2, instead of z = 1.96. Therefore, an approximate 95% confidence interval estimate for the population mean is given by

$$\bar{x} \pm 2 s_{\bar{x}}$$

In Part a of this exercise, $s_{\bar{x}}$ was found to be 4.86, and the sample mean was given as $\bar{x} = 150$. Hence, the 95% confidence interval for the population mean is computed as

$$\bar{x} \pm 2 s_{\bar{x}} = 150 \pm (2)(4.86) = 150 \pm 9.72$$

which results in an interval of $140.28 to $159.72. The above interval indicates that we are 95% confident that the actual mean of the population is between $140.28 to $159.72.

2. Simple random sampling has been used to obtain a sample of size 60 from a population of size 700. The sample mean was 500 with a standard deviation of 60.

(a) Estimate the standard error of the mean.

(b) Develop an approximate 95% confidence interval for the population mean.

3. Assume simple random sampling has been used, and the following information was obtained.

Population size	$N = 900$
Sample size	$n = 36$
Sample mean	$\bar{x} = 300$
Sample standard deviation	$s = 72$

(a) Estimate the standard error of the mean.

(b) Develop an approximate 95% confidence interval for the population mean.

***4.** The accounting firm of Shannon Lipscomb and Associates (SLA) was commissioned to audit a population of 500 accounts. For this audit, SLA selected a simple random sample of 64 accounts. The sample showed an average discrepancy of $120 with a standard deviation of $24.

(a) Estimate the population total discrepancy.

Answer: The point estimator of a population total is computed as

$$\hat{x} = N\bar{x}$$

Since the population consisted of 500 accounts and the sample showed an average discrepancy of $240, we can compute the population total discrepancy as

$$\hat{x} = N\bar{x} = (500)(240) = 60{,}000$$

Therefore, the population total discrepancy is estimated to be $60,000.

(b) Develop an approximate 95% confidence interval for the population total discrepancy.

Answer: The approximate 95% confidence interval estimate for the population total is given by

$$N\bar{x} \pm 2s_{\hat{x}}$$

where the estimate of the standard error of $\hat{x}$ is given by

$$s_{\hat{x}} = Ns_{\bar{x}}$$

First, let us compute the standard error of the mean $s_{\bar{x}}$ as

$$s_{\bar{x}} = \sqrt{\frac{N-n}{N}}\left(\frac{s}{\sqrt{n}}\right) = \sqrt{\frac{500-64}{500}}\left(\frac{24}{\sqrt{64}}\right) = 2.80 \text{ (rounded)}$$

Now, we can compute the standard error of $\hat{x}$ as

$$s_{\hat{x}} = Ns_{\bar{x}} = (500)(2.8) = 1{,}400$$

Finally, the interval can be computed as

$$N\bar{x} \pm 2s_{\hat{x}} = (500)(240) \pm (2)(1{,}400)$$

$$= 60{,}000 \pm 2{,}800$$

or 57,200 to 62,800. This interval indicates that we are 95% confident that the total discrepancy in the population is between \$57,200 to \$62,800.

5. A university has 5,000 students. The manager of the food services at the university is interested in determining the total lunch expenditure. A simple random sample of 121 lunch receipts was selected. The sample showed an average of \$3 with a standard deviation of \$0.40.

(a) Estimate the standard error of the sample mean.

(b) Develop an approximate 95% confidence interval for the population mean.

(c) Estimate the population's total expenditure for lunch.

(d) Develop an approximate 95% confidence interval for the total expenditure for lunch.

6. A simple random sample of size 144 is taken from a population of size 1,000. The sample mean is 250, and the standard deviation of the sample is 48.

(a) Estimate the population total.

(b) Determine the standard error of the estimate of the population.

(c) Develop an approximate 95% confidence interval for the population total.

7. Simple random sampling has been used to obtain a sample of $n = 49$ elements from a population of $N = 500$. The sample mean was 120, and the sample standard deviation was found to be 21.

(a) Estimate the population total.

(b) Determine the standard error of the estimate of the population.

(c) Develop an approximate 95% confidence interval for the population total.

***8.** A company has 3,000 employees. The management of the company is interested in determining what percentage of the employees would be interested in enrolling in a new dental care plan. A random sample of 200 employees was selected, and 40 employees indicated that they were interested in enrolling in the plan. Develop an approximate 95% confidence interval for the proportion of the population of the employees who were interested in the plan.

Answer: The sample proportion is determined as

$$\bar{p} = \frac{40}{200} = 0.2$$

Then, we can compute the standard error of the proportion:

$$s_{\bar{p}} = \sqrt{\left(\frac{N-n}{N}\right)\left(\frac{\bar{p}(1-\bar{p})}{n-1}\right)} = \sqrt{\frac{3000-200}{3000}\left(\frac{0.2(1-0.2)}{200-1}\right)} = 0.0274$$

Finally, we can compute an approximate 95% confidence interval as

$$\bar{p} \pm 2s_{\bar{p}} = 0.2 \pm (2)(0.0274) = 0.2 \pm 0.055$$

or 0.145 to 0.255

9. A small department store has 850 customers with charge cards. In a sample of 100 customers, 45 indicated they are satisfied with the store's credit policies. Develop an approximate 95% confidence interval for the proportion of the population of customers who were satisfied.

***10.** A bank has 4,000 customers with checking accounts. We are interested in selecting a sample of checking accounts in order to develop an approximate 95% confidence interval. A pilot study resulted in a standard deviation of $300.

(a) How large must the sample be if we want to develop an approximate 95% confidence interval with a width of at most $80?

Answer: The sample size n, when estimating the population mean, is given by

$$n = \frac{Ns^2}{N\left(\frac{B^2}{4}\right) + s^2}$$

In the above equation, N is the size of the population, which in this case is 4,000; s is the standard deviation of the pilot study, which is 300. The desired interval width is 80, which indicates the bound on the sampling error (B) is 40. Substituting these values in the above equation yields

$$n = \frac{Ns^2}{N\left(\frac{B^2}{4}\right) + s^2} = \frac{(4000)(300)^2}{4000\left(\frac{40^2}{4}\right) + (300)^2} = 213.017$$

Rounding up, it is determined that a sample size of 214 will provide an approximate 95% confidence interval with a width of $80.

(b) How large must the sample be if we want to develop an approximate 95% confidence interval for the **total** of the checking accounts with a bound of $150,000?

Answer: The sample size n, when estimating the total, is given by

$$n = \frac{Ns^2}{\frac{B^2}{4N} + s^2}$$

Substituting the values of N = 4,000, s = 300, and B = 150,000 in the above equation yields

$$n = \frac{Ns^2}{\frac{B^2}{4N} + s^2} = \frac{(4000)(300)^2}{\frac{(150,000)^2}{4(4000)} + (300)^2} = 240.6$$

Rounding up, the required sample size is 241. This sample size provides an approximate 95% confidence interval for the total with a bound of $150,000.

11. A sample is to be taken to develop an approximate 95% confidence interval estimate of the population mean. The population consists of 3,000 elements, and a pilot study has resulted in a standard deviation of 600. How large must the sample be if we want to develop an approximate 95% confidence interval with a width of 160?

12. A population consists of 1,500 elements. A sample is to be taken in order to develop an approximate 95% confidence interval for the total. A pilot study resulted in a standard deviation of 40. How large must the sample be if we require a bound of 20,000 for the total?

***13.** A local university has 6,000 students. A pilot study showed that 45% of the students would be interested in taking classes on Saturday mornings. How large of a sample must be taken in order to estimate the proportion of students who are interested in Saturday classes if the desired bound, B = 0.04?

Answer: For estimating the population proportion, the sample size is given by

$$n = \frac{N\bar{p}(1-\bar{p})}{N\left(\frac{B^2}{4}\right)+\bar{p}(1-\bar{p})}$$

In this example, $\bar{p} = 0.45$, N = 6,000, and B = 0.04. Substituting these values in the above equation yields

$$n = \frac{N\bar{p}(1-\bar{p})}{N\left(\frac{B^2}{4}\right)+\bar{p}(1-\bar{p})} = \frac{(6000)(0.45)(1-0.45)}{6000\left(\frac{0.04^2}{4}\right)+(0.45)(1-0.45)} = 560.9$$

Therefore, the desired sample size is 561.

14. A small community has 8,000 registered voters. You are interested in taking a sample in order to determine what percentage of registered voters will vote for the democratic candidate. How large of a sample do you need to take if the desired bound is B = 0.07? (Hint: Since $\bar{p}$ is not known, use $\bar{p} = 0.5$.)

***15.** MNM Corporation has branches in Japan, the United States, Germany, and Switzerland. Of the 4,000 employees (N) of the corporation, 800 are stationed in Germany (N_1), 1,200 in the United States (N_2), 1,600 in Japan (N_3), and 400 in Switzerland (N_4). We are interested in conducting a survey regarding the yearly salaries of the MNM Corporation. Stratified sampling has been adopted, and samples of the employees' records are selected. Table 21.1 shows the results of the samples.

h	Country (h)	$\bar{x}_h$	s_h	N_h	n_h
1	Germany	36	2	800	25
2	United States	33	4	1,200	40
3	Japan	35	3	1,600	50
4	Switzerland	35	5	400	20

Table 21.1

In Table 21.1, the following notations are used.

h = country (Japan = 1, U.S. = 2, Germany = 3, Switzerland = 4)

$\bar{x}_h$ = average salary in country h

s_h = standard deviation of salaries in country h (in $1000)

N_h = number of employees in country h

n_h = sample size in country h

(a) Determine an estimate of the population mean.

Answer: In stratified sampling, the point estimator of the population mean is given by

$$\bar{x}_{st} = \sum_{h=1}^{H} \left(\frac{N_h}{N} \right) \bar{x}_h$$

Using the above equation, the estimate of the population mean is computed as follows.

$$\bar{x}_{st} = \sum_{h=1}^{H}\left(\frac{N_h}{N}\right)\bar{x}_h = \left(\frac{800}{4000}\right)(36)+\left(\frac{1200}{4000}\right)(33)+\left(\frac{1600}{4000}\right)(35)+\left(\frac{400}{4000}\right)(34)$$

$$= 34.5$$

Since the figures were in $1,000s, the above indicates that the point estimator of the population mean is $34,500.

(b) Develop an approximate 95% confidence interval estimate for the population mean.

Answer: The approximate 95% confidence interval is

$$\bar{x}_{st} \pm 2s_{\bar{x}_{st}}$$

In order to compute the interval, first we need to estimate the standard error which is

$$s_{\bar{x}_{st}} = \sqrt{\left(\frac{1}{N^2}\right)\sum_{h=1}^{H} N_h\left(N_h - n_h\right)\frac{s_h^2}{n_h}}$$

To compute the standard error, we first need to compute the value of

$$\sum_{h=1}^{H} N_h\left(N_h - n_h\right)\frac{s_h^2}{n_h}$$

Table 21.2 shows how this value is computed.

Country	h	$N_h(N_h - n_h)\frac{s_h^2}{n_h}$
Germany	1	$800(800-25)\frac{(2)^2}{25} = 99{,}200$
United States	2	$1200(1200-40)\frac{(4)^2}{40} = 556{,}800$
Japan	3	$1600(1600-50)\frac{(3)^2}{50} = 446{,}400$
Switzerland	4	$400(400-20)\frac{(5)^2}{20} = 190{,}000$
		$\sum_{h=1}^{H} N_h(N_h - n_h)\frac{s_h^2}{n_h} = 1{,}292{,}400$

Table 21.2

Now we can compute the standard error as

$$s_{\bar{x}_{st}} = \sqrt{\left(\frac{1}{N^2}\right)\sum_{h=1}^{H} N_h(N_h - n_h)\frac{s_h^2}{n_h}} = \sqrt{\frac{1}{(4000)^2}(1{,}292{,}400)} = 0.2842$$

Finally, an approximate 95% confidence interval of the population mean is computed as

$$\bar{x}_{st} \pm 2s_{\bar{x}_{st}} = 34.5 \pm (2)(0.2842)$$

or 33.932 to 35.068 (rounded). Since the figures were in $1,000s, the interval for the mean is from $33,932 to $35,068.

(c) Determine a point estimate for the total salaries of the MNM Corporation.

Answer: In Part a, the point estimate of the population mean was determined to be 34.5. Now we can determine the point estimate of the population total.

$$\hat{x} = N\bar{x}_{xt} = (4{,}000)(34.5) = 138{,}000$$

Once again, converting the above to dollars, we can indicate an unbiased estimate of the population total earnings to be $138,000,000.

(d) Determine an approximate 95% confidence interval estimate of the population total.

Answer: This interval is given by

$$N\bar{x}_{st} \pm 2s_{\hat{x}}$$

In Part b, the standard error of the mean was determined to be 0.2842. Now, we can estimate the standard error of $\hat{x}$ as

$$s_{\hat{x}} = Ns_{\bar{x}_{st}} = (4{,}000)(0.2842) = 1{,}136.8$$

Using the above figures, the interval for the total is computed as

$$N\bar{x}_{st} \pm 2s_{\hat{x}} = 138{,}000 \pm (2)(1{,}136.8)$$

or 135,726 to 140,274 (rounded), indicating that the interval for total earnings were from $135,726,000 to $140,274,000.

(e) The number of female employees in the sample of 180 is shown below.

Country	nh	Number of Female Employees
Germany	25	10
United States	40	18
Japan	50	20
Switzerland	20	10

Determine an approximate 95% confidence interval for the proportion of female employees.

Answer: The point estimate of the population proportion in stratified simple random sampling is the weighted average of the proportion for each stratum. Thus, the point estimate is computed as

$$\bar{p}_{st} = \sum_{h=1}^{H}\left(\frac{Nh}{N}\right)\bar{p}_h$$

$$= \left(\frac{800}{4000}\right)\left(\frac{10}{25}\right)+\left(\frac{1200}{4000}\right)\left(\frac{18}{40}\right)+\left(\frac{1600}{4000}\right)\left(\frac{20}{50}\right)+\left(\frac{400}{4000}\right)\left(\frac{10}{20}\right) = 0.425$$

Then, we can compute the standard error by

$$s_{\bar{p}_{st}} = \sqrt{\frac{1}{N^2}\sum_{h=1}^{H} N_h\left(N_h - n_h\right)\left[\frac{\bar{p}_h\left(1-\bar{p}_h\right)}{n_h - 1}\right]}$$

Table 21.3 shows a portion of the calculations needed to estimate the standard error.

Country	h	$N_h(N_h - n_h)\left[\frac{\bar{p}_h(1-\bar{p}_h)}{n_h - 1}\right]$	
Germany	1	$800(800-25)\left[\frac{(10/25)(15/25)}{25-1}\right] =$	6,200
United States	2	$1200(1200-40)\left[\frac{(18/40)(22/40)}{40-1}\right] =$	8,833.85
Japan	3	$1600(1600-50)\left[\frac{(20/50)(30/50)}{50-1}\right] =$	12,146.94
Switzerland	4	$400(400-20)\left[\frac{(10/20)(10/20)}{20-1}\right] =$	2,000
		$\sum_{h=1}^{H} N_h(N_h - n_h)\left[\frac{\bar{p}_h(1-\bar{p}_h)}{n_h - 1}\right] =$	29,180.79

Table 21.3

Now, we can compute the standard error as

$$s_{\bar{p}_{st}} = \sqrt{\frac{1}{N^2}\sum_{h=1}^{H} N_h(N_h - n_h)\left[\frac{\bar{p}_h(1-\bar{p}_h)}{n_h - 1}\right]} = \sqrt{\frac{1}{(4000)^2}(29,180.79)} = 0.0427$$

Finally, an approximate 95% confidence is computed:

$$\bar{p}_{st} \pm 2s_{\bar{p}_{st}} = 0.425 \pm (2)(0.0427) \quad \text{or} \quad 0.3395 \text{ to } 0.5104$$

16. A stratified simple random sample has been taken with the following results.

Stratum (h)	$\bar{x}_h$	s_h	$\bar{p}_h$	N_h	n_h
1	50	3	0.40	200	30
2	40	5	0.35	300	40
3	60	2	0.25	400	20
4	90	4	0.40	100	10

(a) Develop an estimate of the population mean.

(b) Compute the standard error of the mean.

(c) Develop an approximate 95% confidence interval of the population mean.

(d) Determine a point estimate of the total.

(e) Determine an approximate 95% confidence interval estimate of the population total.

(f) Compute the point estimate of the population proportion.

(g) Compute the standard error of the proportion.

(h) Develop an approximate 95% confidence interval for the proportion of the population.

17. A stratified simple random sample has been taken with the following results.

Stratum (h)	$\bar{x}_h$	s_h	$\bar{p}_h$	N_h	n_h
1	50	8	0.4	600	50
2	40	9	0.3	400	100
3	60	7	0.5	1000	40

(a) Develop an estimate of the population mean.

(b) Compute the standard error of the mean.

(c) Develop an approximate 95% confidence interval of the population mean.

(d) Determine a point estimate of the total.

(e) Determine an approximate 95% confidence interval estimate of the population total.

(f) Compute the point estimate of the population proportion.

(g) Compute the standard error of the proportion.

(h) Develop an approximate 95% confidence interval for the proportion of the population.

***18.** A population has been divided into three strata. From a past survey, the standard deviations for the three strata are known. The size of each strata and the standard deviations are shown below.

Stratum (h)	N_h	s_h
1	400	50
2	300	40
3	200	30
	N = 900	

We are interested in estimating the population mean.

(a) How large of a sample is needed so that the bound on the error of the estimate of B = 7?

Answer: The sample size when estimating the population mean is given by

$$n = \frac{\left(\sum_{h=1}^{H} N_h s_h\right)^2}{N^2\left(\frac{B^2}{4}\right) + \sum_{h=1}^{H} N_h s_h^2}$$

First, let us compute

$$\sum_{h=1}^{3} N_h s_h = (400)(50) + (300)(40) + (200)(30) = 38{,}000$$

Then we can compute

$$\sum_{h=1}^{H} N_h s_h^2 = (400)(50)^2 + (300)(40)^2 + (200)(30)^2 = 1{,}660{,}000$$

Finally, the sample size is computed as

$$n = \frac{\left(\sum_{h=1}^{H} N_h s_h\right)^2}{N^2\left(\frac{B^2}{4}\right) + \sum_{h=1}^{H} N_h s_h^2} = \frac{(38,000)^2}{\frac{(900)^2(7)^2}{4} + 1,660,000} = 124.6$$

Rounding up, the desired sample size is 125.

(b) How many units should be selected from each stratum?

Answer: Let us use the Neyman allocation procedure to determine the sample size for each stratum. The sample size for each stratum is given by

$$n_h = n\left(\frac{N_h s_h}{\sum_{h=1}^{H} N_h s_h}\right)$$

Hence, the sample sizes are computed as

$$n_1 = 125\left(\frac{(400)(50)}{38,000}\right) = 66$$

$$n_2 = 125\left(\frac{(300)(40)}{38,000}\right) = 39$$

$$n_3 = 125\left(\frac{(200)(30)}{38,000}\right) = 20$$

Note: The figures are rounded.

19. A population has been divided into three strata with $N_1 = 600$, $N_2 = 400$, and $N_3 = 500$. From a past survey, the following estimates for the standard deviations in the three strata are available: $s_1=50$, $s_2=30$, and $s_3=40$.

(a) How large of a sample is needed if an estimate of the population mean with a bound on the error of the estimate of $B = 8$ is required?

(b) How many units should be selected from each stratum?

***20.** There are 100 counties in a state. A cluster of 5 counties is selected in order to obtain information regarding various characteristics of the registered voters. Assume there are 20,000 registered voters in the state. The above information can be written:

N = number of clusters in the population = 100
n = number of clusters selected in the sample = 5
M = number of elements in the population = 20,000
$\overline{M}$ = average number of elements in a cluster = M/N = 20,000/100 = 200

Table 21.4 provides information regarding the five clusters.

County (i)	Registered Voters (M_i)	Total Taxes Paid ($1,000) for Sample ($x_i$)	Democrats (a_i)
1	20	80	16
2	18	60	6
3	30	95	12
4	22	90	18
5	10	40	8
Total	100	365	60

Table 21.4

(a) Determine the point estimate of the population mean.

Answer: The point estimate of the population mean is computed as follows. (The totals are shown in Table 21.4.)

$$\bar{x}_c = \frac{\sum_{i=1}^{n} x_i}{\sum_{i=1}^{n} M_i} = \frac{365}{100} = 3.65$$

Since the tax data are in thousands of dollars, an estimate of the mean tax in the state is $3,650.

(b) Estimate the standard error of the mean.

Answer: The standard error is given by

$$s_{\bar{x}_c} = \sqrt{\left(\frac{N-n}{Nn\overline{M}^2}\right)\frac{\sum_{i=1}^{n}\left(x_i - \bar{x}_c M_i\right)^2}{n-1}}$$

To compute the standard error, first we need to compute $\sum_{i=1}^{5}\left(x_i - \bar{x}_c M_i\right)^2$. Table 21.5 shows the computations necessary for determining this value.

County (i)	M_i	x_i	$(x_i - 3.65M_i)^2$
1	20	80	$[80 - 3.65(20)]^2 =$ 49.00
2	18	60	[60 - 3.65(18)]2 = 32.49
3	30	95	[95 -3.65(30)]2 = 210.25
4	22	90	[90 -3.65(22)]2 = 94.09
5	10	40	[40 -3.65 (10)]2 = 12.25
	100	365	$\sum_{i=1}^{5}\left(x_i - \bar{x}_c M_i\right)^2 = 398.08$

Table 21.5

Then the standard error can be computed:

$$s_{\bar{x}_c} = \sqrt{\left(\frac{N-n}{Nn\overline{M}^2}\right)\frac{\sum_{i=1}^{n}\left(x_i - \bar{x}_c M_i\right)^2}{n-1}} = \sqrt{\left[\frac{100-5}{(100)(5)(200)^2}\right]\left[\frac{398.08}{5-1}\right]} = 0.02174$$

(c) Develop an approximate 95% confidence interval for the mean of the population.

Answer: The approximate 95% confidence interval of the population mean is given by

$$\bar{x}_c \pm 2s_{\bar{x}_c}$$

In Part a, $\bar{x}_c$ was computed to be 3.65; and in Part b, the standard error was computed as 0.02174. Using these figures, the approximate 95% confidence interval for the mean can be computed as

$$\bar{x}_c \pm 2s_{\bar{x}_c} = 3.65 \pm (2)(0.02174) \quad \text{or} \quad 3.606 \text{ to } 3.693$$

Converting the figures to dollars yields an interval of $3,606 to $3,693.

(d) Determine the point estimate of the population total taxes.

Answer: From Part a, we have $\bar{x}_c = 3.65$ and M is given as 200,000. Therefore, the point estimate of the population total is computed as

$$\hat{x} = M\bar{x}_c = (20{,}000)(3.65) = 73{,}000$$

(e) Determine a 95% confidence interval estimate of the population total.

Answer: First, let us compute the standard error of the total.

$$s_{\hat{x}} = Ms_{\bar{x}_c}$$

Now, we can compute the 95% confidence interval for the total as

$$M\bar{x}_c \pm 2s_{\bar{x}_c} = 73{,}000 \pm (2)(434.8) \quad \text{or} \quad 72{,}130.4 \text{ to } 73{,}869.6$$

In terms of dollars, the interval is $72,130,400 to $73,869,600.

(f) Determine a point estimate of the proportion of Democrats in the population.

Answer: The point estimate of the population proportion is computed as

$$\bar{p}_c = \frac{\sum_{i=1}^{n} a_i}{\sum_{i=1}^{n} M_i} = \frac{16+6+12+18+8}{20+18+30+22+10} = 0.6$$

(g) Develop an approximate 95% confidence interval estimate of the population proportion.

Answer: The standard error is given by

$$s_{\bar{p}_c} = \sqrt{\left(\frac{N-n}{Nn\bar{M}^2}\right)\frac{\sum_{i=1}^{n}\left(a_i - \bar{p}_c M_i\right)^2}{n-1}}$$

First, we need to compute $\sum_{i=1}^{n}\left(a_i - \bar{p}_c M_i\right)^2$. Table 21.6 shows the procedure for computing this value.

Country (i)	M_i	a_i	$(a_i - 0.6M_i)^2$
1	20	16	$[16 - 0.6(20)]^2 = 16.00$
2	18	6	$[6 - 0.6(18)]^2 = 23.04$
3	30	12	$[12 - 0.6(30)]^2 = 36.00$
4	22	18	$[18 - 0.6(22)]^2 = 23.04$
5	10	8	$[8 - 0.6(20)]^2 = 4.00$
Total	100	60	$\sum_{i=1}^{n}\left(a_i - \bar{p}_c M_i\right)^2 = 102.08$

Table 21.6

Now, the standard error can be computed as

$$s_{\bar{p}_c} = \sqrt{\left(\frac{N-n}{Nn\bar{M}^2}\right)\frac{\sum_{i=1}^{n}\left(a_i - \bar{p}_c M_i\right)^2}{n-1}} = \sqrt{\left[\frac{(100-5)}{(100)(5)(200)^2}\right]\left[\frac{102.08}{5-1}\right]} = 0.011$$

Finally the approximate 95% confidence interval for the population proportion is computed as

$$\bar{p}_c \pm 2s_{\bar{p}_c} = 0.6 \pm (2)(0.011) = 0.6 \pm 0.022 \text{ or } 0.578 \text{ to } 0.622$$

21. A sample of six clusters is taken from a population with N = 50 clusters and M = 600 elements in the population. The values of M_i, x_i, and a_i for each cluster in the sample are given below.

Cluster	M_i	x_i	a_i
1	6	150	0
2	12	350	5
3	8	200	3
4	12	250	3
5	5	100	1
6	4	60	4

(a) Determine the point estimate of the population mean.

(b) Estimate the standard error of the mean.

(c) Develop an approximate 95% confidence interval of the mean of the population.

(d) Determine the point estimate of the population total.

(e) Determine a 95% confidence interval estimate of the population total.

(f) Determine a point estimate of the population proportion.

(g) Compute the standard error of the proportion.

(h) Develop an approximate 95% confidence for the population proportion.

SELF-TESTING QUESTIONS

In the following multiple choice questions, circle the correct answer. An answer key is provided following the questions.

1. In a sample survey, it is common practice to use a Z value of (when approximating a 95% confidence interval)

a) 4
b) 3
c) 2
d) 1
e) none of the above

2. The population from which the sample is actually selected is

a) always the target population
b) the census
c) the selected sample
d) the sampled population
e) none of the above

3. The target population and the sampled population

a) are always the same
b) are not always the same
c) must be the same for the results to be accurate
d) none of the above

4. The units that are selected for sampling constitute the

a) sample
b) population
c) frame
d) sampling unit
e) none of the above

5. A list of the sampling units for a study is

a) the sampled population
b) called the frame
c) the same as the sample
d) none of the above

6. Probabilistic sampling is any method of sampling in which

a) the probability of the occurrence of various events in the study are known
b) there are only two possible outcomes, such as P(A) and P(A^C)
c) the probability of each possible sample can be computed
d) the probability of each possible sample is one
e) none of the above

7. Convenience sampling is an example of

a) probabilistic sampling
b) sampling where the probabilities are known
c) nonprobabilistic sampling
d) none of the above

8. Which of the following is an example of nonprobabilistic sampling?

a) simple random sampling
b) stratified simple random sampling
c) cluster sampling
d) judgment sampling
e) none of the above

9. The error that occurs because a sample, and not the entire population, is used to estimate a population parameter is a

a) nonsampling error
b) sampling error
c) judgment error
d) standard error
e) none of the above

10. Which of the following sampling methods is a probabilistic sampling method?

a) judgment sampling
b) convenience sampling
c) both a and b
d) cluster sampling
e) none of the above

11. Stratified random sampling is a method of selecting a sample in which

a) the sample is first divided into strata, and then random samples are taken from each stratum
b) various strata are selected from the sample
c) the population is first divided into strata, and then random samples are drawn from each stratum
d) none of the above

12. Cluster sampling is

a) a nonprobability sampling method
b) the same as convenience sampling
c) a probability sampling method
d) none of the above.

13. With nonprobabilistic sampling

a) it is possible to make estimates about the precision of the population parameters.
b) it is not possible to make statements about the precision of estimates made concerning the population parameters.
c) the precision can be estimated if the sample is larger than 30.
d) none of the above

14. Errors such as measurement error, processing error, and interviewer error are

a) sampling errors
b) nonsampling errors
c) could be either sampling or nonsampling errors
d) impossible to detect
e) none of the above

15. Sampling errors

a) can be avoided by increasing the sample size to at least 30
b) can be avoided if the sample is increased so that it will be at least 5% of the population
c) can be avoided by using probabilistic sampling
d) cannot be avoided
e) none of the above

ANSWERS TO THE SELF-TESTING QUESTIONS

1. c
2. d
3. b
4. d
5. b
6. c
7. c
8. d
9. b
10. d
11. c
12. c
13. b
14. b
15. d

ANSWERS TO CHAPTER TWENTY-ONE EXERCISES

2. (a) 7.4 (rounded)
   (b) 458.2 to 514.8

3. (a) 11.76 (rounded)
   (b) 276.48 to 323.52

5. (a) 0.0359
   (b) 2.93 to 3.07 (rounded)
   (c) 15,000
   (d) 14,640.79 to 15, 359.21

6. (a) 250,000
   (b) 3,700.8
   (c) 242,598.4 to 257,401.6

7. (a) 60,000
   (b) 1,424.6
   (c) 57,150.8 to 62,849.2

9. 0.356 to 0.544

11. 210

12. 152

14. 200

16. (a) 55
    (b) 0.3221
    (c) 54.36 to 55.64 (rounded)
    (d) 55,000
    (e) 54,355.8 to 55,644.2
    (f) 0.325
    (g) 0.04976
    (h) 0.2255 to 0.4245

17. (a) 53
(b) 0.6510
(c) 51.70 to 54.30
(d) 106,000
(e) 103,395.7 to 108,604.3
(f) 0.4111
(g) 0.03845
(h) 0.3342 to 0.4880

19. (a) 100

(b) n_1=48

n_2 = 20

n_3 = 32

Note: figures are rounded.

21. (a) 23.617
(b) 0.405
(c) 22.806 to 24.428
(d) 42,510.64
(e) 41,051.6 to 43,969.7
(f) 0.34
(g) 0.0124 (rounded)
(h) 0.3152 to 0.3648

CHAPTER TWENTY-TWO

DECISION ANALYSIS

CHAPTER OUTLINE AND REVIEW

In this final chapter, you have been introduced to the decision analysis approach to decision making. You have studied decision making under uncertainty with the use of probabilities. Some of the new concepts which you have studied are

A. **State of Nature:** The uncontrollable future events that can affect the outcome of a decision.

B. **Payoff:** The outcome measure, such as profit, cost, and time. Each combination of a decision alternative and a state of nature has an associated payoff.

C. **Payoff Table:** A tabular presentation of the payoffs for a decision problem.

D. **Decision Tree:** A graphical representation of the decision-making situation from decision to state of nature to payoff.

E. **Nodes:** The intersection or junction points of the decision tree.

F. **Branches:** Lines or arcs connecting nodes of the decision tree.

G. **Expected Value:** A decision criterion which weights the payoff for each decision by its probability of occurrence.

H.	**Expected Value of Perfect Information (EVPI):**	The expected value of information that would tell the decision maker exactly which state of nature was going to occur (that is, perfect information). EVPI is equal to the expected opportunity loss of the best-decision alternative when no additional information is available.
I.	**Indicators:**	Information about the states of nature obtained by experimentation. An indicator may be the result of a sample.
J.	**Prior Probabilities:**	The probabilities of the states of nature prior to obtaining experimental information.
K.	**Posterior (Revised) Probabilities:**	The probabilities of the states of nature after use of Bayes' theorem to adjust the prior probabilities based upon given indicator information.
L.	**Bayesian Revision:**	The process of adjusting prior probabilities to create the posterior probabilities based upon information obtained by experimentation.
M.	**Expected Value of Sample Information (EVSI):**	The difference between the expected value of an optimal strategy based on new information and the "best" expected value without any new information. It is a measure of the economic value of new information.
N.	**Efficiency:**	The ratio of EVSI to EVPI. Perfect information is 100% efficient.

CHAPTER FORMULAS

The Two Probability Conditions

$$P(s_j) \geq 0 \qquad \text{for all states of nature} \tag{22.1}$$

$$\sum_{j=1}^{N} P(s_j) = P(s_1) + P(s_2) + \cdots + P(s_N) = 1 \tag{22.2}$$

Expected Value of Decision Alternative d_i

$$EV(d_i) = \sum_{j=1}^{N} P(s_j) V_{ij} \tag{22.3}$$

Expected Value of Perfect Information (EVPI)

$$EVPI = |EVwPI - EVwoPI| \tag{22.4}$$

where EVPI = expected value of perfect information
EVwPI = expected value *with* perfect information about the states of nature
EVwoPI = expected value *without* perfect information about the states of nature

Expected Value of Sample Information (EVSI)

$$EVSI = |EVwSI - EVwoSI| \tag{22.5}$$

where EVSI = expected value of sample information
EVwSI = expected value *with* sample information about the states of nature
EVwoSI = expected value *without* sample information about the states of nature

Efficiency of Sample Information (E)

$$E = \frac{EVSI}{EVPI}(100) \tag{22.6}$$

EXERCISES

***1.** Suppose we are interested in investing in one of three investment opportunities: d_1, d_2, or d_3. The following profit payoff table shows the profits (in thousands of dollars) under each of the 3 possible economic conditions: s_1, s_2, and s_3.

Decision Alternative	State of Nature s_1	s_2	s_3
d_1	18	28	30
d_2	19	17	-5
d_3	3	40	16

Assume the states of nature have the following probabilities of occurrence.

$$P(s_1) = 0.2 \qquad P(s_2) = 0.3 \qquad P(s_3) = 0.5$$

(a) Use the expected monetary value criterion to determine the optimal decision.

Answer: The expected monetary value of each decision is determined by

$$EV(d_i) = \sum_{j=1}^{N} P(s_j)V_{ij}$$

Therefore,

$$EV(d_1) = .2(18) + .3(28) + .5(30) = 27.0 \ *$$

$$EV(d_2) = .2(19) + .3(17) + .5(-5) = 6.4$$

$$EV(d_3) = .2(3) + .3(40) + .5(16) = 20.6$$

Hence, the decision alternative d_1 with the highest expected monetary value ($27,000) is recommended.

(b) Construct a decision tree. What is the expected value at each state of nature node?

Answer: The decision tree is shown in Figure 22.1.

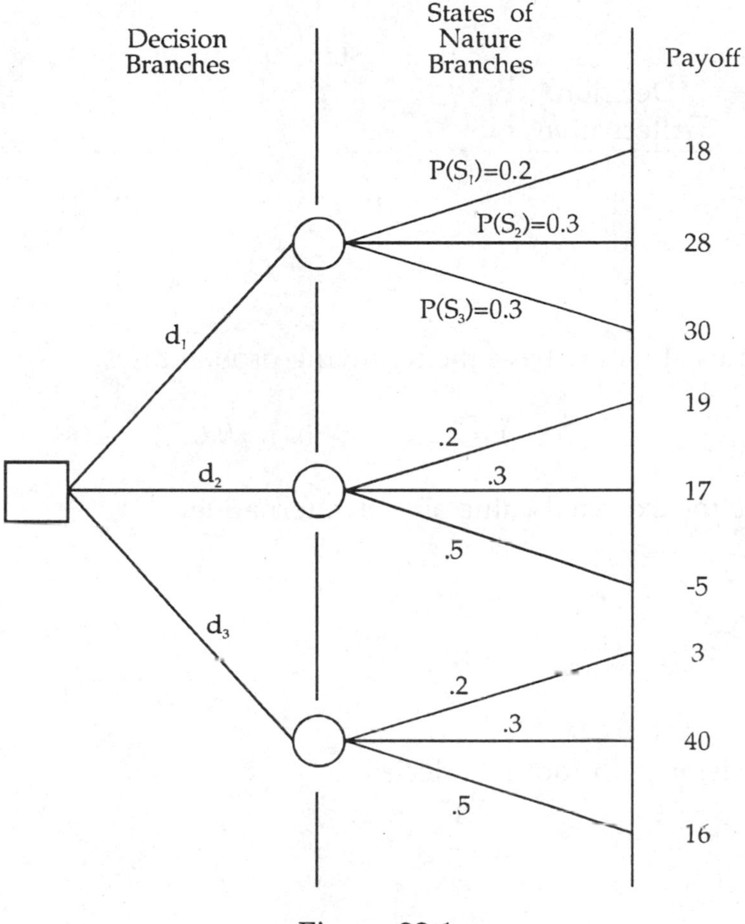

Figure 22.1

The expected value at each state of nature node can be determined by weighting each payoff by its associated probability and showing the results at each node as shown in Figure 22.2. (Note that these are the same as the expected monetary values.)

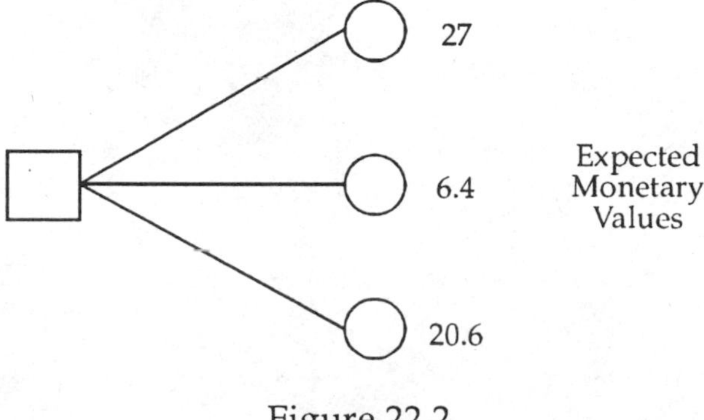

Figure 22.2

2. Assume you are faced with the following decision alternatives and two states of nature. The payoff table is shown below.

Decision Alternative	State of Nature s_1	s_2
d_1	9	18
d_2	0	30
d_3	20	5

Assume the states of nature have the following probabilities.

$P(s_1) = 0.4$ $\qquad$ $P(s_2) = 0.6$

(a) Determine the expected value of each alternative.

(b) Which decision is the optimal decision?

***3.** Refer to exercise 1. Determine the expected value of perfect information.

Answer: The expected value of perfect information (EVPI) is the absolute value of the difference between the expected value *with* perfect information and the expected value *without* perfect information. Thus, EVPI can be determined by

$$EVPI = |EVwPI - EVwoPI|$$

If the decision maker has perfect information and knows that the state of nature s_1 will occur, the selected alternative will be d_2 (the largest payoff of 19). If it is known that s_2 will occur, the best alternative will be d_3 (payoff of 40). Finally, if it is known that s_3 will occur, the selected alternative will be d_1 (payoff of 30). Remember that the probabilities of the states of nature 1, 2, and 3 were given in exercise 1 as 0.2, 0.3, and 0.5. Now assigning probability weights to the payoffs, we can compute the expected value *with* perfect information as

$$EVPI = (0.2)(19) + ().3(40) + (0.5)(30) = 30.8$$

In Part a of exercise 1, the expected value of the best alternative (which is the expected value *without* perfect information) was determined to be 27. Therefore, the expected value of perfect information is determined as

$$EVPI = |EVwPI - EVwoPI| = 30.8 - 27 = 3.8$$

Therefore, the expected value of perfect information is $3,800.

4. Refer to exercise 2. Determine the expected value of perfect information.

***5.** Refer to the following payoff table and probability information.

Decision Alternative	s_1	s_2
d_1	100	200
d_2	50	300
d_3	500	0

$P(s_1) = 0.3$ $\quad P(I_1 \mid s_1) = 0.9$ $\quad P(I_2 \mid s_1) = 0.1$

$P(s_2) = 0.7$ $\quad P(I_1 \mid s_2) = 0.2$ $\quad P(I_2 \mid s_2) = 0.8$

(a) Find the values of $P(I_1)$ and $P(I_2)$.

Answer: Using the tabular approach, we can first find $P(I_1)$ as shown in Table 22.1.

State of Nature	Prior Probability $P(s_j)$	Conditional Probability $P(I_1 \mid s_j)$	Joint Probability $P(I_1\ s_j)$	Posterior Probability $P(s_j \mid I_1)$
s_1	0.3	0.9	0.27	0.6585
s_2	0.7	0.2	0.14	0.3415
			$P(I_1) = 0.41$	

Table 22.1

Now that we have determined $P(I_1) = 0.41$ and we know that $P(I_1) + P(I_2) = 1$, we can simply conclude that $P(I_2) = 1 - 0.41 = 0.59$; or we can compute $P(I_2)$ as shown in Table 22.2.

State of Nature	Prior Probability $P(s_j)$	Conditional Probability $P(I_2 \mid s_j)$	Joint Probability $P(I_2\ s_j)$	Posterior Probability $P(s_j \mid I_2)$
s_1	0.3	0.1	0.03	0.0508
s_2	0.7	0.8	0.56	0.9492
			$P(I_2) = 0.59$	

Table 22.2

(b) What are the values of $P(s_1 | I_1)$, $P(s_2 | I_1)$, $P(s_1 | I_2)$, and $P(s_2 | I_2)$?

Answer: From Table 22.1, we can determine the following probabilities.

$$P(s_1 | I_1) = \frac{0.27}{0.41} = 0.6585$$

$$P(s_2 | I_1) = \frac{0.14}{0.41} = 0.3415$$

And from Table 22.2, the following probabilities can be computed.

$$P(s_1 | I_2) = \frac{0.03}{0.59} = 0.0508$$

$$P(s_2 | I_2) = \frac{0.56}{0.59} = 0.9492$$

The above probabilities are the posterior probabilities and are shown in the last column of Tables 22.1 and 22.2.

(c) Use the decision tree approach and determine the optimal decision strategy. What is the expected value of the solution?

Answer: The decision tree for the above exercise is shown in Figure 22.3.

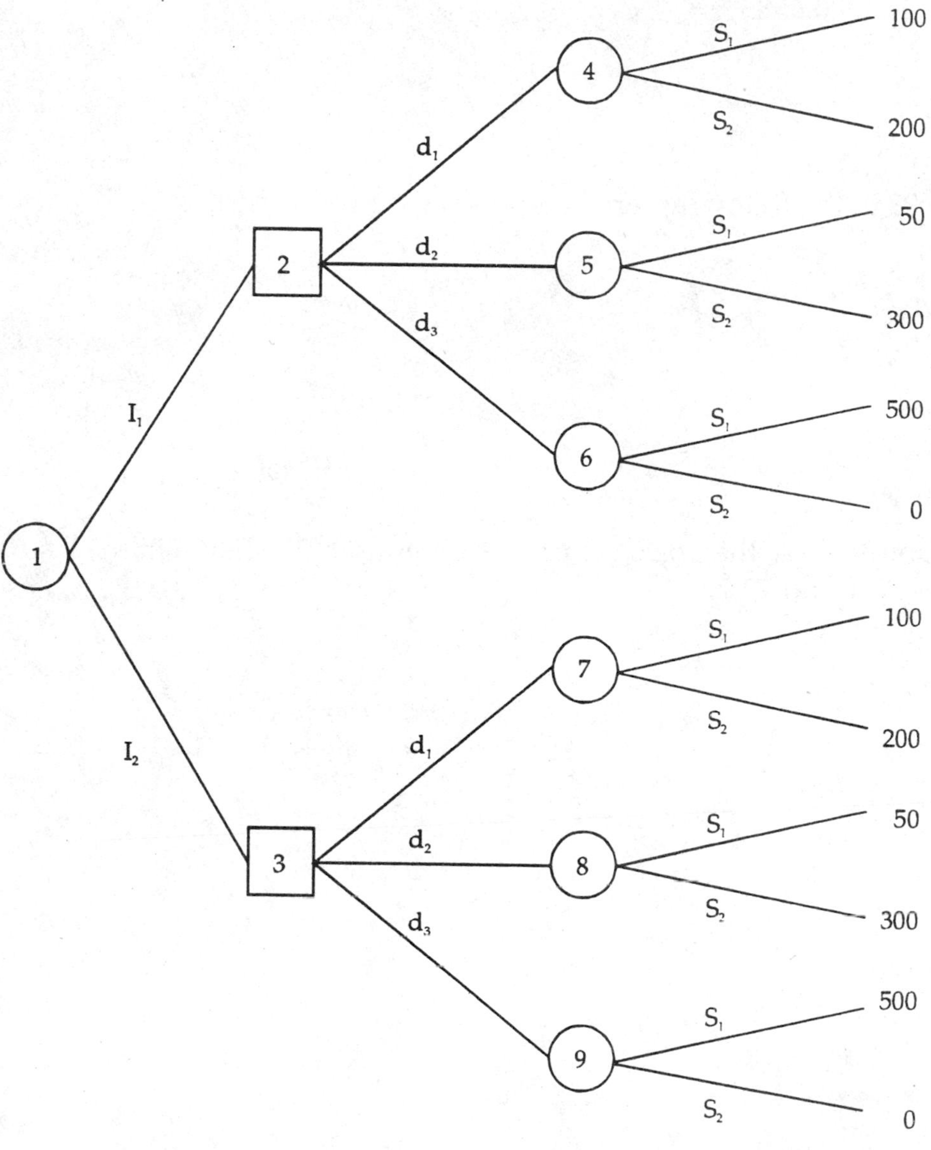

Figure 22.3

In Part b of this exercise, we determined the posterior probabilities. Using those probabilities, we can determine the expected monetary values for nodes 4 through 9 as follows.

$$\begin{aligned}
EV(\text{Node } 4) &= (0.6585)(100) + (0.3415)(200) = 134.15\\
EV(\text{Node } 5) &= (0.6585)(50) + (0.3415)(300) = 135.38\\
EV(\text{Node } 6) &= (0.6585)(500) + (0.3415)(0) = 329.25\\
EV(\text{Node } 7) &= (0.0508)(100) + (0.9492)(200) = 194.92\\
EV(\text{Node } 8) &= (0.0508)(50) + (0.9492)(300) = 287.30\\
EV(\text{Node } 9) &= (0.0508)(500) + (0.9492)(0) = 25.40
\end{aligned}$$

Now comparing the EV's of nodes 4, 5, and 6, we note the highest EV is 329.25 (for node 6); and comparing the EV's of nodes 7, 8, and 9, the highest EV is 287.3 (for node 8). Thus, the decision tree can be reduced to and presented as shown in Figure 22.4.

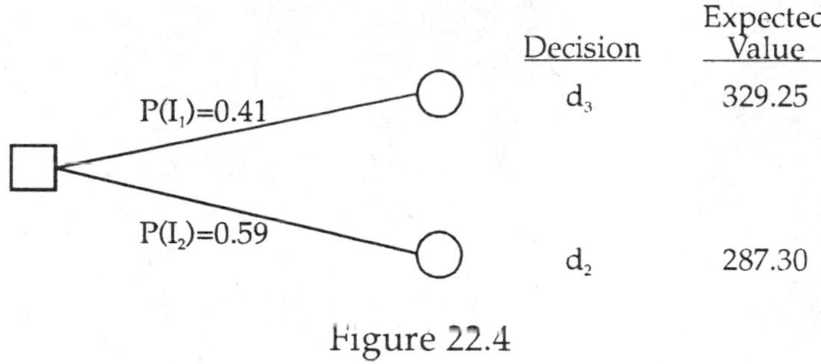

Figure 22.4

Figure 22.4 indicates that if I_1 occurs, the optimal decision is d_3; and if I_2 occurs, the optimal decision is d_2. Therefore, the expected value of the optimal strategy or the expected value of node 1 is

$$EV(\text{Node } 1) = (0.41)(329.25) + (0.59)(287.30) = 304.50$$

6. Refer to the following payoff table and probability information.

Decision Alternative	s_1	s_2
d_1	1000	3000
d_2	4000	500

$P(s_1) = 0.45$ $\quad P(I_1 \mid s_1) = 0.7$ $\quad P(I_2 \mid s_1) = 0.3$

$P(s_2) = 0.55$ $\quad P(I_1 \mid s_2) = 0.6$ $\quad P(I_2 \mid s_2) = 0.4$

(a) Find the values of $P(I_1)$ and $P(I_2)$.

(b) Determine the values of $P(s_1 \mid I_1)$, $P(s_2 \mid I_1)$, $P(s_1 \mid I_2)$, and $P(s_2 \mid I_2)$.

(c) Use the decision tree approach and determine the optimal strategy. What is the expected value of your solution?

***7.** Refer to exercise 5. Determine the expected value of the sample information.

Answer: The expected value of the sample information is

$$EVSI = |EVwSI - EVwoSI|$$

The expected value of the optimal decision with the sample information was determined in Part c of exercise 5 as 304.50. The optimal decision without the sample information can be determined as follows.

$$EV(d_1) = (.3)(100) + (.7)(200) = 170$$

$$EV(d_2) = (.3)(50) + (.7)(300) = 225 \, *$$

$$EV(d_3) = (.3)(500) + (.7)(0) = 150$$

Thus, the optimal decision is d_2 with an expected value of 225. Now we can determine the expected value of the sample information:

$$EVSI = |EVwSI - EVwoSI| = 304.50 - 225 = 79.50.$$

8. Refer to exercise 6. Determine the expected value of the sample information.

SELF-TESTING QUESTIONS

In the following multiple choice questions, circle the correct answer. An answer key is provided following the questions.

1. The uncontrollable future events that can affect the outcome of a decision are known as

a) alternatives
b) decision outcome
c) payoff
d) state of nature

2. When the state of nature is known, the decision process is said to be under

a) certainty
b) uncertainty
c) risk
d) none of the above

3. A decision criterion which weights the payoff for each decision by its probability of occurrence is known as

a) minimax criterion
b) expected monetary value criterion
c) probability
d) expected value of perfect information

4. When the state of nature is not known, the decision process is said to be under

a) regret
b) uncertainty
c) either a or b
d) none of the above

5. The probability of the states of nature after use of Bayes' theorem to adjust the prior probabilities based upon given indicator information is called

a) marginal probability
b) conditional probability
c) posterior probability
d) none of the above

6. The efficiency of information is the ratio of

a) EOL to EVSI
b) EOL to EVPI
c) EVPI to EVSI
d) EVSI to EVPI

ANSWERS TO THE SELF-TESTING QUESTIONS

1. d
2. a
3. b
4. b
5. c
6. d

ANSWERS TO CHAPTER TWENTY-TWO EXERCISES

2. (a) $EV(d_1) = 10.8$, $EV(d_2) = 18$, $EV(d_3) = 11$
   (b) d_2

4. 8

6. (a) $P(I_1) = 0.645$, $P(I_2) = 0.355$
   (b) $P(s_1 \mid I_1) = 0.488$
   $P(s_2 \mid I_1) = 0.512$
   $P(s_1 \mid I_2) = 0.380$
   $P(s_2 \mid I_2) = 0.620$
   (c) If I_1 occurs, the optimal decision is d_2; if I_2 occurs the optimal decision is d_1. Therefore, the expected value of the optimal strategy is \$2219.36.

8. \$119.36